Stochastic Systems for Engineers

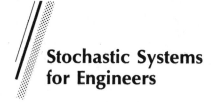

Stochastic Systems for Engineers

Modelling, Estimation and Control

JOHN A. BORRIE

PRENTICE HALL

New York · London · Toronto · Sydney · Tokyo · Singapore

First published 1992 by
Prentice Hall International (UK) Ltd
Campus 400, Maylands Avenue
Hemel Hempstead
Hertfordshire, HP2 7EZ
A division of
Simon & Schuster International Group

Typeset in 10/12 pt Times
by P & R Typesetters

Printed and bound in Great Britain by
Dotesios Ltd, Trowbridge, Wilts

Library of Congress Cataloging-in-Publication Data

Borrie, John A., 1937–
Stochastic systems for engineers : modeling, estimation and control / John A. Borrie.
p. cm.
Includes index.
ISBN 0-13-847351-X
1. Engineering mathematics. 2. Stochastic systems I. Title.
TA340.B65 1992
003′.76′02462–dc20 92–10927
CIP

British Library Cataloguing in Publication Data

A catalogue record for this book is available from the British Library

ISBN 0-13-847351-X (pbk)

1 2 3 4 5 96 95 94 93 92

Contents

Preface

This book describes the main properties of stochastic dynamical systems and practical techniques for their modelling, estimation and control. The material, which is suitable for practising engineers, postgraduate students and advanced undergraduates, is largely self-contained and develops the subject with definitions, formulae and explanations, but without detailed mathematical proofs – the interested reader being directed for these to a few carefully chosen references.

The problems of estimating and controlling stochastic systems are far from solved, and a considerable amount of research is under way. Thus, in addition to describing established methods, topics giving the reader access to this research are introduced. These include mathematical tools such as continuous time models using Itô stochastic integrals and engineering topics such as H_∞ control techniques.

Chapter 1 establishes the necessary probability theory. Chapter 2 deals with properties of continuous time stochastic processes and dynamical systems, and Chapter 3 with the discrete time equivalents. Simple exercises are provided.

Chapters 4, 5 and 6 amount to a manual of useful methods of computer modelling, state estimation (Kalman filtering) and control techniques, illustrated in most cases with worked examples. Design exercises, which inevitably require computing and computer-aided design facilities, are provided at the ends of these chapters.

The author is grateful to the Cranfield students who have investigated these methods, and to the many others who have provided problems, made suggestions and otherwise contributed to this work.

The author is also grateful to the literary executor of the late Sir Ronald A. Fisher, FRS, to Dr Frank Yates, FRS, and the Longman Group Ltd, London, for permission to reprint part of Table XXXIII from their book *Statistical Tables for Biological, Agricultural and Medical Research* (6th edn, 1974).

J. BORRIE

1 Probability and statistics

1.1 Introduction

The definitions and properties of random variables which are needed in the study of engineering stochastic systems are set out in this chapter. The underlying ideas of probability are first explored in their natural context of discrete events, and then developed to deal with continuous random variables. This opens the door to the topics considered in the remainder of the book.

Section 1.2 deals with basic definitions, Sections 1.3–1.5 with discrete events, and Sections 1.6–1.11 with continuous random variables. Sections 1.12–1.14 deal with functions of random variables, correlation and dependence.

1.2 Definitions of probability

The probability of an event, X, occurring as the result of some experiment or trial is defined by considering many trials:

$$P(X) = \lim_{n \to \infty} \left(\frac{n_X}{n} \right) \tag{1.1}$$

where

n is the number of trials conducted,
n_X is the number of trials resulting in X.

This 'relative frequency' definition of probability involves conducting a large number of trials, and is usually impractical as a basis for calculation.

A more practical, but fundamentally unsatisfactory, definition is

$$P(X) = \frac{N_X}{N} \tag{1.2}$$

where

N is the number of possible, 'equally probable', outcomes of one trial (the 'sample space'),
N_X is the number of outcomes 'favourable' to X.

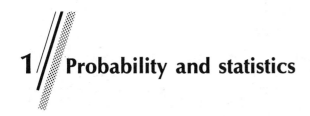

This 'classical' definition of probability fails as a logical statement in depending on the term 'equally probable'. It does, however, allow practical calculation of probabilities, and, with common sense, yields results consistent with those based on Eqn (1.1). Simple examples illustrate these points.

Examples

(a) Consider the probability of throwing a 3 when an unbiased die is tossed.

In 600 tosses, a 3 might occur 95 times, and $\dfrac{n_3}{n} = \dfrac{95}{600} = 0.158$

In 6000 tosses, a 3 might occur 990 times, and $\dfrac{n_3}{n} = \dfrac{990}{6000} = 0.165$

It is not difficult to believe that $P(3) = \lim\limits_{n \to \infty} \left(\dfrac{n_3}{n}\right) = \dfrac{1}{6} = 0.167$

More conveniently, using Eqn (1.2), there are six equally probable outcomes of the experiment, one of these being favourable to the event '3':

$$P(3) = \frac{N_3}{N} = \frac{1}{6} = 0.167$$

(b) Consider the probability that a new baby will be male. From Eqn (1.2),

$P(\text{male}) = 0.5$

However, it is common knowledge that the number of male births slightly exceeds that of females; from Eqn (1.1) and the relevant data,

$P(\text{male}) > 0.5$

This illustrates the lack of rigour in Eqn (1.2) and points up the need for caution. Unfortunately, the definition of Eqn (1.1) also lacks rigour–in particular, there is no clear reason why $P(X)$ should approach a limit as $n \to \infty$. This has given rise to a modern axiomatic definition [refs. 1, 2] which is, strictly, necessary as a sound basis for the description of probability and stochastic systems. From an engineering point of view, however, this is of rather philosophical interest, and is not pursued here.

Returning to Eqn (1.2), it is evident that the practical calculation of probabilities involves counting the possible outcomes of a trial, the possible outcomes favourable to the event in question, and dividing one count by the other, taking care that all the outcomes are 'equally probable'. The counting process frequently involves permutations and combinations (Appendix C, Section C1).

Examples

(a)′ If thirteen cards are dealt at random from a standard pack, what is the probability that all thirteen are red?

In this case, X is the event that all thirteen cards are red (Eqn (1.2)).

There are $\binom{52}{13}$ ways of selecting, or 'combining' any thirteen red cards from the pack (Appendix C, Section C1) and there are $\binom{26}{13}$ ways of selecting any thirteen cards from the twenty-six red cards in the pack. The probability of X is therefore (Eqn (C1.2))

$$P(X) = \frac{\binom{26}{13}}{\binom{52}{13}} = \frac{26!}{13!13!} \frac{13!39!}{52!} = 1.638 \times 10^{-5}$$

(b)′ If thirteen cards are dealt at random from a standard pack, what is the probability that all thirteen have the same suit?

In this case, X is the event that all thirteen cards are of one of the four suits.

There are $\binom{52}{13}$ ways of dealing any thirteen cards. There are four ways of dealing thirteen cards all of the same suit. The probability of X is therefore (Eqn (C1.2))

$$P(X) = \frac{4.13!39!}{52!} = 6.299 \times 10^{-12}$$

1.3 Probabilities of multiple events

1.3.1 Independent events

Two events, X and Y, are 'independent' if the occurrence (or non-occurrence) of X does not affect the probability of Y, and vice versa. In this case the probability of both events occurring is

$$P(XY) = P(X) \cdot P(Y) \tag{1.3}$$

where

$P(XY)$ is the probability of (X and Y),
$P(X)$ is the probability of X,
$P(Y)$ is the probability of Y.

Naturally, this can be extended to any number of independent events, $X_1, X_2, \ldots, X_n$:

$$P(X_1, X_2, \ldots, X_n) = P(X_1) \cdot P(X_2) \cdot P(X_3) \ldots P(X_n) \tag{1.4}$$

Example

Consider the probability of two 6s occurring when two unbiased dice are thrown.

The two events are independent since the occurrence of one 6 has no effect on the probability of the other:

$$P(66) = P(6) \cdot P(6) = \frac{1}{6} \cdot \frac{1}{6} = 0.0278$$

1.3.2 Dependent events

The event Y is 'dependent' on X if the probability of Y, $P(Y)$, is affected by the occurrence (or non-occurrence) of X:

$$P(XY) = P(X) \cdot P(Y|X) \tag{1.5}$$

where

$P(XY)$ is the probability of (X and Y),
$P(X)$ is the probability of X,
$P(Y|X)$ is the probability of Y given that X has occurred.

Example

Consider the probability of randomly drawing a king followed by a second king from a standard pack of cards:

$$P(K_1K_2) = P(K_1)P(K_2|K_1) = \frac{4}{52} \cdot \frac{3}{51} = 0.004\,52$$

If, on the first draw, a king had not been selected, the probability of selecting a king on the second draw would have been different; the two events are dependent.

1.3.3 Mutually exclusive events

Two events, X and Y, are mutually exclusive if the occurrence of X precludes that of Y and vice versa.

In this case,

$$P(X + Y) = P(X) + P(Y) \tag{1.6}$$

where

$P(X + Y)$ is the probability of (X or Y),
$P(X)$ is the probability of X,
$P(Y)$ is the probability of Y.

Mutual exclusivity can be regarded (with some provisos) as an extreme case of dependence; if X, Y are mutually exclusive,

$$P(X|Y) = P(Y|X) = 0$$

Example

Consider the probability of throwing a 2 or 3 when a die is tossed once:

$$P(2+3) = P(2) + P(3) = \frac{1}{6} + \frac{1}{6} = 0.333$$

In this case, the occurrence of a 2 precludes a 3 and vice versa; the two events are mutually exclusive.

1.3.4 Non-mutually-exclusive events

If the events X and Y are not mutually exclusive, Eqn (1.2) yields

$$P(X+Y) = \frac{N_X}{N} + \frac{N_Y}{N} - \frac{N_{XY}}{N} = P(X) + P(Y) - P(XY) \tag{1.7}$$

where

$P(X+Y)$ is the probability of (X or Y),
$P(XY)$ is the probability of (X and Y).

This is easily seen from the 'Venn' diagram (Figure 1.1) in which the sample space is represented by a set containing N elements (or outcomes of trials) and the outcomes favourable to X and Y by appropriate subsets containing $N_X N_Y$ elements respectively. These overlap to give a set containing N_{XY} elements – the number of results favourable to both X and Y.

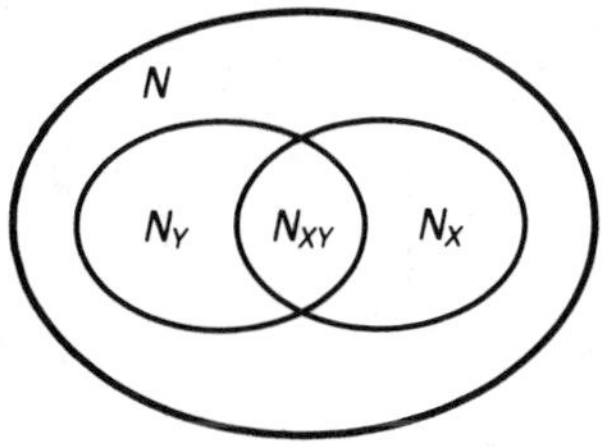

Figure 1.1 Venn diagram.

Example

The probability of throwing a 3 if two dice are tossed is (Eqn (1.7)):

$$P(3 + 3) = \frac{6}{36} + \frac{6}{36} - \frac{1}{36} = 0.306$$

The relevant Venn diagram is shown in Figure 1.2, where eleven of the thirty-six possible results contain a 3. This logic can, with some complication, be extended to more than two events.

Applying the procedures outlined in this section can be surprisingly difficult; indeed in many apparently simple cases the difficulty is sufficient to preclude successful calculation of probabilities altogether. An example illustrates this point.

Example

A standard pack of fifty-two cards is dealt to four players. What are the probabilities that

(a) a particular player has only red cards,
(b) any player has only red cards,
(c) each player has a complete suit,
(d) any player has a complete suit?

(a) This problem is solved in Section 1.2, Example (a)′. P (a particular player has only red cards) $= 1.638 \times 10^{-5}$.

(b) The fact that one player has only red cards is not mutually exclusive to any of the other three players having only red cards. Moreover, the fact that one player has only red cards does affect the probability of another player having only red cards.

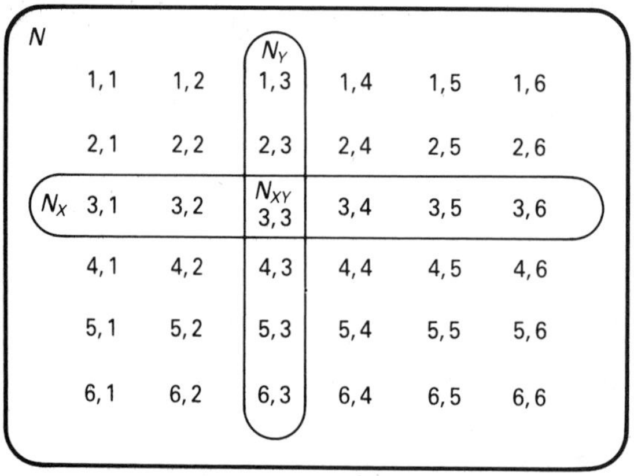

Figure 1.2 Venn diagram.

These events are neither mutually exclusive nor independent and the solution of this apparently innocuous problem involves an almost impossibly complicated calculation.

(c) Considering the players in order, there are four ways in which the first player can be dealt a complete suit. Since there are $\binom{52}{13}$ ways in which he can be dealt any hand, the probability of his having a complete suit is

$$P\text{ (first player has a complete suit)} = \frac{4.13!39!}{52!} = 6.299 \times 10^{-12}$$

Given that the first player has such a hand, and applying the same logic, with the appropriately reduced number of cards, to the second, third and fourth players,

$$P\text{ (second player has a complete suit and first player has a complete suit)}$$
$$= \frac{3.13!26!}{39!} = 3.693 \times 10^{-10}$$

$$P\text{(third player has a complete suit and first two players have a complete suit)}$$

$$= \frac{2.13!13!}{26!} = 1.923 \times 10^{-7}$$

$$P\text{(fourth player has a complete suit and first three players have a complete suit)}$$
$$= \frac{1.13!0!}{13!} = 1$$

Hence

$$P\text{ (each player has a complete suit)}$$
$$= 6.299 \times 3.693 \times 1.923 \times 10^{-29} = 4.473 \times 10^{-28}$$

(d) As in (b), the fact that one player has a complete suit is neither mutually exclusive to, nor independent of, any other player having a complete suit. The solution of this problem is almost impossibly complicated.

1.4 The binomial distribution

Suppose the probability of an event X occurring is p, and that of it not occurring is q:

$$P(X) = p, \quad P(\bar{X}) = q = (1 - p)$$

The probability of X happening r times in n experiments is easily determined. The result of a particular sequence of n trials, in which X occurs r times is

$$X, X, \bar{X}, X, \bar{X}, \ldots, \bar{X}$$

and the probability of this result is (Eqn (1.4))

$$ppqpq \ldots q = q^{n-r}p^r$$

Since there are $\binom{n}{r}$ such results, the probability of the event 'any result with r occurrences of X' is

$$P(r \text{ occurrences of } X) = \binom{n}{r} q^{n-r} p^r \tag{1.8}$$

From the binomial theorem (Appendix C, Section C2),

$$(q + p)^n = \sum_{r=0}^{n} \binom{n}{r} q^{n-r} p^r$$

Thus the probability of r occurrences of X in n experiments is equal to the corresponding coefficient in the binomial expansion of $(q + p)^n$.

Example

Consider the probabilities of throwing 0, 1, 2, 3, 4 fives when four dice are tossed. Here,

$$p = \frac{1}{6}, q = \frac{5}{6}, n = 4, r = 0, 1, 2, 3, 4$$

$$P(r = 0) = \binom{4}{0}\left(\frac{5}{6}\right)^4\left(\frac{1}{6}\right)^0 = 0.482$$

$$P(r = 1) = \binom{4}{1}\left(\frac{5}{6}\right)^3\left(\frac{1}{6}\right)^1 = 0.386$$

$$P(r = 2) = \binom{4}{2}\left(\frac{5}{6}\right)^2\left(\frac{1}{6}\right)^2 = 0.116$$

$$P(r = 3) = \binom{4}{3}\left(\frac{5}{6}\right)^1\left(\frac{1}{6}\right)^3 = 0.0154$$

$$P(r = 4) = \binom{4}{4}\left(\frac{5}{6}\right)^0\left(\frac{1}{6}\right)^4 = 0.000\,772$$

Since the events $r = 0, 1, 2, 3, 4$ are mutually exclusive, these probabilities are summed directly (Eqn (1.6)) to give the probability that r has any value 0, 1, ..., 4:

$$P(r \text{ any value}) = 0.482 + 0.386 + \cdots = 1$$

1.5 Variates, discrete variate distributions and probabilities

It is appropriate at this stage to examine and refine the symbols being used.

$P(X)$ is the probability of an event X, which, in the last example, is the number of 5s when four dice are tossed; one such event, X, is $(r = 0)$, another is $(r = 1)$, and so on.

A value x can be assigned to each event X and generally it seems natural, as in this case, to assign the numerical value of r to x, i.e. $x = 0$ to $r = 0$, $x = 1$ to $r = 1$, etc. This procedure, which defines a 'variate' or 'random variable', x, also covers cases where X does not take natural values (e.g. if a coin is tossed, the events $(X = \text{head})$, $(X = \text{tail})$ might have assigned to them $x = 1$, $x = 0$ respectively).

Thus, a variate x is defined as a number assigned to an experimental result, usually equal to that result in numerical cases.

The values of x constitute a finite population of N numbers, the distribution of which is characterized by some commonly accepted, though rather coarse, properties. Each value of x occurs $F(x)$ times in the population.

Example

Consider a set of four dice thrown 100 times, and let x represent the number of 5s resulting from each throw. The results are tabulated in Figure 1.3(a), where $F(x)$ represents the frequency of occurrence of the discrete variate x in the 100 trials

x	$F(x)$	$xF(x)$	$x^2F(x)$
0	48	0	0
1	39	39	39
2	11	22	44
3	2	6	18
4	0	0	0
	$\Sigma F(x) = 100$	$\Sigma xF(x) = 67$	$\Sigma x^2F(x) = 101$

(a)

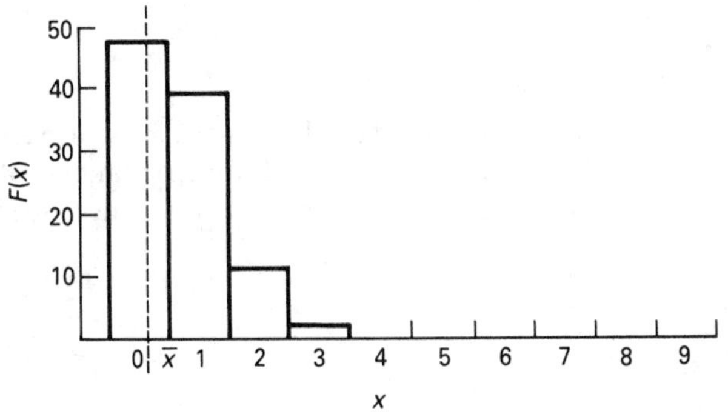

(b)

Figure 1.3 Data table and histogram.

performed. Clearly,

$$\sum F(x) = N = 100$$

The relationship of $F(x)$ with x is conveniently illustrated in the form of a 'histogram', Figure 1.3(b), in which the discrete values of x are represented by sections of (not points on) the horizontal axis.

The mean value, or 'first moment', of x is defined:

$$\bar{x} = \frac{\sum xF(x)}{\sum F(x)} = \frac{\sum xF(x)}{N} \tag{1.9}$$

Since x is unlikely to equal one of the discrete values of x, it can be represented on the histogram only in an informal manner. The 'shape' of the histogram is described very roughly by the 'central moments' of x.

The 'first central moment' is zero:

$$\overline{(x - \bar{x})} = \frac{\sum (x - \bar{x})F(x)}{\sum F(x)} = \frac{\sum xF(x)}{\sum F(x)} - \frac{\bar{x}\sum F(x)}{\sum F(x)}$$
$$= 0$$

The 'second central moment', or 'variance' describes the spread of the distribution about $\bar{x}$:

$$\sigma_x^2 = \overline{(x - \bar{x})^2} = \frac{\sum (x - \bar{x})^2 F(x)}{\sum F(x)} \tag{1.10}$$

This can be rewritten as the 'mean of the squares minus the square of the means':

$$\sigma_x^2 = \frac{\sum x^2 F(x)}{\sum F(x)} - \frac{2\bar{x}\sum xF(x)}{\sum F(x)} + \frac{\bar{x}^2 \sum F(x)}{\sum F(x)}$$
$$= \overline{(x^2)} - (\bar{x})^2 \tag{1.11}$$

The square root of the variance, or 'standard derivation', σ_x, is also a measure of the 'spread' of x about $\bar{x}$, and has the, sometimes worthwhile, advantage of having the same dimension as x. The skewness of the histogram about $\bar{x}$ is described by the 'third central moment':

$$\overline{(x - \bar{x})^3} = \frac{\sum (x - \bar{x})^3 F(x)}{\sum F(x)} \tag{1.12}$$

If this is positive, a 'tail' to the right of $\bar{x}$ is indicated; if negative, a tail to the left; and if zero, symmetry about $\bar{x}$.

Higher order moments are not of much practical interest.

Example

Returning to the last example concerning the number of 5s occurring when four dice are tossed 100 times (Figure 1.3), the mean is calculated (Eqn (1.9)):

$$\bar{x} = \frac{67}{100} = 0.670 \text{ (5s)}$$

This is indicated informally on the histogram of Figure 1.3(a). The variance, σ_x^2, a measure of the 'spread' around $\bar{x}$, is (Eqn (1.10) or (1.11))

$$\sigma_x^2 = \frac{101}{100} - (0.670)^2 = 0.571 \text{ (5s)}^2$$

The third central moment (Eqn (1.12)), being positive, indicates a 'tail' to the right of $\bar{x}$, as shown in Figure 1.3(b):

$$\overline{(x - \bar{x})^3} = 0.670 \text{ (5s)}^3$$

The formulae given above can be used with any set of data – including those from an apparently continuous variate – to yield distributions, histograms, means, and variances. The data is simply divided into non-overlapping groups, usually between seven and fifteen in number, the discrete variate, x, assigned to the groups, and the formulae applied.

Example

The weights (kg) of fifty items in a consignment of goods are 10.1 11.5 12.2 12.5 13.0 13.2 13.9 14.0 14.2 14.3 14.3 14.4 14.7 15.1 15.3 15.4 15.4 15.9 16.1 16.4 16.4 16.7 16.9 16.9 16.9 17.0 17.2 17.4 17.7 17.7 17.9 17.9 18.0 18.1 18.2 18.3 18.4 18.5 18.5 19.2 19.3 19.5 19.6 20.0 20.4 21.1 21.1 21.6 21.8 23.8. The data is divided into seven non-overlapping groups and the variate x is defined as the 'central' value of each group; $F(x)$, $xF(x)$ and $x^2F(x)$ are tabulated and a histogram constructed, as shown in Figure 1.4(a, b).

The mean and variance are readily calculated (Eqns (1.9) and (1.11)):

$$\bar{x} = \frac{\sum xF(x)}{\sum F(x)} = \frac{848}{50} = 16.960 \text{ (kg)}$$

$$\sigma_n^2 = \frac{\sum x^2F(x)}{\sum F(x)} - (\bar{x})^2 = \frac{14\,754}{50} - (16.960)^2 = 7.438 \text{ (kg}^2\text{)}$$

$$\sigma_x = \sqrt{7.438} = 2.727 \text{ (kg)}$$

Recalling the definition of Eqn (1.1), the probability of the event, X, that the discrete variate x occurs when one trial is conducted is found by considering the limiting values of the formula, $N \to \infty$:

$$P(X) = P_d(x) = \lim_{N\to\infty}\left(\frac{F(x)}{N}\right) = \lim_{N\to\infty}\left(\frac{F(x)}{\sum F(x)}\right) \tag{1.13}$$

where $P_d(x)$ is the 'probability function' of the discrete variable, x. The mean, or 'expected value' of x, usually written $\varepsilon(x)$, and its variance, σ_x^2, in terms of $P_d(x)$, are easily developed:

$$\varepsilon(x) = \lim_{N\to\infty}(\bar{x}) = \lim_{N\to\infty}\left(\frac{\sum xF(x)}{\sum F(x)}\right)$$

$$= \sum xP_d(x) \tag{1.14}$$

$$\sigma_x^2 = \varepsilon(x - \varepsilon(x))^2 = \varepsilon(x^2) - (\varepsilon(x))^2 \tag{1.15}$$

In practice, the value of the probability function, $P_d(x)$ can be calculated using the definition of Eqn (1.2) which concerns the probability of the event X which corresponds to x, namely $P(X)$.

Group	Group limits	x	$F(x)$	$xF(x)$	$x^2F(x)$
1	10.00–11.99	11	2	22	242
2	12.00–13.99	13	5	65	845
3	14.00–15.99	15	11	165	2475
4	16.00–17.99	17	14	238	4046
5	18.00–19.99	19	11	209	3971
6	20.00–21.99	21	6	126	2646
7	22.00–23.99	23	1	23	529
			50	848	14754

(a)

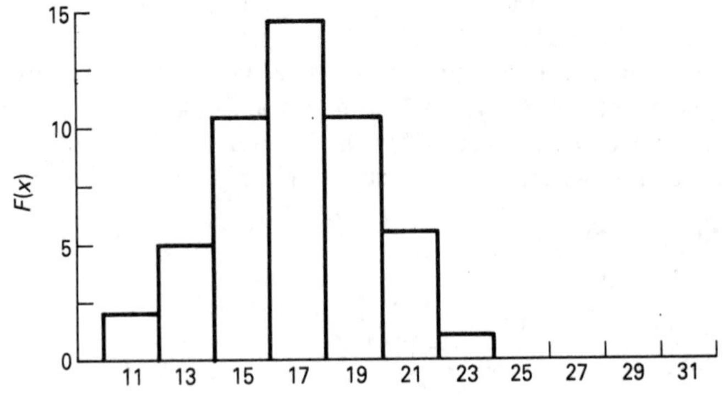

(b)

Figure 1.4 Data table and histogram.

Thus, in the case of the binomial distribution, for example (Eqn (1.8)),

$$P_d(x) = P(r = x) = \binom{n}{x} q^{n-x} p^x \tag{1.16}$$

(where n, p and q are as defined in Section 1.4). It is worth noting that, for the binomial distribution,

$$\varepsilon(x) = np \tag{1.17}$$

$$\sigma_x^2 = npq \text{ or } \sigma_x = \sqrt{npq} \tag{1.18}$$

Example

Consider the example concerning the number of 5s occurring when four dice are tossed. The table and histogram of Figure 1.3 illustrate the distribution of the variate x where $\sum F(x) = N = 100$.

$P_d(x)$ concerns the variate x as $N \to \infty$; it is most readily calculated using Eqn (1.16) (based on Eqn (1.2)):

$$P_d(0) = 0.482, \quad P_d(1) = 0.385, \quad P_d(2) = 0.116,$$

$$P_d(3) = 0.0154, \quad P_d(4) = 0.000\,772$$

Also (Eqns (1.17) and (1.18)),

$$\varepsilon(x) = 4 \times \frac{1}{6} = 0.667; \quad \sigma_x^2 = 4 \times \frac{1}{6} \times \frac{5}{6} = 0.556; \quad \sigma_x = 0.745$$

A histogram of $P_d(x)$, on which $\varepsilon(x)$ and σ_x are informally represented, is shown in Figure 1.5.

The histogram of Figure 1.3(b) would approach the shape of that in Figure 1.5 if $N = \sum F(x)$ were increased indefinitely.

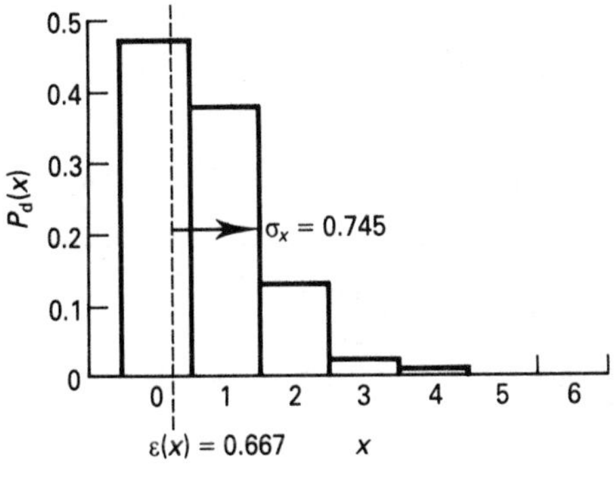

Figure 1.5 Histogram of $P_d(x)$.

A similar exercise could be undertaken, with somewhat more difficulty, for the variate illustrated in Figure 1.4.

1.6 Continuous variable probability functions

Having established the meanings of probability, and probability functions of discrete variables, it is now appropriate to consider probability functions of continuous variables.

Assigning a discrete value x to the event X (Section 1.5) yields the probability function $P_d(x)$ (Eqn (1.13)). In contrast, a probability function of the continuous variable, ξ, can be defined, again after assigning a value x to X, in such a way that x is also a continuous variable:

$$P(\xi) = \text{Prob}(x \leqslant \xi) \qquad (1.19)$$

where ξ is a real number, and the right-hand side of Eqn (1.19) corresponds to the definition of Eqn (1.1), the event X being that $x \leqslant \xi$. This proves to be a useful and flexible representation of probability.

A typical probability function is illustrated in Figure 1.6. The main properties of $P(\xi)$ are

(i) $P(\xi) \geqslant 0, \quad -\infty \leqslant \xi \leqslant +\infty,$

(ii) $P(-\infty) = 0, \quad P(+\infty) = 1,$

(iii) $\dfrac{dP(\xi)}{d\xi} \geqslant 0, \quad -\infty \leqslant \xi \leqslant +\infty,$

(iv) It is easily shown that

$$\text{Prob}(\xi_1 < x \leqslant \xi_2) = P(\xi_2) - P(\xi_1).$$

Simple examples illustrate these points.

Examples

(a) If x is the (value assigned to the) result of tossing a die, $P(\xi)$ is as shown in

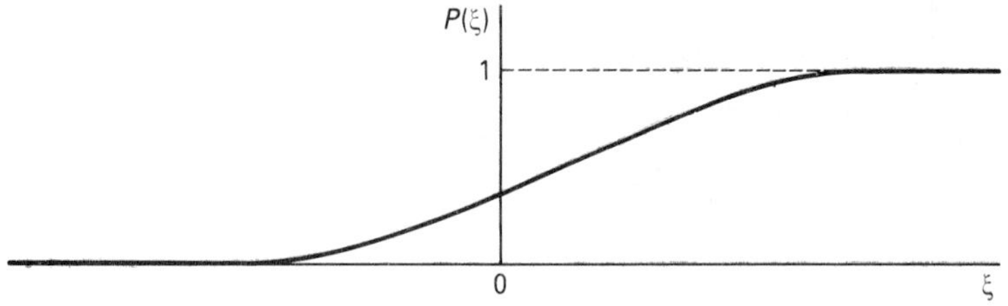

Figure 1.6 Typical probability function.

Figure 1.7(a), where it can be seen that

$$\text{Prob}(x \leqslant 0) = P(0) = 0$$
$$\text{Prob}(x \leqslant 9) = P(9) = 1$$
$$\text{Prob}(x \leqslant 2.5) = P(2.5) = 2/6 = 0.333$$
$$\text{Prob}(x \leqslant -\infty) = P(-\infty) = 0$$
$$\text{Prob}(0.3 < x < 3.6) = P(3.6) - P(0.3) = 3/6 = 0.5$$

(b) If x is a random variable evenly distributed between -1 and $+3$, $P(\xi)$ is as shown in Figure 1.7(b). In this case,

$$\text{Prob}(x \leqslant -1) = P(-1) = 0$$
$$\text{Prob}(x \leqslant -\infty) = P(-\infty) = 0$$
$$\text{Prob}(x \leqslant 0) = P(0) = 0.25$$

$$P(\xi) = \begin{cases} 0, \xi < +1 \\ \frac{1}{6}, 1 \leqslant \xi < 2 \\ \frac{2}{6}, 2 \leqslant \xi < 3 \\ \vdots \\ 1, \xi \geqslant 6 \end{cases}$$

(a)

$$P(\xi) = \begin{cases} 0, \xi < -1 \\ (\frac{\xi}{4} + 0.25), -1 \leqslant \xi < +3 \\ 1, \xi \geqslant +3 \end{cases}$$

(b)

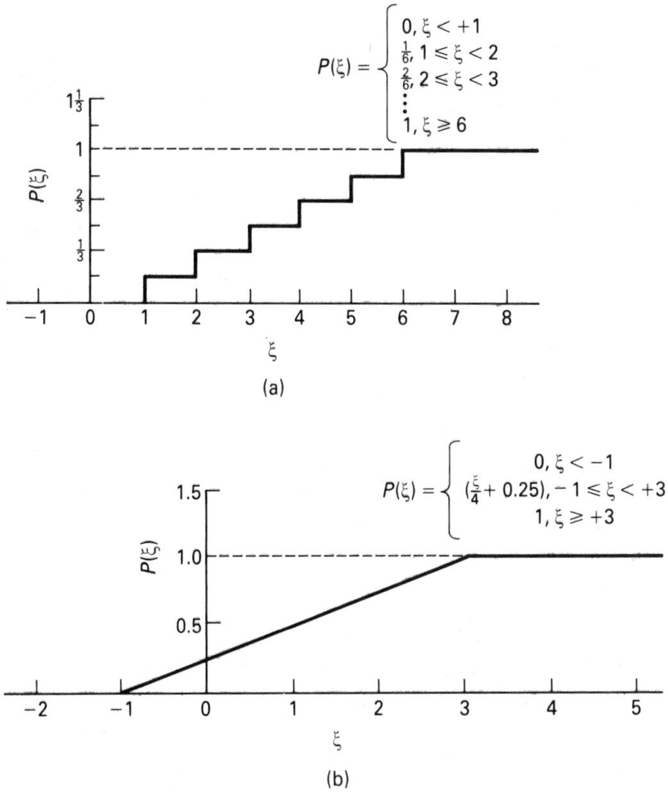

Figure 1.7 Probability function.

$$\text{Prob}(x \leqslant 5.6) = P(5.6) = 1$$

$$\text{Prob}(2.5 < x \leqslant 3.9) = P(3.9) - P(2.5) = 1 - 0.875 = 0.125$$

The precise relationship between $P(\xi)$ (Eqn (1.19)) and $P_d(x)$ (Eqn (1.13), in which x adopts discrete values), can be expressed, with some complication, in terms of the Stieltjes integral (Appendix E, Section E3).

It turns out that the straightforward probability function, $P(\xi)$, is less useful than its derivative, the probability density function, described in Section 1.7.

1.7 Probability density functions

The probability density function, $p(\xi)$, which describes the distribution of a random variable x (using the terminology explained in Section 1.5), is defined:

$$p(\xi) = \frac{\mathrm{d}P(\xi)}{\mathrm{d}\xi} \qquad (1.20)$$

where $P(\xi)$ is defined by Eqn (1.19).

The main properties of $p(\xi)$ are

(i) $p(\xi) \geqslant 0, \quad -\infty \leqslant \xi \leqslant +\infty$

(ii) $p(-\infty) = p(+\infty) = 0$

(iii) $$\text{Prob}(x \leqslant \xi_1) = \int_{-\infty}^{\xi_1} p(\xi)\,\mathrm{d}\xi \qquad (1.21)$$

(iv) $$\text{Prob}(\xi_1 < x \leqslant \xi_2) = \int_{\xi_1}^{\xi_2} p(\xi)\,\mathrm{d}\xi \qquad (1.22)$$

(v) $$p(\xi) = \lim_{\delta\xi \to 0}\left[\frac{\text{Prob}(\xi < x \leqslant \xi + \delta\xi)}{\delta\xi}\right]$$

(vi) $$\text{Prob}(-\infty < x \leqslant +\infty) = \int_{-\infty}^{+\infty} p(\xi)\,\mathrm{d}\xi = 1 \qquad (1.23)$$

Examples

(a) Where x is the result of tossing a die, $p(\xi)$ is as shown in Figure 1.8 (cf. Figure 1.7(a)). Then $p(\xi)$ is found by differentiating $P(\xi)$ WRT ξ, which in this case results in Dirac δ functions (Appendix D, Section D1):

From Figure 1.8, and the properties of $p(\xi)$,

$$\text{Prob}(x \leqslant 0) = \int_{-\infty}^{0} \frac{1}{6}\sum_{i=1}^{6} \delta(\xi - i)\,\mathrm{d}\xi = 0$$

$$\text{Prob}(x \leqslant 3.5) = \int_{-\infty}^{3.5} \frac{1}{6} \sum_{i=1}^{6} \delta(\xi - i)\,\mathrm{d}\xi = \frac{3}{6}$$

(b) If x is a random variable evenly distributed between -1 and $+3$, $p(\xi)$ is as shown in Figure 1.9 (cf. Figure 1.7(b)). From Figure 1.9 and Eqns (1.21) and (1.22),

$$\text{Prob}(x \leqslant 0) = \int_{-\infty}^{-1} 0\,\mathrm{d}\xi + \int_{-1}^{0} 0.25\,\mathrm{d}\xi = 0.25$$

$$\text{Prob}(1 < x \leqslant 5) = \int_{1}^{3} 0.25\,\mathrm{d}\xi + \int_{3}^{5} 0\,\mathrm{d}\xi = 0.5$$

(c) Consider a random variable, x, with the exponential probability density function shown in Figure 1.10. C can be evaluated using Eqn (1.23):

$$\int_{0}^{\infty} C\,\mathrm{e}^{-\xi}\,\mathrm{d}\xi = C = 1$$

Thus

$$p(\xi) = \begin{cases} 0, & \xi < 0 \\ \mathrm{e}^{-\xi}, & \xi \geqslant 0 \end{cases}$$

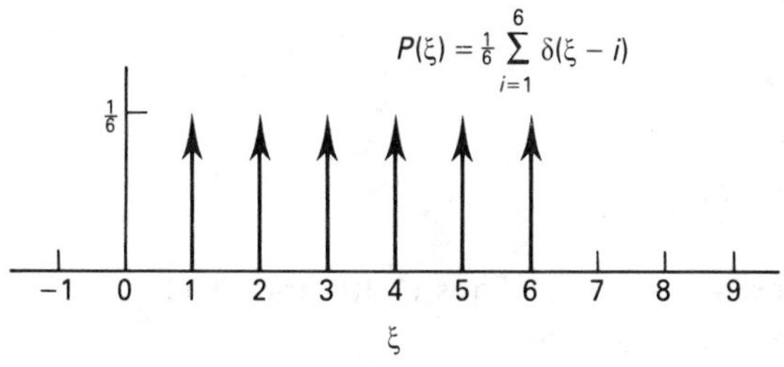

Figure 1.8 Probability density function.

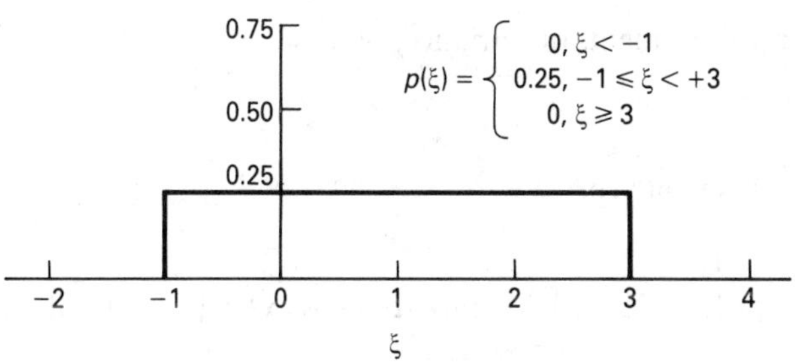

Figure 1.9 Probability density function.

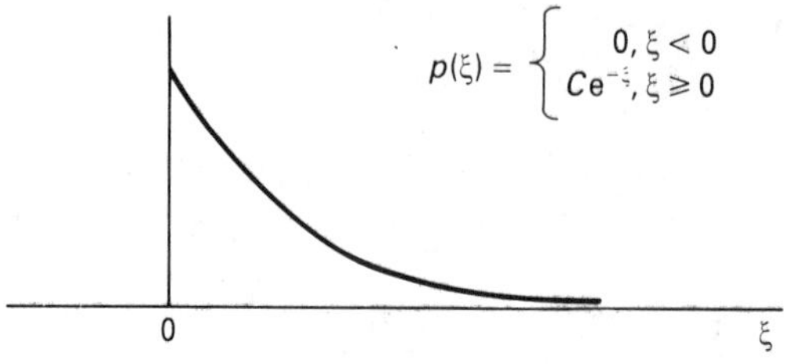

Figure 1.10 Probability density function.

and

$$\text{Prob}(x \leqslant 0.5) = \int_0^{0.5} e^{-\xi}\, d\xi = 0.393$$

$$\text{Prob}(-1 < x \leqslant 0.25) = \int_0^{0.25} e^{-\xi}\, d\xi = 0.221$$

1.8 Properties of random variables

The main properties of random variables can now be defined in terms of their probability density functions.

1.8.1 The mean, or expected value (cf. Eqn (1.9)

The mean or expected value of x is

$$\varepsilon(x) = \int_{-\infty}^{+\infty} \xi p(\xi)\, d\xi \tag{1.24}$$

$\varepsilon(x)$ is also the 'first moment' of x.

1.8.2 Second moment and variance (cf. Eqns (1.10) and (1.11))

The second moment of x is

$$\varepsilon(x^2) = \int_{-\infty}^{+\infty} \xi^2 p(\xi)\, d\xi$$

The second central moment, or variance, of x is

$$\begin{aligned}\sigma_x^2 &= \varepsilon(x - \varepsilon(x))^2 \\ &= \int_{-\infty}^{+\infty} (\xi - \varepsilon(x))^2 p(\xi)\, d\xi \\ &= \int_{-\infty}^{+\infty} \xi^2 p(\xi)\, d\xi - 2\varepsilon(x)\int_{-\infty}^{+\infty} \xi p(\xi)\, d\xi + (\varepsilon(x))^2 \int_{-\infty}^{+\infty} p(\xi)\, d\xi \\ &= \varepsilon(x^2) - (\varepsilon(x))^2 \end{aligned} \tag{1.25}$$

The variance, σ_x^2, and its square root, the standard deviation, are measures of the spread of x about $\varepsilon(x)$.

1.8.3 Other moments

The nth moment and central moment are defined:

$$\left.\begin{aligned}\varepsilon(x^n) &= \int_{-\infty}^{+\infty} \xi^n p(\xi)\,\mathrm{d}\xi \\ \varepsilon(x-\varepsilon(x))^n &= \int_{-\infty}^{+\infty} (\xi-\varepsilon(x))^n p(\xi)\,\mathrm{d}\xi\end{aligned}\right\} \tag{1.26}$$

The third central moment provides some indication of the skewness of a distribution about $\varepsilon(x)$; a positive value indicates a 'tail' to the right, a negative value a tail to the left, and zero, symmetry about $\varepsilon(x)$.

Higher moments are not of much practical interest.

1.8.4 Expected value of a function of a random variable

The mean of a function of $x, f(x)$, is

$$\varepsilon(f(x)) = \int_{-\infty}^{+\infty} f(\xi)p(\xi)\,\mathrm{d}\xi \tag{1.27}$$

Example

Consider a random variable, x, with PDF $p(\xi) = \mathrm{e}^{-\xi}, \xi \geqslant 0$, as shown in Figure 1.10.

The mean (Eqn (1.24)) is

$$\varepsilon(x) = \int_0^\infty \xi\,\mathrm{e}^{-\xi}\,\mathrm{d}\xi = 1$$

The variance (Eqn (1.25)) is

$$\sigma_x^2 = \int_0^\infty \xi^2\,\mathrm{e}^{-\xi}\,\mathrm{d}\xi - (1)^2 = 1$$

The third central moment (Eqn (1.26), $n = 3$) is

$$\varepsilon(x-\varepsilon(x))^3 = \int_0^\infty (\xi-1)^3\,\mathrm{e}^{-\xi}\,\mathrm{d}\xi = 2$$

Being positive, this indicates a skewed PDF, with a 'tail' to the right of $\varepsilon(x)$.

The means and variances of functions of x may also be found (Eqn (1.27)).

Consider $y = e^{-x}$:

$$\varepsilon(y) = \int_0^\infty e^{-\xi} e^{-\xi}\, d\xi = 0.5$$

$$\varepsilon(y^2) = \int_0^\infty e^{-2\xi} e^{-\xi}\, d\xi = 0.333$$

Hence,

$$\sigma_y^2 = 0.333 - (0.5)^2 = 0.0833; \quad \sigma_y = 0.289$$

1.9 Some common random variables

1.9.1 An evenly distributed random variable

Consider a random variable x, evenly distributed between a, b, as indicated in Figure 1.11. The mean, variance and standard deviation of x are easily found (Eqns (1.24), (1.25)):

$$\varepsilon(x) = \int_a^b \frac{\xi}{b-a}\, d\xi = \frac{a+b}{2}$$

$$\varepsilon(x^2) = \int_a^b \frac{\xi^2}{(b-a)}\, d\xi = \frac{1}{3}(a^2 + ab + b^2)$$

$$\sigma_x^2 = \frac{1}{3}(a^2 + ab + b^2) - \frac{(a+b)^2}{4} = \frac{(b-a)^2}{12}$$

$$\sigma_x = \frac{(b-a)}{\sqrt{12}}$$

If $a = 1, b = 3$, and if x represents the radius (m) of a circle, the mean and variance of its area are readily found (Eqn (1.27)):

$$\varepsilon(A) = \varepsilon(\pi x^2) = \int_1^3 \pi\xi^2 \frac{1}{2}\, d\xi = 13.615\ (\text{m}^2)$$

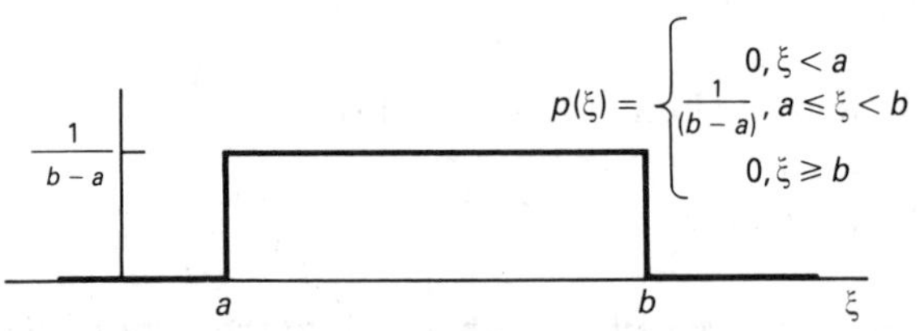

Figure 1.11 PDF of evenly distributed random variable.

$$\varepsilon(A^2) = \varepsilon(\pi^2 x^4) = \int_1^3 \pi^2 \xi^4 \frac{1}{2} \mathrm{d}\xi = 238.906 \text{ (m}^4\text{)}$$

Hence (Eqn (1.25)),

$$\sigma_A^2 = \varepsilon(A^2) - (\varepsilon(A))^2 = 53.538\,(m^4)$$

$$\sigma_A = 7.317\ (m^2)$$

1.9.2 The Gaussian, or normal probability density function

A random variable, x, with a Gaussian or 'normal' distribution has the PDF (Figure 1.12(a))

$$p(\xi) = \frac{1}{\sigma_x\sqrt{2\pi}} \exp\left[-\frac{(\xi - \varepsilon(x))^2}{2\sigma_x^2}\right] \tag{1.28}$$

This can be derived from Eqn (1.16), the binominal distribution, for the limiting case, $n \to \infty$, using the Stieltjes integral relationship (Appendix E, Section E2) to bridge the gap between the discrete and continuous variables, x, in the two equations. The Gaussian distribution is common in the natural world for the reasons outlined in Section 1.13.

Since the Gaussian distribution is common, and since probabilities are calculated using Eqn (1.28) only with some difficulty, a 'standard normal form' variate, q, with zero mean and unity standard deviation, is useful. The PDF of q is illustrated in Figure 1.12(b):

$$q = \frac{x - \varepsilon(x)}{\sigma_x}; \quad u = \frac{\xi - \varepsilon(x)}{\sigma_x}$$

A table of areas under $p(u)$ is shown in Appendix C, Figure C3.2.

Most problems involving the calculation of Gaussian probabilities can be reduced to equivalent problems involving this standard normal form PDF, $p(u)$.

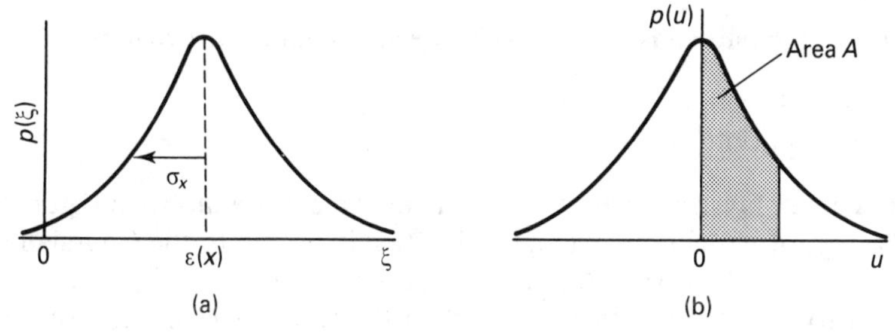

Figure 1.12 Gaussian probability density function.

Example

The thickness, x, of metal strip leaving a mill, which can reasonably be expected to be normally distributed, has a mean of 5.0 (mm) and standard deviation of 0.9 (mm).

The standard normal form variate, q, is defined so that $\varepsilon(q) = 0$, $\sigma_q = 1$:

$$q = \frac{x - 5}{0.9}$$

Consider the probability that the metal thickness at some point lies between 4.3 and 5.6 mm:

$$u_{4.3} = \frac{4.3 - 5.0}{0.9} = -0.78$$

$$u_{5.6} = \frac{5.6 - 5.0}{0.9} = 0.67$$

From the table (Appendix C, Figure C3.2), the areas under the curve which represent the corresponding probabilities are

$A_{4.3} = 0.2177$ (left-hand tail)

$A_{5.6} = 0.2514$ (right-hand tail)

Thus the probability that the thickness lies in the specified range is $(1 - 0.2177 - 0.2514)$, or 0.53.

Figure 1.12 shows the 'clustering' of the Gaussian PDF around the mean value, a feature summarized in terms of the 'σ-limits' of the distribution:

68% of the area lies within the σ limits, $\varepsilon(x) \pm \sigma_x$

95% of the area lies within the 2σ limits, $\varepsilon(x) \pm 2\sigma_x$

99.7% of the area lies within the 3σ limits, $\varepsilon(x) \pm 3\sigma_x$

For the steel strip in the above example:

68% of the steel has a thickness between 4.10 mm and 5.90 mm,

95% of the steel has a thickness between 3.20 mm and 6.80 mm,

99.7% of the steel has a thickness between 2.30 mm and 7.70 mm.

1.9.3 Other distributions

Some PDFs commonly used to model natural phenomena are shown in Figure 1.13.

1. The lognormal variate, x (Figure 1.13(a)) is related to the Gaussian variate, y, by $x = e^y$, where y is Gaussian. All values of x are positive.

If a variate is the product, rather than the sum (Section 1.13), of a number of independent variates, its distribution is lognormal.

2. The gamma distribution (Figure 1.13(b)) is commonly associated with the waiting times between events in random processes; $p(\xi)$ involves the gamma function

$$\Gamma(\alpha) = \int_0^{\infty} q^{\alpha-1}\, e^{-q}\, dq$$

This has the property that $\Gamma(\alpha + 1) = \alpha\Gamma(\alpha)$ and is the continuous equivalent of the factorial function of an integer:

$$(\alpha + 1)! = (\alpha + 1)\alpha!$$

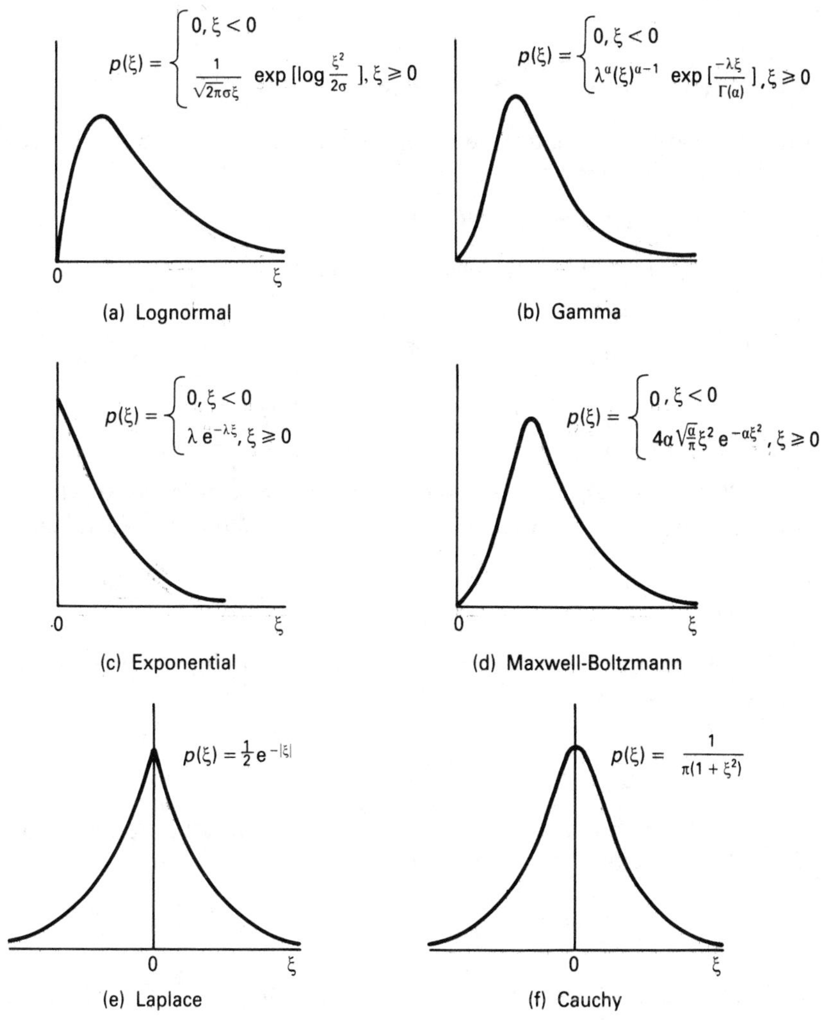

Figure 1.13 Some well-known PDFs.

If $\alpha = 1$, $p(\xi)$ becomes the exponential function of Figure 1.13(c), while if α is an integer, $p(\xi)$ becomes the Erlang function, associated with queueing theory:

$$p(\xi) = \begin{cases} 0, & \xi < 0 \\ \dfrac{\lambda\,\mathrm{e}^{-\lambda\xi}(\lambda\xi)^{\alpha-1}}{(\alpha-1)!}, & \xi \geqslant 0 \end{cases}$$

3. The Maxwell–Boltzmann distribution (Figure 1.13(d)) describes the velocity of gas molecules as a function of their mass and temperature.

4. The symmetrical Laplace and Cauchy distributions, illustrated in Figure 1.13(e, f), are also common in scientific literature.

1.10 Conditional probability functions

If a random variable, x, is dependent on another random variable, y, a conditional probability function can be defined (cf. Section 1.3.2):

$$\text{Prob}(x \leqslant \xi; \quad y = \chi) = P(\xi|\chi)$$

The corresponding probability density function is

$$p(\xi|\chi) = \frac{\mathrm{d}P(\xi|\chi)}{\mathrm{d}\xi}$$

$$= \lim_{\delta\xi \to 0}\left[\frac{\text{Prob}(\xi < x \leqslant \xi + \delta\xi; y = \chi)}{\delta\xi}\right]$$

Clearly, $P(\xi|\chi)$, $p(\xi|\chi)$ have essentially the same properties as the unconditional functions $P(\xi)$, $p(\xi)$ (Sections 1.7 and 1.8) and require no special explanation.

Example

Consider a standard pack of fifty-two cards. The probability of selecting a $1, 2, \ldots, 13$ (king) at random is described by the unconditional PDF

$$p(\xi) = \frac{4}{52}\sum_{i=1}^{13} \delta(\xi - i)$$

If, however, a card is removed from the pack – say a 2 – then the probability of selecting a $1, 2, \ldots, 13$ is described by the conditional PDF

$$p(\xi|2) = \frac{4}{51}\sum_{i=1}^{13} \delta(\xi - i) - \frac{1}{51}\delta(\xi - 2)$$

These functions are illustrated in Figure 1.14(a, b). Clearly, the PDF of Figure 1.14(b) is conditional upon a 2 having been chosen first; had another face value been selected, this function would have been different.

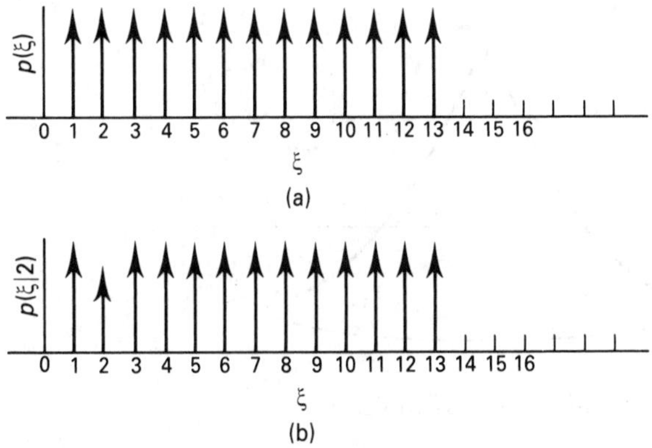

Figure 1.14 Unconditional and conditional PDFs.

1.11 Joint probability functions

If m random variables, $x_1, x_2, \ldots, x_m$, are considered together, a joint probability function can be defined (cf. Eqn (1.19)):

$$\text{Prob}(x_1 \leqslant \xi_1; x_2 \leqslant \xi_2; \ldots; x_m \leqslant \xi_m) = P(\xi_1, \xi_2, \ldots, \xi_m)$$

This is the probability that $x_1 \leqslant \xi_1$ *and* $x_2 \leqslant \xi_2, x_3 \leqslant \xi_3$, etc.

The corresponding joint PDF is (cf. Eqn (1.20))

$$p(\xi_1, \xi_2, \ldots, \xi_m) = \frac{\partial^m P(\xi_1, \xi_2, \ldots, \xi_m)}{\partial\xi_1 . \partial\xi_2 . \ldots . \partial\xi_m} \tag{1.29}$$

The properties of this function parallel those of the single variable PDF:

(i) $p(\xi_1, \xi_2, \ldots, \xi_m) \geqslant 0$ for all $\xi_1, \xi_2, \ldots, \xi_m$

(ii) $p(-\infty, -\infty, \ldots, -\infty) = p(+\infty, +\infty, \ldots, +\infty) = 0$

(iii) $\text{Prob}(\xi_{11} < x_1 \leqslant \xi_{12}, \xi_{21} < x_2 \leqslant \xi_{22}, \ldots, \xi_{m1} < x_m \leqslant \xi_{m2})$

$$= \int_{\xi_{11}}^{\xi_{12}} \int_{\xi_{21}}^{\xi_{22}} \cdots \int_{\xi_{m1}}^{\xi_{m2}} p(\xi_1, \xi_2, \ldots, \xi_m)\, d\xi_1\, d\xi_2 \ldots d\xi_m$$

(iv) $\displaystyle\int_{-\infty}^{+\infty} \int_{-\infty}^{+\infty} \cdots \int_{-\infty}^{+\infty} p(\xi_1, \xi_2, \ldots, \xi_m)\, d\xi_1\, d\xi_2 \ldots d\xi_m = 1$

(v) The PDF of any one variable, x_i, regardless of the others, is

$$p_i(\xi_i) = \int_{-\infty}^{+\infty} \int_{-\infty}^{+\infty} \cdots \int_{-\infty}^{+\infty} p(\xi_1, \xi_2, \ldots, \xi_m)\, d\xi_1\, d\xi_2 \ldots d\xi_{i-1}\, d\xi_{i+1} \ldots d\xi_m \tag{1.30}$$

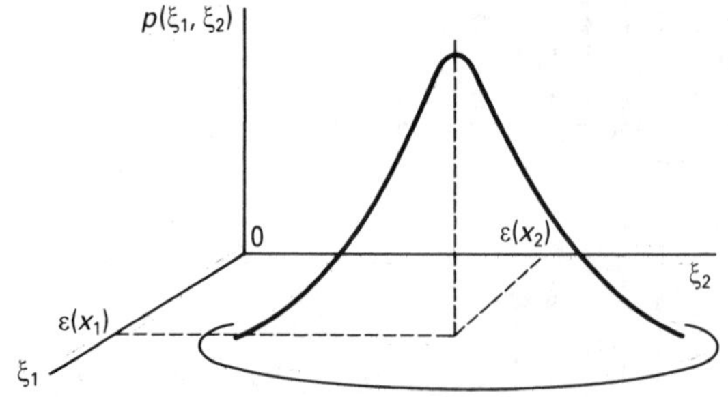

Figure 1.15 Two-dimensional Gaussian PDF.

(vi) The probability that the single variable x_i lies in the range $\xi_{i1} < x_i \leqslant \xi_{i2}$ is

$$\text{Prob}(\xi_{i1} < x_i \leqslant \xi_{i2}) = \int_{-\infty}^{+\infty} \int_{-\infty}^{+\infty} \cdots \int_{\xi_{i1}}^{\xi_{i2}} \cdots \int_{-\infty}^{+\infty} p(\xi_1, \xi_2, \ldots, \xi_m)\, \mathrm{d}\xi_1\, \mathrm{d}\xi_2 \ldots \mathrm{d}\xi_m \tag{1.31}$$

(vii) The expected or mean value of a function of the variables $x_1, x_2, \ldots, x_m$ is

$$\varepsilon[f(x_1, x_2, \ldots, x_m)] = \int_{-\infty}^{+\infty} \int_{-\infty}^{+\infty} \cdots \int_{-\infty}^{+\infty} f(\xi_1, \xi_2, \ldots, \xi_m) \times p(\xi_1, \xi_2, \ldots, \xi_m)\, \mathrm{d}\xi_1\, \mathrm{d}\xi_2 \ldots \mathrm{d}\xi_m \tag{1.32}$$

(viii) The mean and variance of any one variable, x_i, regardless of the others, are

$$\varepsilon(x_i) = \int_{-\infty}^{+\infty} \int_{-\infty}^{+\infty} \cdots \int_{-\infty}^{+\infty} \xi_i p(\xi_1, \xi_2, \ldots, \xi_m)\, \mathrm{d}\xi_1\, \mathrm{d}\xi_2 \ldots \mathrm{d}\xi_m$$

$$\sigma_{x_i}^2 = \int_{-\infty}^{+\infty} \int_{-\infty}^{+\infty} \cdots \int_{-\infty}^{+\infty} (\xi_i - \varepsilon(x_i))^2 p(\xi_1, \xi_2, \ldots, \xi_m)\, \mathrm{d}\xi_1\, \mathrm{d}\xi_2 \ldots \mathrm{d}\xi_m$$

(ix) If the variables $x_1, x_2, \ldots, x_m$ are independent, it is not difficult to show that

$$p(\xi_1, \xi_2, \ldots, \xi_m) = p(\xi_1)p(\xi_2) \ldots p(\xi_m) \tag{1.33}$$

The converse is also true; if Eqn (1.33) holds, then $x_1, x_2, \ldots, x_m$ are independent.

Example

Consider the multivariate Gaussian joint PDF

$$p(\xi_1, \xi_2, \ldots, \xi_m) = p(\boldsymbol{\xi}) = \frac{1}{(2\pi)^{m/2}|\mathbf{R}|^{1/2}} \exp[-\frac{1}{2}(\boldsymbol{\xi} - \varepsilon(\mathbf{x}))^T \mathbf{R}^{-1}(\boldsymbol{\xi} - \varepsilon(\mathbf{x})] \tag{1.34}$$

where, using vector format for convenience,

$$\mathbf{x} = (x_1 \ x_2 \ \dots \ x_m)^T,$$

$$\boldsymbol{\xi} = (\xi_1 \ \xi_2 \ \dots \ \xi_m),$$

$$\mathbf{R} = \ \varepsilon[(\mathbf{x} - \varepsilon(\mathbf{x}))(\mathbf{x} - \varepsilon(\mathbf{x}))^T],$$

$|\mathbf{R}|$ denotes the determinant of $\mathbf{R}$.

$p(\boldsymbol{\xi})$ is shown for $m = 2$ in Figure 1.15. If $x_1, x_2, \dots, x_m$ are independent, $\mathbf{R}$ is diagonal, and it is easy to show that Eqn (1.33) holds true for Eqn (1.34).

1.12 Functions of a random variable

Consider a random variable x, PDF $p_x(\xi)$, and another (dependent) random variable y, a function of x:

$$y = f(x)$$

Given $f(x)$ and $p_x(\xi)$, it is of interest to find the PDF of y, $p_y(\chi)$. This turns out to be possible only if

(i) y exists for every possible value of x,
(ii) the values $y = \pm\infty$ have zero probability,
(iii) the inverse function $x = f^{-1}(y)$ has a finite number of roots.

(Strictly, $f(x)$ must be a 'Baire' function [ref. 2].)
It can be shown [ref. 1] that

$$p_y(\chi) = p_x(\xi_1)\left[\left|\frac{\mathrm{d}f(\xi)}{\mathrm{d}\xi}\right|_{\xi_1}\right]^{-1} + p_x(\xi_2)\left[\left|\frac{\mathrm{d}f(\xi)}{\mathrm{d}\xi}\right|_{\xi_2}\right]^{-1} + \dots p_x(\xi_m)\left[\left|\frac{\mathrm{d}f(\xi)}{\mathrm{d}\xi}\right|_{\xi_m}\right]^{-1} \tag{1.35}$$

where $\xi_1, \xi_2, \dots, \xi_m$ are the m roots of the equation

$$\chi = f(\xi)$$

Eqn (1.35) is easy to evaluate if one value of x corresponds to one value of y, i.e. if $x = f^{-1}(y)$ has only one root. Otherwise, the relationships can be complicated.

Examples

(a) Consider a random variable x with uniform PDF:

$$p_x(\xi) = \begin{cases} 0, & \xi < 1 \\ 0.5, & 1 \leqslant \xi < 3 \\ 0, & \xi \geqslant 3 \end{cases} \tag{1.36}$$

Consider $y = ax + b$ where a, b are constants, $a \geqslant 0, b \geqslant 0$. These functions are shown in Figure 1.16(a, b).

In this case, $p_y(\chi)$ can be calculated from first principles by considering the probability functions $P_x(\xi), P_y(\chi)$:

$$\begin{aligned} P_y(\chi) &= \text{Prob}(y \leqslant \chi) \\ &= \text{Prob}(x \leqslant f^{-1}(\chi)) \\ &= \text{Prob}\left(x \leqslant \frac{\chi - b}{a}\right) \\ &= P_x\left(\frac{\chi - b}{a}\right) \qquad \text{(Figure 1.16(c))} \end{aligned}$$

Recalling that

$$p_y(\chi) = \frac{\mathrm{d}P_y(\chi)}{\mathrm{d}\chi}$$

$$p_y(\chi) = \begin{cases} 0, & \chi < a + b \\ \dfrac{1}{2a}, & a + b \leqslant \chi < 3a + b \\ 0, & \chi \geqslant 3a + b \end{cases}$$

This is shown in Figure 1.16(d).

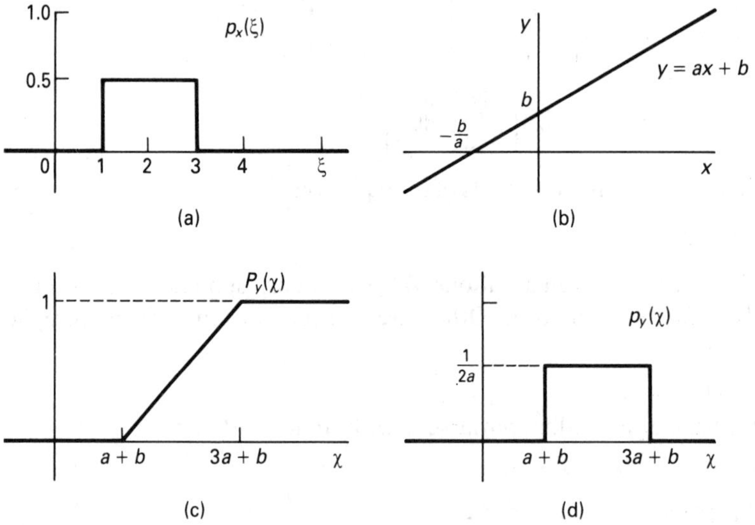

Figure 1.16 Function of a random variable.

Alternatively, Eqn (1.35):

$$\xi_1 = \frac{\chi - b}{a}$$

$$\left.\frac{\mathrm{d}f(\xi)}{\mathrm{d}\xi}\right|_{\xi_1} = a$$

$$p_y(\chi) = \frac{1}{a} p_x\left(\frac{\chi - b}{a}\right)$$

$$= \begin{cases} 0, & \chi < a + b \\ \dfrac{1}{2a}, & a + b \leqslant \chi < 3a + b \\ 0, & \chi \geqslant 3a + b \end{cases}$$

If a or b is negative, the problem is only slightly more complicated.

(b) Consider the more difficult case where x has the PDF of Eqn (1.36) and $y = x^2/4$, as shown in Figure 1.17(a). It is required to find the PDF of y, $p_y(\chi)$.

The inverse function has two roots:

$$x = \pm 2\sqrt{y}$$

Applying straightforward logic is complicated:

(i) For $\chi \leqslant 0$, $P_y(\chi) = \text{Prob}(y \leqslant \chi) = 0$

(ii) For $\chi > 0$, $P_y(\chi) = \text{Prob}(y \leqslant \chi)$
$= \text{Prob}(0 < x \leqslant 2\sqrt{\chi}) + \text{Prob}(-2\sqrt{\chi} < x \leqslant 0)$

Hence:

$$P_y(\chi) = \begin{cases} 0, & \chi < 0.25 \\ \displaystyle\int_1^{2\sqrt{\chi}} 0.5\,\mathrm{d}\xi = \sqrt{\chi} - 0.5, & 0.25 \leqslant \chi < 2.25 \\ 0, & \chi \geqslant 2.25 \end{cases}$$

This is shown in Figure 1.17(b).

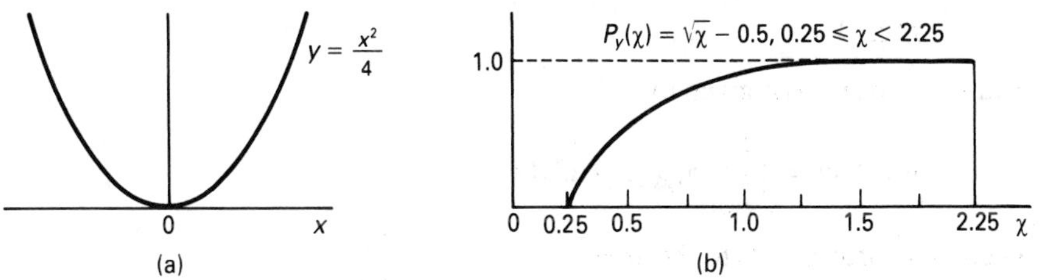

Figure 1.17 Function of a random variable.

Alternatively, and much more easily, using Eqn (1.35):

$$\xi_1 = -2\sqrt{\chi}, \quad \xi_2 = 2\sqrt{\chi}, \quad \chi \geqslant 0$$

$$p_x(\xi_1)\left[\left|\frac{\mathrm{d}f(\xi_1)}{\mathrm{d}\xi_1}\right|_{\xi_1}\right]^{-1} = 0, \quad p_x(\xi_2)\left[\left|\frac{\mathrm{d}f(\xi)}{\mathrm{d}\xi}\right|_{\xi_2}\right]^{-1} = \frac{1}{2\sqrt{\chi}}$$

Hence,

$$p_y(\chi) = \begin{cases} 0, & \chi < 0.25 \\ \dfrac{1}{2\sqrt{\chi}}, & 0.25 \leqslant \chi < 4.25 \\ 0, & \chi \geqslant 2.25 \end{cases}$$

(c) It is easy to show that any linear function of a Gaussian random variable, x, is also Gaussian. Consider $y = ax$, where the PDF of x is given by Eqn (1.28). The PDF of y is given by Eqn (1.35), or simply by considering the meaning of the terms in Eqn (1.28) and substituting $x = y/a$ and $\xi = \dfrac{\chi}{a}$ where appropriate to give the distribution:

$$p_y(\chi) = \frac{a}{\sigma_y\sqrt{2\pi}}\exp\left[-\frac{(a\chi - \varepsilon(y^2))^2}{2\sigma_y^2}\right]$$

Thus y, any linear function of x, is Gaussian.

1.13 Functions of two random variables

Formulae can be developed fairly readily to express the PDF of a random variable, z, a function of two random variables, x, y, whose joint PDF, $p_{xy}(\xi, \chi)$, is known. Suppose $z = f(x, y)$; then the probability function of z, $P_z(\gamma)$ is

$$P_z(\gamma) = \mathrm{Prob}(z \leqslant \gamma)$$

$$= \iint_D p_{xy}(\xi, \chi)\,\mathrm{d}\xi\,\mathrm{d}\chi \tag{1.37}$$

where the double integral is taken over the area(s), D, in the plane corresponding to

$$f(\xi, \chi) = \gamma$$

Equivalently, $p_z(\gamma)$ is given by

$$p_z(\gamma)\,\mathrm{d}\gamma = \iint_{\Delta D} p_{xy}(\xi, \chi)\,\mathrm{d}\xi\,\mathrm{d}\chi \tag{1.38}$$

where ΔD corresponds to the area

$$\gamma < f(\xi, \chi) \leqslant \gamma + \mathrm{d}\gamma$$

Where $z = x + y$, Eqns (1.37) and (1.38) simplify to yield

$$P_z(\gamma) = \int_{-\infty}^{+\infty} \int_{-\infty}^{\gamma-\chi} p_{xy}(\xi, \chi)\, d\xi\, d\chi$$
$$= \int_{-\infty}^{+\infty} \int_{-\infty}^{\gamma-\xi} p_{xy}(\xi, \chi)\, d\xi\, d\chi \quad (1.39)$$

and

$$p_z(\gamma)\, d\gamma = \int_{-\infty}^{+\infty} p_{xy}((\gamma - \chi), \chi)\, d\chi\, d\gamma$$
$$= \int_{-\infty}^{+\infty} p_{xy}(\xi, \gamma - \xi)\, d\xi\, d\gamma \quad (1.40)$$

A further simplification is possible if x, y are independent:

$$p_z(\gamma) = \int_{-\infty}^{+\infty} p_x(\gamma - \chi) p_y(\chi)\, d\chi$$
$$= \int_{-\infty}^{+\infty} p_x(\xi) p_y(\gamma - \xi)\, d\xi \quad (1.41)$$

Example

Consider the random variable $z = x + y$ where x, y are independent, with PDFs

$$p_x(\xi) = \begin{cases} 0, & \xi < a \\ \dfrac{1}{b-a}, & a \leqslant \xi < b \\ 0, & \xi \geqslant b \end{cases}$$

$$p_y(\chi) = \begin{cases} 0, & \chi < c \\ \dfrac{1}{d-c}, & c \leqslant \chi < d \\ 0, & \gamma \geqslant d \end{cases}$$

where a, b, c, d are positive constants. $p_x(\xi)$, $p_y(\chi)$ are illustrated in Figure 1.18(a, b), and $p_z(\gamma)$ in Figure 1.18(c).

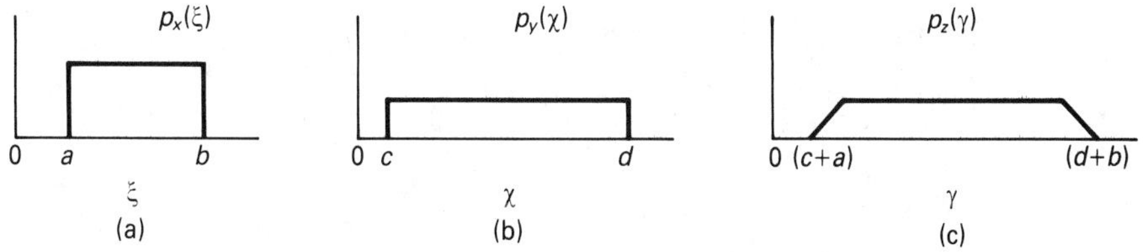

Figure 1.18 Function of two random variables.

Eqn (1.41) can be used to find the PDF of a random variable, z, which is a linear function of many independent random variables $x_1, x_2, \ldots, x_m$:

$$z = a_1 x_1 + a_2 x_2 + \cdots + a_m x_m$$

where $a_1, a_2, \ldots, a_m$ are constants. If m is 'large', it can be shown that z is Gaussian. This is the 'Central Limit Theorem', which explains why many natural random variables are Gaussian; they are linear combinations of many independent (not necessarily Gaussian) random variables.

1.14 Correlation and dependence

The joint probability functions described in Section 1.13 are useful in defining correlation and dependence (or the absence of these) between random variables.

1.14.1 Correlation

The correlation function, $\bar{r}_{xy}$, of two random variables, x, y, is a measure of correlation, or relationship, between them:

$$\bar{r}_{xy} = \varepsilon(xy) \tag{1.42}$$

If x and y are uncorrelated,

$$\bar{r}_{xy} = 0$$

The converse is not invariably true: x, y can be related although $\bar{r}_{xy} = 0$. Formally, if $\bar{r}_{xy} = 0$, x, y are described as 'statistically orthogonal'.

Any 'mean value' content of both x and y does constitute a correlation, so if

$\varepsilon(x) \neq 0$ and $\varepsilon(y) \neq 0$, $r_{xy} \neq 0$

Finally, variables which consist of a weighted sum of uncorrelated variables are worth examination. Consider

$$z = a_1 x_1 + a_2 x_2 + \cdots + a_m x_m$$

where $a_1, a_2, \ldots, a_m$ are constants,

$x_1, x_2, \ldots, x_m$ are uncorrelated random variables.

Then it is not difficult to show that the mean and variance of z are

$$\varepsilon(z) = a_1 \varepsilon(x_1) + a_2 \varepsilon(x_2) + \cdots + a_m \varepsilon(x_m) \tag{1.43}$$

$$\sigma_z^2 = a_1^2 \sigma_{x_1}^2 + a_2^2 \sigma_{x_2}^2 + \cdots + a_m^2 \sigma_{x_m}^2 \tag{1.44}$$

1.14.2 Covariance

The covariance function, r_{xy}, of two random variables, x, y, is a measure of correlation between the deviations of x, y, from their expected values:

$$r_{xy} = \text{cov}(x, y) = \varepsilon[(x - \varepsilon(x))(y - \varepsilon(y))] \tag{1.45}$$

$$= \varepsilon(xy) - \varepsilon(x) \cdot \varepsilon(y)$$

$$= \bar{r}_{xy} - \varepsilon(x) \cdot \varepsilon(y) \tag{1.46}$$

If x, y have joint PDF, $p_{xy}(\xi, \chi)$, then their covariance is (Eqn (1.32))

$$r_{xy} = \int_{-\infty}^{+\infty} \int_{-\infty}^{+\infty} (\xi - \varepsilon(x))(\chi - \varepsilon(y)) p_{xy}(\xi, \chi)\, \mathrm{d}\xi\, \mathrm{d}\chi$$

$$= \int_{-\infty}^{+\infty} \int_{-\infty}^{+\infty} \xi \chi p_{xy}(\xi, \chi)\, \mathrm{d}\xi\, \mathrm{d}\chi - \varepsilon(x)\varepsilon(y) \tag{1.47}$$

where

$$\varepsilon(x) = \int_{-\infty}^{+\infty} \int_{-\infty}^{+\infty} \xi p_{xy}(\xi, \chi)\, \mathrm{d}\xi\, \mathrm{d}\chi$$

$$\varepsilon(y) = \int_{-\infty}^{+\infty} \int_{-\infty}^{+\infty} \chi p_{xy}(\xi, \chi)\, \mathrm{d}\xi\, \mathrm{d}\chi$$

1.14.3 Independence

Two random variables x and y are independent if their joint PDF can be expressed as the product of their individual PDFs:

$$p_{xy}(\xi, \chi) = p_x(\xi) p_y(\chi) \tag{1.48}$$

Naturally, independent processes have zero correlation (and covariance) functions, i.e. they are statistically orthogonal. The inverse is not necessarily true. However, for Gaussian variables, zero correlation does imply independence.

A variable which consists of the weighted sum of a large number of independent (and therefore uncorrelated) variables has the mean and variance given by Eqns (1.43) and (1.44), and is Gaussian in accordance with the Central Limit Theorem (Section 1.13).

Examples

Three artificially simple, but illustrative, cases are considered here.

(a) Consider the variables x, y whose joint PDF (Figure 1.19) is

$$p_{xy}(\xi, \chi) = \begin{cases} 0, & \xi < -a, \quad \chi < -a \\ \dfrac{1}{4a^2}, & -a \leqslant \xi < +a, \quad -a \leqslant \chi < +a \\ 0, & \xi \geqslant +a, \quad \chi \geqslant +a \end{cases} \tag{1.49}$$

Substitution in Eqn (1.49) shows that

$$p_{xy}(\xi, \chi) = p_x(\xi) \cdot p_y(\chi)$$

where

$$p_x(\xi) = \begin{cases} 0, & \xi < -a \\ \dfrac{1}{2a}, & -a \leqslant \xi < +a \\ 0, & \xi \geqslant +a \end{cases}$$

$$p_y(\chi) = \begin{cases} 0, & \xi < -a \\ \dfrac{1}{2a}, & -a \leqslant \chi < +a \\ 0, & \chi \geqslant +a \end{cases}$$

The random variables x, y are thus independent. They are therefore uncorrelated.

It is also easy to show that $\varepsilon(x) = \varepsilon(y) = 0$, and that $\varepsilon(xy) = 0$, confirming that their correlation, $\bar{r}_{xy}$, and covariance, r_{xy}, functions are zero.

(b) Consider the variables x, y, $\varepsilon(x) = q$, $\varepsilon(y) = r$, and

$$p_{xy}(\xi, \chi) = \begin{cases} 0, & \xi < q - a, \quad \chi < r - a \\ \dfrac{1}{4a^2}, & q - a \leqslant \xi < q + a, \quad r - a \leqslant \chi < r + a \\ 0, & \xi \geqslant q + a, \quad \chi \geqslant r + a \end{cases}$$

In this case $p(\xi, \chi)$ cannot be represented as the product of two PDFs, so x and y are not formally independent, and $\bar{r}_{xy} \neq 0$. However, the deviations of x and y from $q, r, (x - q), (y - r)$, are independent, and it is easy to show that

$$\bar{r}_{xy} = \varepsilon(xy) = qr = \varepsilon(x) \cdot \varepsilon(y)$$

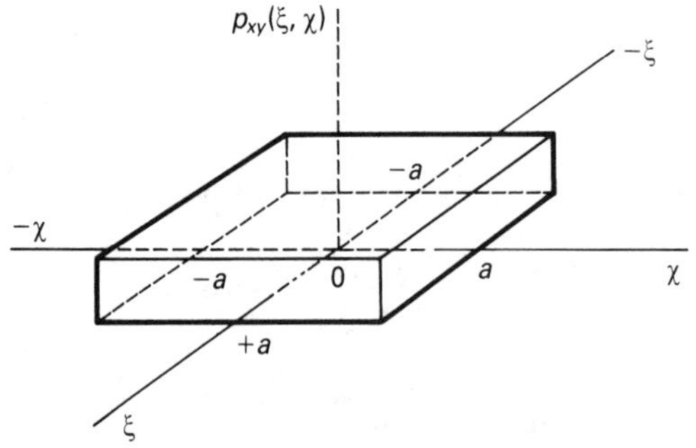

Figure 1.19 Joint PDF.

Consequently, (Eqn (1.46))

$$r_{xy} = 0$$

In this case, though their covariance is zero, x and y are not orthogonal or independent; their mutual dependence results from their both having non-zero means.

(c) Consider the joint PDF

$$p_{xy}(\xi, \chi) = \begin{cases} \dfrac{1}{2a^2}, & -a \leqslant \xi < +a, \quad -a \leqslant \chi < +a, \quad \chi \geqslant \xi \\ 0, & \text{otherwise} \end{cases}$$

Since $p_{xy}(\xi, \chi)$ cannot be rewritten as the product of two independent PDFs, x and y are dependent.

The means of x and y are

$$\varepsilon(x) = \int_{-a}^{+a} \left[\int_{-a}^{\chi} \xi \frac{1}{2a^2} \mathrm{d}\xi \right] \mathrm{d}\chi = \frac{-a}{3}$$

$$\varepsilon(y) = \int_{-a}^{+a} \left[\int_{\xi}^{a} \chi \frac{1}{2a^2} \mathrm{d}\chi \right] \mathrm{d}\xi = \frac{+a}{3}$$

Their correlation function is (Eqns (1.42))

$$\begin{aligned} \bar{r}_{xy} &= \int_{-\infty}^{+\infty} \int_{-\infty}^{+\infty} \xi \chi p_{xy}(\xi, \chi) \, \mathrm{d}\xi \, \mathrm{d}\chi \\ &= \int_{-a}^{+a} \int_{-a}^{+a} \xi \chi \frac{1}{2a^2} \mathrm{d}\xi \, \mathrm{d}\chi = 0 \end{aligned}$$

Thus (Eqn (1.46))

$$r_{xy} = \frac{a^2}{9}$$

In this case x and y are dependent and $r_{xy} \neq 0$. Unusually, however, they have zero correlation, $\bar{r}_{xy} = 0$, and are thus statistically orthogonal.

EXERCISES

1 A bag contains six red and five blue balls. Four are drawn out at random, without being replaced. What is the probability that they are alternately of different colours?

2 What is the probability of the four players at a game of bridge each being dealt a complete suit?

3 How often must a pair of dice be tossed to make it more likely than not that a double 6 appears?

4 If, on average, rain falls in ten days out of thirty, find the probabilities that:
(a) the first three days of a given week are wet,
(b) the first three days are wet and the remaining four are fine,
(c) any three days are wet.

5 A chain consists of ten links. The tensile strength of each link is equally likely to have any value between 500 and 600 newtons. What are the probabilities that the chain will break under:
(a) 510 newtons,
(b) 540 newtons?

6 Verify that (np) and (npq) give the mean and variance of the binomial distribution described by $(0.3 + 0.7)^4$.

7 A random variable, x, has a PDF:

$$p(\xi) = \begin{cases} 0, & \xi < 0 \\ Ke^{-\xi}, & \xi \geqslant 0 \end{cases}$$

What is the value of K, and what are the probabilities that:
(a) $x < 0.5$,
(b) $0.2 < x < 0.7$?
Evaluate $\varepsilon(x)$, $\varepsilon(x^2)$, and σ_x^2.

8 A random variable, x, has a PDF:

$$p(\xi) = \begin{cases} 3\xi^2, & 0 < \xi \leqslant 1 \\ 0, & \text{elsewhere} \end{cases}$$

Calculate $\varepsilon(x)$, $\varepsilon(x^2)$ and σ_x^2, and find a number, m, such that $P(x > m) = P(x \leqslant m)$.

9 The radius of a sphere is uniformly distributed between 0 and 2 m. What are the mean volume and surface area of the sphere? What are the probabilities that the volume and surface area exceed half their maximum values?

10 A stick of length L is broken at random, and the larger piece, length l, is then broken at random, thus producing three shorter sticks. What is the probability that they can form the sides of a triangle?

11 The joint PDF of two variables x, y, is:

$$p(\xi, \chi) = \begin{cases} K(\xi + \chi), & 0 < \xi \leqslant 1, \quad 0 < \chi \leqslant 1 \\ 0, & \text{elsewhere} \end{cases}$$

Sketch $p(\xi, \chi)$, calculate K, and find $\varepsilon(x)$, σ_x^2, $\varepsilon(y^2)$, σ_y^2.

12 A random variable, x, has uniform distribution between -1 and $+1$. Find $y = f(x)$ such that the PDF of y is:

$$p(\chi) = \begin{cases} 0, & \chi < 0 \\ 2e^{-2\chi}, & \chi \geqslant 0 \end{cases}$$

13 A random variable, x, has uniform distribution between -3 and $+3$. Find the PDF of y if:

(a) $y = x^3$,

(b) $y = \sin\left(\frac{\pi}{6}x\right)$.

14 Two independent random variables, x, y, have PDFs:

$$p(\xi) = \begin{cases} 0.5, & 0 < x < 2 \\ 0, & \text{elsewhere,} \end{cases}$$

$$p(\chi) = \begin{cases} 1 - |\chi|, & -1 < \chi \leqslant 1 \\ 0, & \text{elsewhere} \end{cases}$$

Find the probability that $x \geqslant y$.

15 x and y are uniformly distributed, jointly, over a square of side 2, and their joint PDF is:

$$p(\xi, \chi) = \begin{cases} C, & -1 < \xi \leqslant +1, \quad -1 < \chi \leqslant +1 \\ 0, & \text{elsewhere} \end{cases}$$

Find the PDF of z where $z = x + y$

16 The variables x, y have joint PDF

$$p(\xi, \chi) = \begin{cases} C(1 - \xi^2 - \chi^2), & 0 \leqslant \xi^2 + \chi^2 \leqslant 1 \\ 0, & \text{elsewhere} \end{cases}$$

Find C and the covariance of x and y.

REFERENCES

[1] Miller, I. and Freund, J. E., *Probability and Statistics for Engineers*, Prentice Hall, Englewood Cliffs, NJ, 1985.

[2] Papoulis, A., *Probability, Random Variables and Stochastic Processes*, McGraw-Hill, New York, 1984.

2 Continuous time stochastic processes and systems

2.1 Introduction

This chapter deals with the properties of continuous time stochastic processes and with the structure and analysis of standard form mathematical models of dynamical systems. Details of the construction of such models, involving as they do such subjects as Lagrangian mechanics, thermodynamics and flight dynamics, are beyond the scope of this book. However, the main constructional tools required are described in sufficient depth to be useful, to promote understanding and to allow access to advanced literature on the subject.

Definitions and properties of stochastic processes are considered in Sections 2.2–2.5, differentiation and integration in Sections 2.6–2.8, and vector processes in Section 2.9. Mathematical models of the behaviour of dynamical systems are explored in Sections 2.10–2.12.

2.2 Stochastic processes

Stochastic processes, or signals, are those affected by random noise. All practical processes are stochastic, but where noise content is negligible, processes are commonly regarded as deterministic.

Consider the random output signal of a radio receiver tuned to no station. The properties of this stochastic process are studied by considering the outputs of an infinite set of such radios simultaneously. Each signal, $\bar{x}_i(t)$, is a 'sample function', while the infinite set of signals is an 'ensemble'. *Any* sample function is represented by the symbol $x(t)$. The ensemble is illustrated in Figure 2.1.

Formally, $x(t)$ is a time-dependent value assigned to any of the sample functions of the ensemble, and is thus a time-dependent random variable in accordance with the definition of Section 1.5. The amplitude of $x(t)$ is therefore described by a probability density function (Section 1.7) which is in general a function of time, $p(\xi, t)$.

Function $p(\xi, t)$ describes the dynamic behaviour of the statistics of the amplitude of $x(t)$ and amounts to a very sparse description of the behaviour of $x(t)$. The fullest possible description of this would be given by the infinite dimensional joint PDF (Eqn (1.29)):

$$p(\xi_1, \xi_2, \ldots, \xi_i, \ldots; t_1, t_2, \ldots, t_i \ldots) \tag{2.1}$$

where ξ_1 refers to $x(t_1)$, ξ_2 to $x(t_2), \ldots, \xi_i$ to $x(t_i)$, etc. The complexity of this renders it of little practical use, and simpler, much coarser, terms are invariably employed to describe the behaviour of $x(t)$; these are examined in Section 2.3.

The variable $x(t)$ is 'stationary' or, more formally, 'strictly stationary', if its statistics do not vary with time, that is, if the joint PDF of Eqn (2.1) is time invariant:

$$p(\xi_1, \xi_2, \ldots; t_1, t_2, \ldots) = p(\xi_1, \xi_2, \ldots; t_1 + \tau, t_2 + \tau, \ldots) \tag{2.2}$$

for any τ. This implies that $p(\xi, t)$ is not a function of time, and also that $p(\xi_1, \xi_2; t_1, t_2)$ is time invariant:

$$p(\xi, t) = p(\xi) \text{ and } p(\xi_1, \xi_2; t_1, t_2) = p(\xi_1, \xi_2; t_1 + \tau, t_2 + \tau) \tag{2.3}$$

Considerably weaker, but more easily fulfilled, criteria are that the mean and autocorrelation of $x(t)$, to be defined in Section 2.3, are time invariant:

$$\varepsilon(x(t)) = \varepsilon(x(t + \tau)) = m_x$$

$$\varepsilon(x(t) \cdot x(t + \tau)) = r_x(\tau)$$

If $x(t)$ meets these conditions, it is 'wide sense stationary'. Even weaker, though still useful, criteria are simply that the mean and variance of $x(t)$ are time invariant:

$$\varepsilon(x(t)) = \varepsilon(x(t + \tau)) = m_x$$

$$\varepsilon(x(t) - m_x)^2 = \varepsilon(x(t + \tau) - m_x)^2$$

If $x(t)$ meets these conditions, it is 'weakly stationary'.

Clearly, any strictly stationary process is wide sense and weakly stationary, and any wide sense stationary process is weakly stationary, but the reverse statements are not necessarily true.

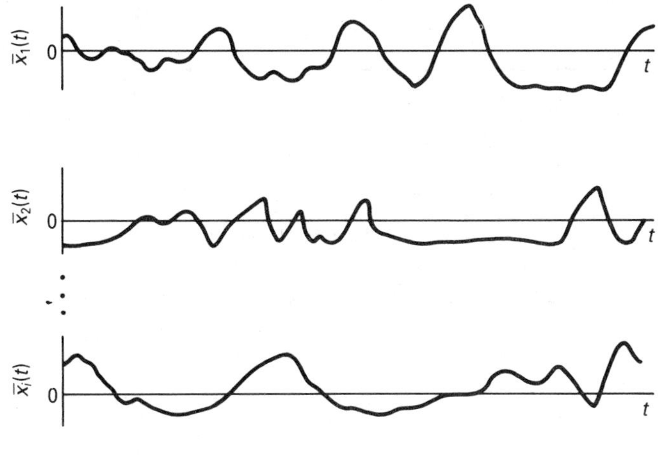

Figure 2.1 Sample functions.

Function $x(t)$ is 'ergodic' if the statistics of numbers sampled from any sample function, $\bar{x}_i(t_1), \bar{x}_i(t_2), \ldots, \bar{x}_i(t_m) \ldots$, are identical to the statistics of the numbers collected 'across' the ensemble at any instant, $\bar{x}_1(t), \bar{x}_2(t), \ldots, \bar{x}_m(t), \ldots$. This is possible – and indeed makes sense – only if $x(t)$ is strictly stationary.

Ergodicity can be demonstrated properly only by collecting sets of data from many individual sample functions, and from across the ensemble, and equating the statistics of the two sets. In practice, processes are frequently assumed to be ergodic simply when it seems reasonable to do so.

It is worth noting that while all ergodic processes are strictly stationary, most, but not all, strictly stationary processes are ergodic.

Examples

(a) Consider the process of throwing a die once at $t = 0$, yielding the stochastic result $x(t)$. The ensemble of $x(t)$ comprises the results, $\bar{x}_1(t), \bar{x}_2(t), \ldots, \bar{x}_i(t), \ldots$, of an infinite number of dice thrown once at $t = 0$, as indicated in Figure 2.2(a).

Some sample functions are:

$$\bar{x}_1(t) = \begin{cases} 0, & t < 0 \\ 4, & t \geqslant 0 \end{cases}$$

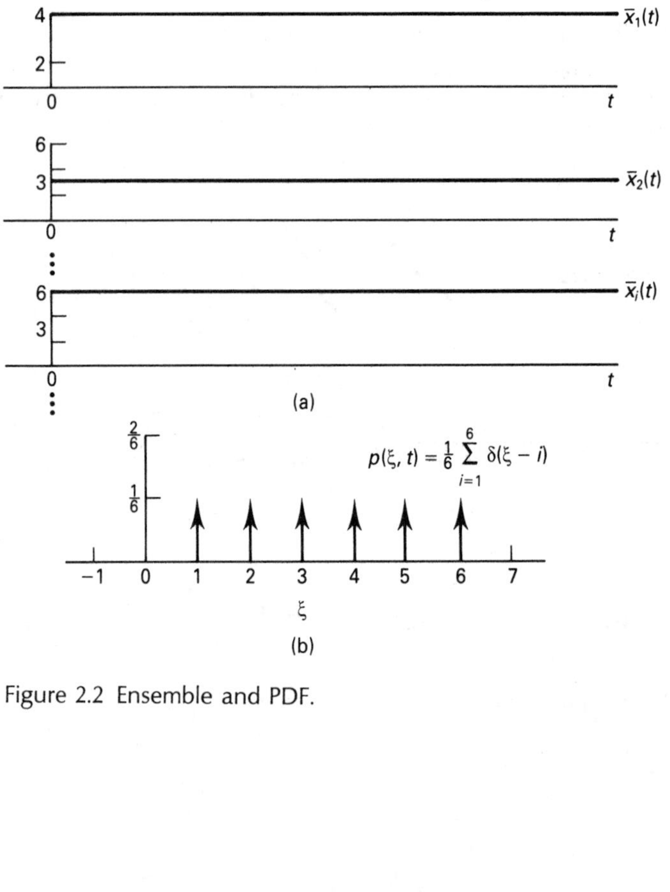

Figure 2.2 Ensemble and PDF.

$$\bar{x}_2(t) = \begin{cases} 0, & t < 0 \\ 3, & t \geqslant 0 \end{cases}$$

$$\vdots$$

$$\bar{x}_i(t) = \begin{cases} 0, & t < 0 \\ 6, & t \geqslant 0 \end{cases}$$

The variable $x(t)$ represents any one of these, and its PDF (Figure 2.2(b)) is obviously time invariant:

$$p(\xi, t) = p(\xi) = \frac{1}{6} \sum_{i=1}^{6} \delta(\xi - i)$$

In view of this, $x(t)$ is taken to be stationary, though, strictly, a joint PDF should be developed and checked for stationarity using Eqn (2.2) – a very considerable task.

The PDF of numbers collected from any sample function over time is clearly very different from $p(\xi, t)$. For example, $\bar{x}_2(t)$ has the PDF

$$\bar{p}_2(\xi) = \delta(\xi - 3)$$

Therefore, $x(t)$ is not ergodic (though it is stationary).

(b) Consider a die thrown repeatedly, once every second, yielding $x(t)$. The ensemble is indicated in Figure 2.3. Across the ensemble, $x(t)$ has the PDF

$$p(\xi, t) = p(\xi) = \frac{1}{6} \sum_{i=1}^{6} \delta(\xi - i)$$

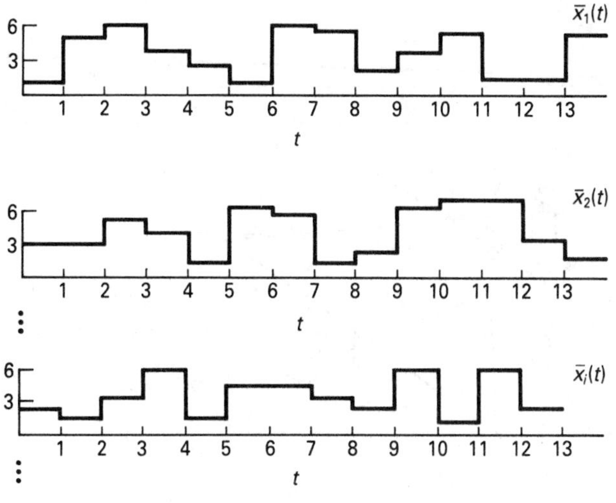

Figure 2.3 Ensemble.

Since $p(\xi, t)$ is independent of time, $x(t)$ is taken to be stationary. The same PDF describes $\bar{x}_i(t)$, at least for a data sampling period greater than one second. Thus, ignoring the complexities associated with short sampling periods, $x(t)$ is ergodic.

(c) Consider the stochastic process

$$x(t) = a + bt$$

where a, b are independent, evenly distributed, random variables:

$$p_a(\xi) = p_b(\xi) = \begin{cases} 0, & \xi < -2 \\ 0.25, & -2 \leqslant \xi < +2 \\ 0, & +2 \leqslant \xi \end{cases}$$

Some sample functions are shown in Figure 2.4(a).

Since $x(t)$ is the sum of two independent random variables, a and bt, its PDF, $p(\xi, t)$, is found using Eqn (1.41). This turns out to be quite involved, even for this apparently simple case:

$$p_a(\xi) = \begin{cases} 0, & \xi < -2 \\ 0.25, & -2 \leqslant \xi < +2 \\ 0, & +2 \leqslant \xi \end{cases}$$

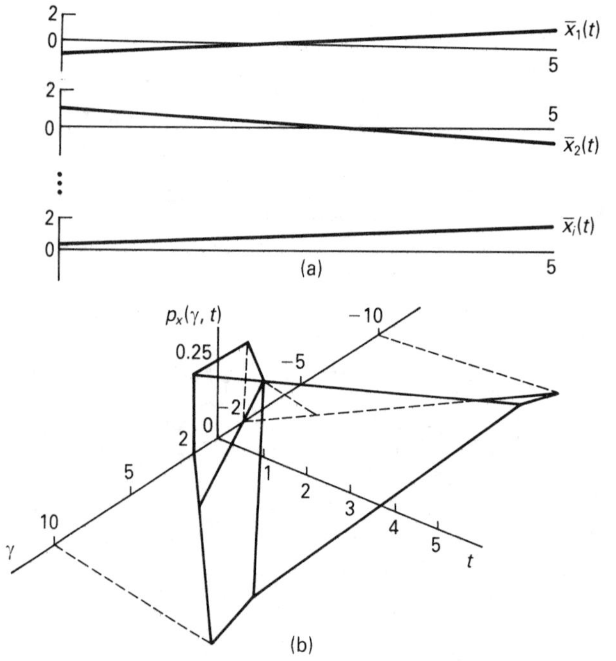

Figure 2.4 Ensemble and PDF.

and

$$p_{bt}(\gamma - \xi) = \begin{cases} 0, & \xi \leqslant \gamma - 2t \\ \dfrac{0.25}{t}, & \gamma - 2t < \xi \leqslant \gamma + 2t \\ 0, & \gamma + 2t < \xi \end{cases}$$

Multiplying and integrating, for $t \leqslant 1$,

$$p_x(\gamma) = \begin{cases} 0, & \gamma \leqslant -2 - 2t \\ \dfrac{(0.25)^2}{t}(\gamma + 2t + 2), & -2 - 2t < \gamma \leqslant -2 + 2t \\ 0.25, & -2 + 2t < \gamma \leqslant 2 - 2t \\ \dfrac{(0.25)^2}{t}(\gamma - 2t - 2), & 2 - 2t < \gamma \leqslant 2 + 2t \\ 0, & 2 + 2t \leqslant \gamma \end{cases}$$

For $t \geqslant 1$,

$$p_x(\gamma) = \begin{cases} 0, & \gamma \leqslant -2t - 2 \\ \dfrac{(0.25)^2}{t}(\gamma + 2t + 2), & -2t - 2 < \gamma \leqslant -2t + 2 \\ \dfrac{0.25}{t}, & -2t + 2 < \gamma \leqslant 2t - 2 \\ \dfrac{(0.25)^2}{t}(\gamma - 2t - 2), & 2t - 2 < \gamma \leqslant 2t + 2 \\ 0, & 2t + 2 < \gamma \end{cases}$$

The propagation of $p(\xi, t)$ with time is shown in Figure 2.4(b) where it can be seen, for example, that at $t = 0$, $-2 \leqslant x(0) < +2$ and $p(\xi, 0)$ is rectangular, while at $t = 4$, $-10 \leqslant x(4) < +10$, and $p(\xi, 4)$ is trapezoidal.

2.3 Ensemble properties of stochastic processes

As indicated in Section 2.2, the joint probability density function (Eqn (2.1)), which fully characterizes a stochastic process, is far too complicated to constitute a useful description, and in practice the rather crude functions which describe random variables, namely mean, variance, correlation and covariance (Sections 1.8 and 1.14) are used to describe stochastic processes.

2.3.1 Mean and variance (cf. Section 1.8)

The mean, or expected value (Eqn (1.24)) of a stochastic process, $x(t)$, taken across the ensemble at time t, is

$$\begin{aligned}\varepsilon(x(t)) &= m_x(t)\\ &= \int_{-\infty}^{+\infty} \xi p(\xi, t)\,\mathrm{d}\xi \qquad (2.4)\end{aligned}$$

The variance, or 'spread', of $x(t)$ (Eqn (1.25)) is a crude measure of its amplitude:

$$\begin{aligned}\sigma_x^2(t) &= \varepsilon[(x(t) - \varepsilon(x(t))^2]\\ &= \int_{-\infty}^{+\infty} (\xi - \varepsilon(x(t)))^2 p(\xi, t)\,\mathrm{d}\xi \qquad (2.5)\end{aligned}$$

Functions $m_x(t)$ and $\sigma_x(t)$ are deterministic (not stochastic) functions of time; if $x(t)$ is stationary in any sense, both are constant:

$$m_x(t) = m_x$$

$$\sigma_x^2(t) = \sigma_x^2$$

2.3.2 Autocorrelation and covariance (cf. Section 1.14)

These functions are measures of the rate of change, or dynamics, of a stochastic process, and also, incidentally, of its amplitude.

The autocorrelation of $x(t)$ is (cf. Eqn (1.42))

$$\begin{aligned}\bar{r}(t, s) &= \varepsilon(x(t)\cdot x(s))\\ &= \int_{-\infty}^{+\infty}\int_{-\infty}^{+\infty} \xi_t \xi_s \mathrm{p}(\xi_t, \xi_s; t, s)\,\mathrm{d}\xi_t\,\mathrm{d}\xi_s \qquad (2.6)\end{aligned}$$

where $p(\xi_t, \xi_s; t, s)$ is the joint PDF describing the distribution of $x(t), x(s)$.

The same property, but centralized about the means $\varepsilon(x(t)), \varepsilon(x(s))$, is given by the autocovariance of $x(t)$ (cf. Eqn (1.45)), usually referred to simply as the covariance of $x(t)$:

$$\begin{aligned}r(t, s) &= \varepsilon[(x(t) - \varepsilon(x(t)))(x(s) - \varepsilon(x(s)))]\\ &= \int_{-\infty}^{+\infty}\int_{-\infty}^{+\infty} (\xi_t - \varepsilon(x(t)))(\xi_s - \varepsilon(x(s)))p(\xi_t, \xi_s; t, s)\,\mathrm{d}\xi_t\,\mathrm{d}\xi_s \qquad (2.7)\end{aligned}$$

It is easy to show that (cf. Eqn (1.46))

$$r(t, s) = \bar{r}(t, s) - \varepsilon(x(t))\cdot\varepsilon(x(s)) \qquad (2.8)$$

Functions $\bar{r}(t, s)$ and $r(t, s)$ are deterministic functions of the two independent time variables, t, s. If $x(t)$ is wide sense stationary, both are independent of the absolute time and are even functions of the time difference, $\tau = (s - t)$:

$$\bar{r}(t, s) = \bar{r}(s - t) = \bar{r}(\tau) = \bar{r}(-\tau) = \bar{r}(t - s)$$

$$r(t, s) = r(s - t) = r(\tau) = r(-\tau) = r(t - s) \tag{2.9}$$

It is also worth noting that the variance of $x(t)$ is

$$\sigma_x^2(t) = r(t, t)$$

If $x(t)$ is stationary (in any sense),

$$\sigma_x^2(t) = \sigma_x^2 = r(\tau)|_{\tau=0} = r(0)$$

Examples

(a) Consider a die thrown once at $t = 0$. The relevant ensemble is shown in Figure 2.2(a), and the process is stationary, though not ergodic (Section 2.2). The PDF of $x(t)$ (Figure 2.2(b)) is

$$p(\xi, t) = p(\xi) = \frac{1}{6}\sum_{i=1}^{6} \delta(\xi - i)$$

The mean (Eqn (2.4)) is

$$\begin{aligned} \varepsilon(x(t)) &= m_x(t) = m_x \\ &= \frac{1}{6}\int_{-\infty}^{+\infty} \sum_{i=1}^{6} \delta(\xi - i)\,\mathrm{d}\xi \\ &= 3.500 \end{aligned}$$

The variance (Eqn (2.5)) is

$$\begin{aligned} \sigma_x^2(t) &= \sigma_x^2 \\ &= \frac{1}{6}\int_{-\infty}^{+\infty} (\xi - 3.5)^2 \sum_{i=1}^{6} \delta(\xi - i)\,\mathrm{d}\xi \\ &= 2.917 \end{aligned}$$

$p(\xi)$, m_x and σ_x are shown in Figure 2.5(a).

The joint PDF describing $x(t)$, $x(s)$ is (Figure 2.5(b))

$$\begin{aligned} p(\xi_t, \xi_s; t, s) &= p(\xi_t, \xi_s) \\ &= \frac{1}{6}\sum_{i=1}^{6} \delta(\xi_t - i)\delta(\xi_s - i) \end{aligned}$$

The autocorrelation (Eqn (2.6)) and covariance (Eqn (2.8)) are

$$\bar{r}(t, s) = \bar{r}(s - t)$$
$$= \int_{-\infty}^{+\infty}\int_{-\infty}^{+\infty} \xi_t \xi_s \sum_{i=1}^{6} \delta(\xi_t - i)\delta(\xi_s - i)\,\mathrm{d}\xi_t\,\mathrm{d}\xi_s$$
$$= 15.167$$
$$r(t, s) = 15.167 - (3.5)^2$$
$$= 2.917$$

These figures agree with those calculated directly from the ensemble using discrete distribution formulae (Eqns (1.14) and (1.15)) under the assumption of Eqn (1.2):

$$m_x = \frac{1}{6}(1 + 2 + 3 + \cdots + 6) = 3.5$$

$$\sigma_x^2 = \frac{1}{6}(1^2 + 2^2 + 3^2 + \cdots + 6^2) - (3.5)^2 = 2.917$$

$$\bar{r}(t, s) = \bar{r}(\tau) = \frac{1}{6}(1 \times 1 + 2 \times 2 + \cdots + 6 \times 6) = 15.167$$

$$r(t, s) = \sigma_x^2 = 2.917$$

(b) A die thrown at regular one-second intervals contrasts interestingly with the preceding example. The PDF and ensemble are shown in Figures 2.2(b) and 2.3.

$$\varepsilon(x(t)) = m_x$$
$$= \frac{1}{6}\int_{-\infty}^{+\infty} \xi \sum_{i=1}^{6} \delta(\xi - i)\,\mathrm{d}\xi$$
$$= 3.500$$
$$\sigma_x^2 \frac{1}{6}\int_{-\infty}^{+\infty} (\xi - 3.5)^2 \sum_{i=1}^{6} \delta(\xi - i)\,\mathrm{d}\xi$$
$$= 2.917$$

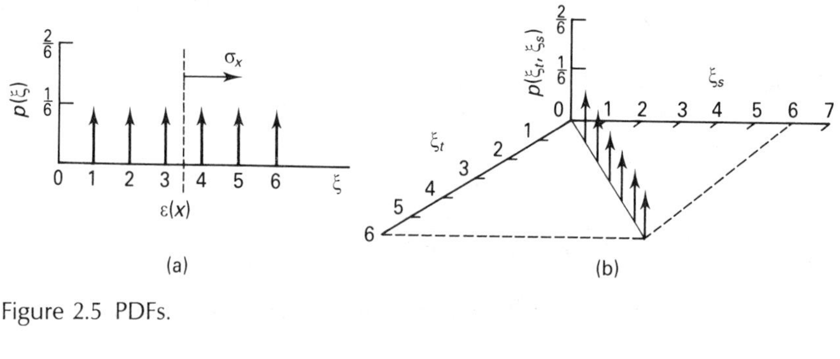

Figure 2.5 PDFs.

Regarding the dynamic behaviour of $x(t)$, if $(s - t) < 1$, $p(\xi_t, \xi_s)$ is complicated and not perhaps of great interest, but if $(s - t) > 1$:

$$p(\xi_t, \xi_s) = \frac{1}{36} \sum_{j=1}^{6} \sum_{i=1}^{6} \delta(\xi_t - i)\delta(\xi_s - j)$$

This is shown in Figure 2.6(a).

The autocorrelation and covariance (Figure 2.6(b)) are found from Eqns (2.6) and (2.8):

$$\bar{r}(t, s) = \int_{-\infty}^{+\infty} \int_{-\infty}^{+\infty} \xi_t \xi_s \frac{1}{36} \sum_{j=1}^{6} \sum_{i=1}^{6} \delta(\xi_t - i)\delta(\xi_s - j)\, d\xi_t\, d\xi_s$$

$$= 12.250$$

$$r(t, s) = r(\tau) = 12.250 - (3.5)^2 = 0$$

For $(s - t) < 1$, the situation is complicated, but clearly,

$$\bar{r}(t, t) = \bar{r}(\tau)|_{\tau=0} = \bar{r}(0) = 15.167$$

$$r(t, t) = r(\tau)|_{\tau=0} = r(0) = 2.917$$

As in the preceding example, these figures agree with those found using discrete distribution formulae under the assumption of Eqn (1.2).

(c) Consider the non-stationary process (Figure 2.4):

$$x(t) = a + bt$$

where a, b are independent random numbers, $\varepsilon(ab) = 0$, and

$$p_a(\xi) = p_b(\xi) = \begin{cases} 0, & \xi < -2 \\ 0.25, & -2 \leqslant \xi < +2 \\ 0, & +2 \leqslant \xi \end{cases}$$

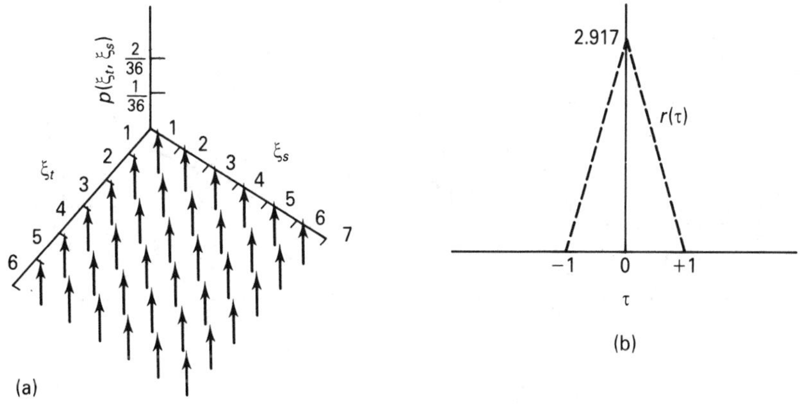

Figure 2.6 Joint PDF and covariance.

The PDF of $x(t)$, $p(\gamma, t)$ is shown in Figure 2.4. It would clearly be possible, if rather complicated, to use this to find $\varepsilon(x(t))$, $\bar{r}(t,s)$ and $r(t,s)$, but in this case it is easier to proceed directly:

$$\varepsilon(x(t)) = \varepsilon(a + bt) = \varepsilon(a) + t \cdot \varepsilon(b) = 0$$

$$\begin{aligned}\bar{r}(t,s) &= \varepsilon[(a + bt)(a + bs)] \\ &= \varepsilon(a^2) + (t + s)\varepsilon(ab) + ts\varepsilon(b^2) \\ &= \frac{4}{3}(1 + ts)\end{aligned}$$

Also,

$$r(t,s) = \frac{4}{3}(1 + ts)$$

and

$$\sigma_x^2(t) = r(t,t) = \frac{4}{3}(1 + t^2)$$

$\varepsilon(x(t))$, $r(t,s)$ *and* $\sigma_x^2(t)$ are shown in Figure 2.7; $\sigma_x^2(t)$ lies on the diagonal $s = t$ in Figure 2.7(b).

2.3.3 Cross-correlation and cross-covariance

A measure of relationship between two stochastic processes, $x(t)$ and $y(t)$, evaluated at times t, s, is given by the correlation function of $x(t)$, $y(s)$ (Eqn (1.42)), usually

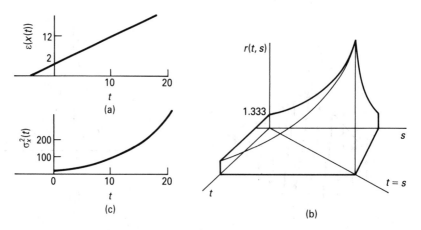

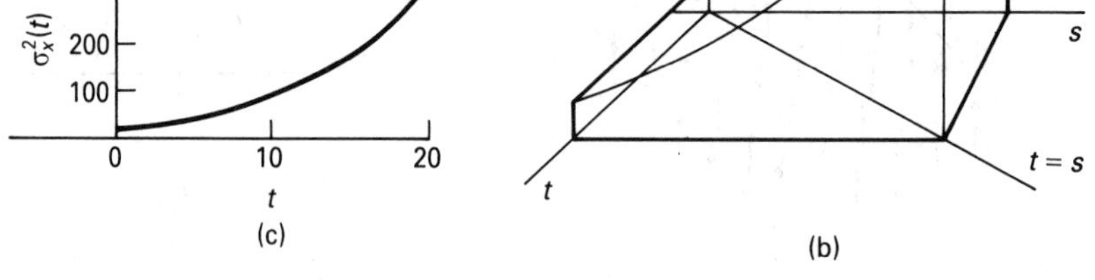

Figure 2.7 Non-stationary process.

termed the 'cross-correlation' of $x(t), y(t)$ to distinguish it from the autocorrelation (Eqn (2.6))

$$\bar{r}_{xy}(t, s) = \varepsilon(x(t) \cdot y(s))$$
$$= \int_{-\infty}^{+\infty} \int_{-\infty}^{+\infty} \xi_t \chi_s p(\xi_t, \chi_s; t, s) \, d\xi_t \, d\chi_s \quad (2.10)$$

where $p(\xi_t, \chi_s; t, s)$ is the joint PDF describing $x(t), y(s)$, as indicated in Figure 2.8(a).

If $x(t), y(t)$ are independent $\bar{r}_{xy}(t, s) = 0$, in which case $\varepsilon(x(t)) = 0$, or $\varepsilon(y(t)) = 0$, or both, and a possible joint PDF is as shown in Figure 2.8(b).

As in Section 1.14.3, if $r_{xy}(t, s) = 0$, $x(t), y(t)$ are statistically orthogonal, which usually, but not necessarily, implies that they are independent.

The 'cross-covariance' of $x(t), y(t)$ is simply the covariance function of $x(t), y(s)$ (cf. Eqn (2.8)):

$$r_{xy}(t, s) = \varepsilon[(x(t) - \varepsilon(x(t)))(y(s) - \varepsilon(y(s)))]$$
$$= \bar{r}_{xy}(t, s) - \varepsilon(x(t)). \, \varepsilon(y(s)) \quad (2.11)$$

If $x(t), y(t)$ are wide sense stationary, these functions are 'symmetrical' and depend only on the time difference, $\tau = (s - t)$:

$$\bar{r}_{xy}(t, s) = \bar{r}_{xy}(s - t) = \bar{r}_{xy}(\tau) = \bar{r}_{yx}(-\tau) = \bar{r}_{yx}(s, t)$$

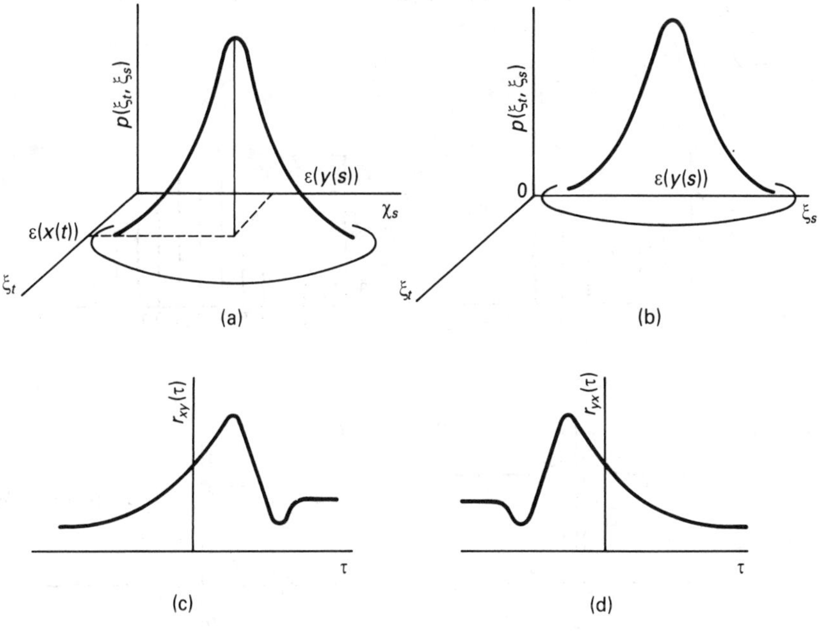

Figure 2.8 PDFs and cross-correlation functions.

Similarly,

$$r_{xy}(t, s) = r_{xy}(s - t) = r_{xy}(\tau) = r_{yx}(-\tau) = r_{yx}(s, t)$$

Example

Two dice are thrown simultaneously at regular one-second intervals, $x(t)$ being the result on one die and $y(t)$ the sum of the two results. Then $x(t), y(t)$ are clearly stationary and dependent.

Probability density functions are shown in Figure 2.9(a, b). Direct use of the various formulae involving PDFs is cumbersome, but by treating values from the ensembles as discrete variables (Section 1.5), it is not difficult to calculate the mean, cross-correlation and cross-covariance functions.

The means are

$$\varepsilon(x(t)) = \frac{1}{6}(1 + 2 + \cdots + 6) = 3.5$$

$$\varepsilon(y(s)) = \frac{1}{36}(2 + 3 \times 2 + 4 \times 3 + 5 \times 4 + \cdots + 12 \times 1) = 7.0$$

The cross-correlation is also easy to calculate, at least for $\tau = 0$ and $\tau \geqslant 1$:

$$\bar{r}_{xy}(0) = \frac{1}{36}[1 \times (2 + 3 + \cdots + 7) + 2(3 + 4 + \cdots + 8) + \cdots + 6(7 + 8 + \cdots + 12)]$$

$$= 27.417$$

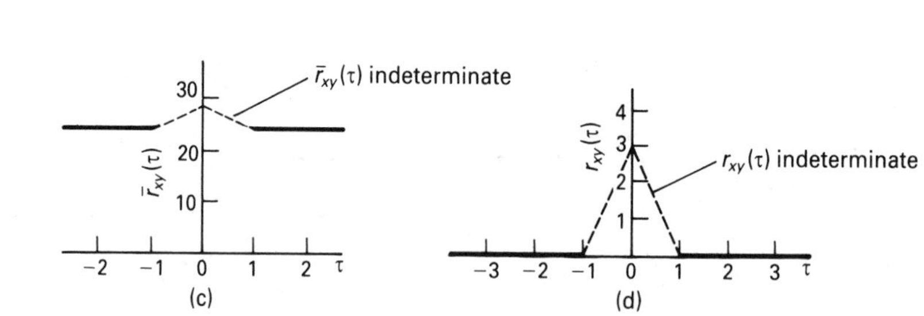

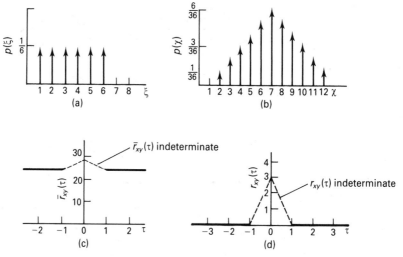

Figure 2.9 PDFs, cross-correlation and cross-covariance functions.

Hence (Eqn (2.11)),

$$r_{xy}(0) = 27.417 - (3.500)(7.000) = 2.917 \tag{2.12}$$

For $\tau \leqslant -1$, $\tau \geqslant 1$:

$$\bar{r}_{xy}(\tau) = \frac{1}{6 \times 36}[1 \times 252 + 2 \times 252 + \cdots + 6 \times 252]$$

$$= 24.500$$

$$r_{xy}(\tau) = 24.500 - (3.500)(7.000) = 0$$

For $-1 < \tau < +1$, $r_{xy}(\tau)$ depends on t and is effectively indeterminate. $\bar{r}_{xy}(\tau)$ and $r_{xy}(\tau)$ are shown in Figure 2.9(c, d).

2.4 Time-averaged properties of ergodic processes

The properties of a general stochastic process are defined in Section 2.3 in terms of data collected across its ensemble. For an ergodic process, the same properties can clearly be defined in terms of data collected from any single sample function, making them much easier to calculate in practice.

2.4.1 Time-averaged mean

Considering one sample function, $\bar{x}_i(t)$, of the ergodic process $x(t)$, the time-averaged mean is

$$m_t = \lim_{T\to\infty} \left(\frac{1}{2T}\int_{-T}^{+T} \bar{x}_i(t)\,\mathrm{d}t\right)$$

Since $x(t)$ is ergodic, m_t is the same for any sample function, and since any sample function is represented by $x(t)$:

$$m_t = \lim_{T\to\infty} \left(\frac{1}{2T}\int_{-T}^{+T} x(t)\,\mathrm{d}t\right) \tag{2.13}$$

Eqn (2.13) assumes that the integral of a stochastic function WRT time is meaningful – a workable definition is established in Section 2.6.2 – and that m_t reaches a limit as $T \to \infty$. Naturally, since $x(t)$ is ergodic, the time-averaged mean is equal to the ensemble mean (Eqn (2.4)):

$$m_t = m_x$$

2.4.2 Time-averaged autocorrelation

The time-averaged autocorrelation of a sample function $\bar{x}_i(t)$ is

$$\bar{r}_t(\tau) = \lim_{T\to\infty}\left(\frac{1}{2T}\int_{-T}^{+T} \bar{x}_i(t)\cdot\bar{x}_i(t+\tau)\,\mathrm{d}t\right)$$

$\bar{r}_t(\tau)$ is the average value of $\bar{x}_i(t)$ multiplied by itself advanced in time by τ, a definition which is sensible only if $x(t)$ is wide sense stationary. If $x(t)$ is also ergodic, $\bar{r}_t(\tau)$ is the same for any sample function, and, interpreting the integral according to Section 2.6.2:

$$\bar{r}_t(\tau) = \lim_{T\to\infty}\left(\frac{1}{2T}\int_{-T}^{+T} x(t)x(t+\tau)\,\mathrm{d}t\right) \tag{2.14}$$

Since $x(t)$ is ergodic, $\bar{r}_t(\tau)$ is equal to the ensemble-defined autocorrelation, $\bar{r}(\tau)$ (Eqn (2.6)).

It would be easy to define a time-averaged covariance function, $r_t(\tau)$, to correspond to the ensemble-defined covariance (Eqns (2.7) and (2.9)), but this proves unnecessary for most practical purposes.

Examples

(a) Consider a die tossed at regular one-second intervals. This is ergodic, at least for $|\tau| \geqslant 1$. Collecting numbers from any sample function and using the methods of Section 1.5:

$$m_t = \frac{1}{6}(1 + 2 + \cdots + 6) = 3.5$$

$$\begin{aligned}\bar{r}_t(\tau) &= \frac{1}{36}[(1\times 1 + 1\times 2 + \cdots + 1\times 6) + (2\times 1 + 2\times 2 + \cdots + 2\times 6)\\ &\quad + \cdots(6\times 1 + 6\times 2 + \cdots + 6\times 6)]\\ &= 15.167\end{aligned}$$

These results agree with the ensemble results (Section 2.3.2, Example (b)).

(b) Consider the periodic stochastic process:

$$x(t) = \cos(\omega_0 t + \alpha)$$

where ω_0 is constant and α is a random variable evenly distributed between $\pm\pi$.

The ensemble (Figure 2.10(a)) shows $x(t)$ to be ergodic with zero mean:

$$m_x = m_t = 0$$

The ensemble autocorrelation (Eqn (2.6))

$$\bar{r}(\tau) = \varepsilon(x(t)\cdot x(t+\tau))$$
$$= \varepsilon(\cos(\omega_0 t + \alpha)\cdot\cos(\omega_0(t+\tau)+\alpha))$$
$$= \cos(\omega_0\tau)$$

This is illustrated in Figure 2.10(b).

Since $m_x = 0$, the covariance is the same (Eqn (2.8)):

$$r(\tau) = \tfrac{1}{2}\cos(\omega_0\tau)$$

The time-averaged autocorrelation (Eqn (2.14)) is

$$\bar{r}_t(\tau) = \frac{1}{2\pi}\int_{-\pi}^{+\pi}\cos(\omega_0 t+\alpha)\cos(\omega_0(t+\tau)+\alpha)\,\mathrm{d}t$$
$$= \tfrac{1}{2}\cos(\omega_0\tau)$$
$$= \bar{r}(\tau)$$

2.4.3 Time-averaged cross-correlation

The cross-correlation between the sample functions $\bar{x}_i(t), \bar{y}_j(t)$ of two stochastic processes, $x(t), y(t)$ is

$$\bar{r}_{txy}(\tau) = \lim_{T\to\infty}\left(\frac{1}{2T}\int_{-T}^{+T}\bar{x}_i(t)\bar{y}_j(t+\tau)\,\mathrm{d}t\right)$$

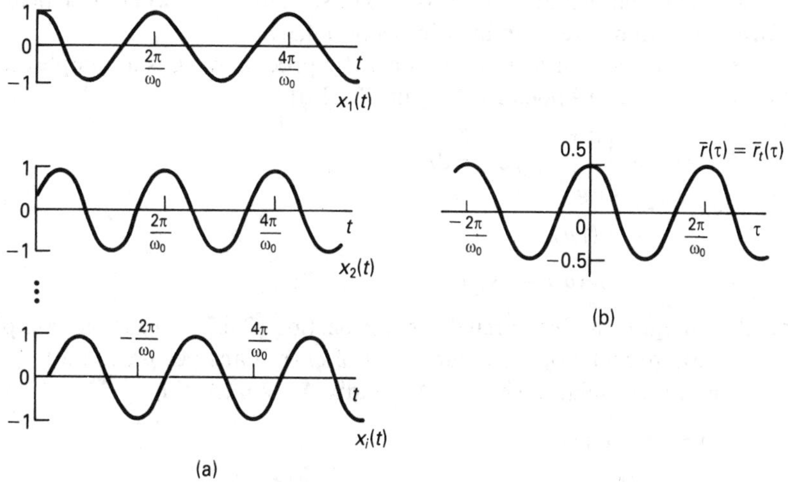

Figure 2.10 Stochastic process.

$\bar{r}_{txy}(\tau)$ is meaningful – or at least useful – only for ergodic processes, where the definition holds for any pair of sample functions. Thus, with the integral interpreted according to Section 2.6.2:

$$\bar{r}_{txy}(\tau) = \lim_{T\to\infty}\left(\frac{1}{2T}\int_{-T}^{+T} x(t)y(t+\tau)\,\mathrm{d}t\right) \tag{2.15}$$

In this case, $r_{txy}(\tau) = r_{xy}(\tau)$, the ensemble cross-correlation function (Eqn (2.10)).

Example

Consider two dice, tossed once per second, where $x(t)$ is the result on one die, and $y(t)$ the sum of the results on both dice. From the statistics of any two sample functions (Figure 2.9), or, more conveniently, using the methods of Section 1.5:

$$\begin{aligned}\bar{r}_{txy}(\tau) &= \frac{1}{36}[(1\times 2 + 1\times 3 + \cdots + 1\times 7) + (2\times 3 + 2\times 4 + \cdots + 2\times 8)\\ &\quad + \cdots(6\times 7 + 6\times 8 + \cdots + 6\times 12)]\\ &= 24.5, \quad |\tau| \geqslant 1\end{aligned}$$

This agrees with the ensemble cross-correlation found for the same example in Section 2.3.3.

2.5 Frequency domain considerations

The Fourier transform (Appendix A, Section A1.3) is a mathematical tool whose application to stationary stochastic processes yields important analytical and design techniques. The mathematics of this is considered here, while practical Fourier transform algorithms are examined in Section 4.3.2.

The Fourier transform of a stochastic process, $x(t)$, is a complex stochastic function of frequency (Appendix A, Eqn (A1.20)):

$$\begin{aligned}\mathcal{F}(x(t)) &= \int_{-\infty}^{+\infty} x(t)\,\mathrm{e}^{-\mathrm{j}\omega t}\,\mathrm{d}t\\ &= X(\mathrm{j}\omega)\\ &= X_R(\omega) - \mathrm{j}X_I(\omega)\end{aligned}$$

where the integral is interpreted as in Section 2.6.2, $X(\mathrm{j}\omega)$ is complex, and $X_R(\omega)$, $X_I(\omega)$ are real stochastic functions of ω (radians per unit time).

The inverse relationship is (Appendix A, Eqn (A1.21))

$$\begin{aligned}x(t) &= \mathcal{F}^{-1}(X(\mathrm{j}\omega))\\ &= \frac{1}{2\pi}\int_{-\infty}^{+\infty} X(\mathrm{j}\omega)\,\mathrm{e}^{\mathrm{j}\omega t}\,\mathrm{d}\omega\end{aligned}$$

Being itself stochastic, $X(j\omega)$ is not of significantly greater analytical value than $x(t)$, though it does give some indication of the frequency content of $x(t)$.

Example

Consider the ergodic (and therefore wide sense stationary) process

$$x(t) = \cos(\omega_0 t + \alpha)$$

where ω_0 is constant and α is evenly distributed between $\pm\pi$. From Appendix A, Figure A1.6:

$$\begin{aligned} X(j\omega) &= \mathscr{F}\left[\cos\left(\omega_0\left(t + \frac{\alpha}{\omega_0}\right)\right)\right] \\ &= X_R(\omega) - jX_I(\omega) \\ &= \pi(\delta(\omega - \omega_0) + \delta(\omega + \omega_0))\left[\cos\left(\frac{\alpha\omega}{\omega_0}\right) - j\sin\left(\frac{\alpha\omega}{\omega_0}\right)\right] \end{aligned}$$

Some sample functions of $x(t)$ and $X(j\omega)$ are shown in Figure 2.11.

A much more practical, deterministic, measure of the frequency content of $x(t)$ is provided by its 'power spectral density', or PSD, $S(\omega)$, describing the power per unit bandwidth in $x(t)$ (typically W rad^{-1} s^{-1}). Clearly, $S(\omega)$ is meaningful only for wide sense stationary processes and, since it describes the frequency content

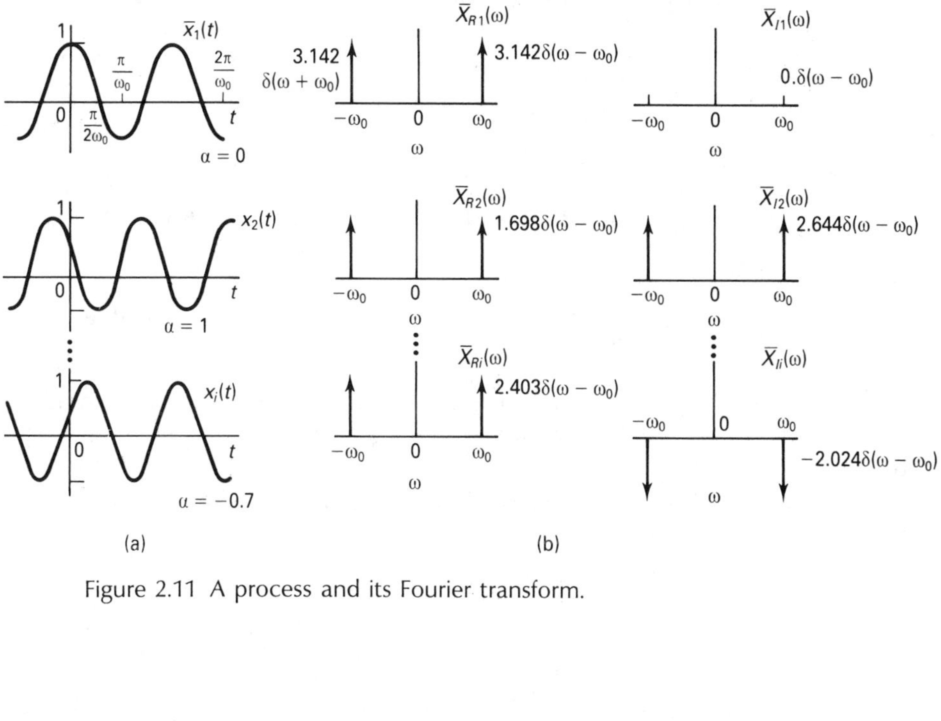

Figure 2.11 A process and its Fourier transform.

of $x(t)$, it is related to its autocorrelation function, $r(\tau)$ (Eqn (2.6)), describing its rate of change. In fact, $S(\omega)$ and $r(\tau)$ form a Fourier transform pair:

$$S(\omega) = \mathscr{F}(\bar{r}(\tau)) = \int_{-\infty}^{+\infty} \bar{r}(\tau)\,\mathrm{e}^{-\mathrm{j}\omega\tau}\,\mathrm{d}\tau \tag{2.16}$$

$$\bar{r}(\tau) = \mathscr{F}^{-1}(S(\omega)) = \frac{1}{2\pi}\int_{-\infty}^{+\infty} S(\omega)\,\mathrm{e}^{\mathrm{j}\omega\tau}\,\mathrm{d}\omega \tag{2.17}$$

This relationship is known as the Wiener–Khinchine theorem [refs 1,2].

Since $\bar{r}(\tau)$ is even and real for a wide sense stationary process, $S(\omega)$ is real and even (Appendix A, Section A1.3).

Examples

Correlation functions and PSDs typically involve substantial arrays of data which are dealt with by computer algorithms (Section 4.3). Simple closed-form analytical functions are rather artificial in this context, but are useful for illustrative purposes.

(a) Consider the ergodic process

$$x(t) = \cos(\omega_0 t + \alpha)$$

ω_0 is constant, α is evenly distributed between $\pm\pi$ (Figure 2.10(a)):

$$\begin{aligned} r(\tau) = \bar{r}(t, s) &= \varepsilon(x(t)\cdot x(s)) \\ &= \varepsilon(\cos(\omega_0 t + \alpha)\cdot\cos(\omega_0 s + \alpha)) \\ &= \frac{1}{2}\varepsilon(\cos\omega_0(s - t) + \cos(\omega_0(s + t) + 2\alpha)) \\ &= \frac{1}{2}\cos\omega_0\tau \end{aligned}$$

The PSD is given by the Fourier transform of this (Appendix A, Figure A1.6):

$$\begin{aligned} S(\omega) &= \mathscr{F}(\bar{r}_t(\tau)) \\ &= \frac{\pi}{2}[\delta(\omega - \omega_0) + \delta(\omega + \omega_0)] \end{aligned}$$

These functions are illustrated in Figure 2.12(a), where it can be seen that all the power in any sample function occurs at $\omega = \omega_0$, corresponding to the fact that $S(\omega) = 0, \omega \neq \omega_0$.

(b) A commonly quoted example is the process with autocorrelation:

$$\bar{r}(\tau) = \frac{a}{2}\mathrm{e}^{-a|\tau|}$$

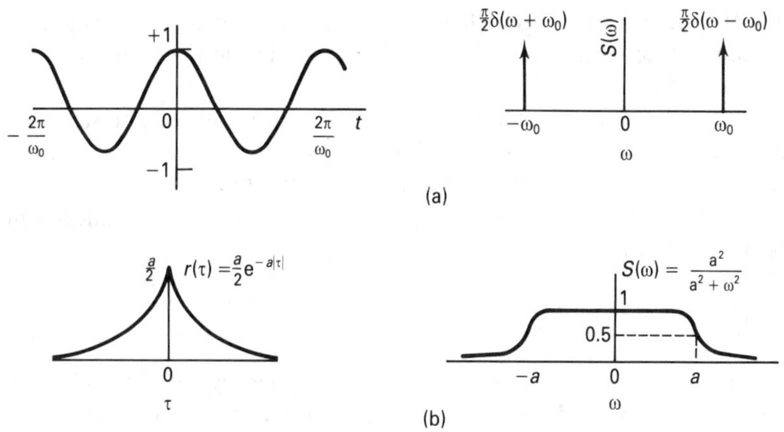

Figure 2.12 Autocorrelation and PSD.

The Fourier transform of this yields (Appendix A, Figure A1.6)

$$S(\omega) = \int_{-\infty}^{+\infty} \frac{a}{2} e^{-a|\tau|} e^{-j\omega\tau} d\tau$$

$$= \frac{a^2}{a^2 + \omega^2}$$

These functions are illustrated in Figure 2.12(b). In this case $S(\omega)$ indicates that the power content is almost constant for $|\omega| < a$, and almost zero for $|\omega| \gg a$.

The 'cross-spectral density', $S_{xy}(j\omega)$, of two wide sense stationary processes, $x(t)$, $y(t)$, is correspondingly defined as the Fourier transform of their cross-correlation function $\bar{r}_{xy}(\tau)$ (Eqn (2.10)):

$$S_{xy}(j\omega) = \mathscr{F}(\bar{r}_{xy}(\tau)) = \int_{-\infty}^{+\infty} \bar{r}_{xy}(\tau) e^{-j\omega\tau} d\tau \tag{2.18}$$

$$\bar{r}_{xy}(\tau) = \mathscr{F}^{-1}(S_{xy}(\omega)) = \frac{1}{2\pi} \int_{-\infty}^{+\infty} S(\omega) e^{j\omega\tau} d\omega \tag{2.19}$$

Since $\bar{r}_{xy}(\tau)$ is not usually even, $S_{xy}(j\omega)$ is usually complex, with the property that since $\bar{r}_{xy}(\tau) = \bar{r}_{xy}(-\tau)$, $S_{xy}(j\omega)$ and $S_{yx}(j\omega)$ are complex conjugates:

$$S_{xy}(j\omega) = S_{yx}^{*}(j\omega); \quad S_{yx}(j\omega) = S_{xy}^{*}(j\omega)$$

2.6 Differentiation and integration of stochastic processes

Having defined the stochastic process, $x(t)$, it is natural to define its differential coefficient and integral WRT time with a view to developing models of dynamical systems. Before proceeding to such definitions, however, it is necessary to give meaning to the concept of 'convergence' as it applies to random variables (and therefore processes).

Consider a set of numbers $(\bar{x}_1, \bar{x}_2, \ldots, \bar{x}_i, \ldots)$. This can be said to 'converge' to the random variable x, as $i \rightarrow \infty$, in various ways, of which perhaps the most straightforward is 'convergence in the mean square' [ref. 2].

The set of numbers $(\bar{x}_1, \bar{x}_2, \ldots, \bar{x}_i, \ldots)$ converges to the random number, x, 'in the mean square' if

$$\lim_{i \rightarrow \infty} [\varepsilon(\bar{x}_i - x)^2] = 0 \tag{2.20}$$

Where convergence occurs, the numbers $\bar{x}_i$ are generally equal to the elements of x, but there is room within this definition for some numbers $(\bar{x}_i - x)$ to be non-zero, provided they are 'few enough' to allow Eqn (2.20) to be satisfied. If Eqn (2.20) is satisfied, the 'limit in the mean square' of the set $(\bar{x}_1, \bar{x}_2, \ldots, \bar{x}_i, \ldots)$ is said to equal x:

$$\mathrm{LIM}(\bar{x}_i) = x \tag{2.21}$$

Other definitions of convergence are possible, notably 'convergence with probability one' [ref. 2], but these are not explored here.

2.6.1 Differentiation WRT time

If the sample functions of a stochastic process, $x(t)$, are differentiated WRT time, they form the sample functions of a new stochastic process. Heuristically, this leads to the definition

$$\dot{x}(t) = \frac{\mathrm{d}x}{\mathrm{d}t} = \lim_{\delta t \rightarrow 0} \left[\frac{x(t + \delta t) - x(t)}{\delta t} \right]$$

If a 'few' sample functions contain 'steps' and cannot be differentiated over all t, a sensible definition is not precluded: the 'mean square derivative', defined using Eqn (2.21), is

$$\dot{x}(t) = \frac{\mathrm{d}x}{\mathrm{d}t} = \mathrm{LIM}\left[\frac{\bar{x}_i(t + \delta t) - \bar{x}_i(t)}{\delta t} \right] \tag{2.22}$$

That is, $\dot{x}(t)$ is such that

$$\lim_{\substack{i \rightarrow \infty \\ \delta t \rightarrow 0}} \left\{ \varepsilon \left[\frac{\bar{x}_i(t + \delta t) - \bar{x}_i(t)}{\delta t} - \frac{\mathrm{d}x}{\mathrm{d}t} \right]^2 \right\} = 0$$

This definition turns out to be substantially more useful than the restricted heuristic one; it is adhered to throughout this book.

Some properties of $\dot{x}(t)$, as defined by Eqn (2.22), are as follows.

1. $\dot{x}(t)$ exists, i.e. $x(t)$ is mean square differentiable, if and only if $\partial^2 r(t,s)/\partial t\,\partial s$ exists for all t, s, where $r(t,s)$ is the covariance of $x(t)$ (Eqns (2.7)).
2. The mean of $\dot{x}(t)$ is

$$\varepsilon\left(\frac{\mathrm{d}x}{\mathrm{d}t}\right)=\frac{\mathrm{d}}{\mathrm{d}t}[\varepsilon(x(t))] \tag{2.23}$$

3. The covariance of $\dot{x}(t)$ is

$$\begin{aligned}\operatorname{cov}\left(\frac{\mathrm{d}x}{\mathrm{d}t}\right)&=\varepsilon\left[\left(\frac{\mathrm{d}x}{\mathrm{d}t}-\varepsilon\left(\frac{\mathrm{d}x}{\mathrm{d}t}\right)\right)\left(\frac{\mathrm{d}x}{\mathrm{d}s}-\varepsilon\left(\frac{\mathrm{d}x}{\mathrm{d}s}\right)\right)\right]\\&=\frac{\partial^2 r(t,s)}{\partial t\,\partial s}\end{aligned} \tag{2.24}$$

4. The cross-covariances of $x(t)$ and $\dot{x}(t)$ are

$$\left.\begin{aligned}r_{x\dot{x}}(t,s)&=\varepsilon\left[(x(t)-\varepsilon x(t))\left(\frac{\mathrm{d}x}{\mathrm{d}t}-\varepsilon\left(\frac{\mathrm{d}x}{\mathrm{d}t}\right)\right)\right]\\&=\frac{\partial r(t,s)}{\partial s}\\r_{\dot{x}x}(t,s)&=\frac{\partial r(t,s)}{\partial t}\end{aligned}\right\} \tag{2.25}$$

5. If $x(t)$ is wide sense stationary, $r(t,s)=r(s-t)=r(\tau)$ and:

$$\begin{aligned}\operatorname{cov}\left(\frac{\mathrm{d}x}{\mathrm{d}t}\right)&=\frac{\partial^2 r(t,s)}{\partial t\,\partial s}\\&=-\frac{\partial}{\partial(s-t)}\left(\frac{\partial r(s-t)}{\partial(s-t)}\right)\\&=-\frac{\partial^2 r}{\partial\tau^2}\end{aligned} \tag{2.26}$$

Example

Consider the process:

$$x(t)=\cos(\omega_0 t+\alpha)$$

ω_0 constant, α a random number evenly distributed between $\pm\pi$. Sample functions are shown in Figure 2.13(a).

$\dot{x}(t)$ is the stochastic process

$$\dot{x}(t)=\frac{\mathrm{d}x}{\mathrm{d}t}=-\omega_0\sin(\omega_0 t+\alpha)$$

Sample functions of $\dot{x}(t)$ are shown in Figure 2.13(b).

It is of interest to check formally that $\dot{x}(t)$ exists (though it obviously does), and to find its properties.

(a) The covariance of $x(t)$ is (Figure 2.10(b))

$$r(t, s) = \tfrac{1}{2}\cos\omega_0(s-t) = \tfrac{1}{2}\cos(\omega_0\tau), \quad \tau = s - t$$

Hence (Eqn (2.24)),

$$\operatorname{cov}\left(\frac{\mathrm{d}x}{\mathrm{d}t}\right) = \frac{\partial^2 r(t,s)}{\partial t\,\partial s} = \omega_0^2\cos(\omega_0(s-t)) = \omega_0^2\cos(\omega_0\tau)$$

Since this exists for all t, s, $\dot{x}(t)$ exists. These functions are illustrated in Figure 2.13(c, d).

(b) The mean of $\dot{x}(t)$ is (Eqn (2.23)):

$$\varepsilon\left(\frac{\mathrm{d}x}{\mathrm{d}t}\right) = \frac{\mathrm{d}}{\mathrm{d}t}(\varepsilon(x(t))) = 0$$

(c) The cross-covariances of $x(t), \dot{x}(t)$ are (Eqn (2.25))

$$\begin{aligned} r_{x\dot{x}}(\tau) &= \operatorname{cov}\left(x(t), \frac{\mathrm{d}x}{\mathrm{d}t}\right) \\ &= \frac{\partial r(t,s)}{\partial s} \\ &= -\tfrac{1}{2}\omega_0\sin(\omega_0\tau) \end{aligned}$$

$$r_{\dot{x}x}(\tau) = \operatorname{cov}\left(\frac{\mathrm{d}x}{\mathrm{d}t}, x(s)\right) = \frac{\partial r(t,s)}{\partial t} = \tfrac{1}{2}\omega_0\sin(\omega_0\tau)$$

As expected (Eqn (2.25)),

$$r_{x\dot{x}}(\tau) = r_{\dot{x}x}(-\tau)$$

These functions are illustrated in Figure 2.13(e, f).

2.6.2 Integration WRT time

If the sample functions of a stochastic process, $x(t)$, are integrated over an interval of time $[a, b]$, a random variable, x_{Iab}, is generated:

$$x_{Iab} = \int_a^b x(t)\,\mathrm{d}t \tag{2.27}$$

Eqn (2.27) is interpreted throughout this book as a straightforward Riemann integral (Appendix E, Section E1), each sample function of $x(t)$ being integrated to produce an element of the random variable x_{Iab}. This incorporates more detailed (and complex) definitions such as 'integration in the mean square' and 'integration with probability one' [ref. 2] and is adequate for most practical models.

Some properties of x_{Iab} are as follows:

1. x_{Iab} exists if and only if

$$\int_a^b \varepsilon(x(t))\,dt \text{ and } \int_a^b \int_a^b r(t,s)\,dt\,ds$$

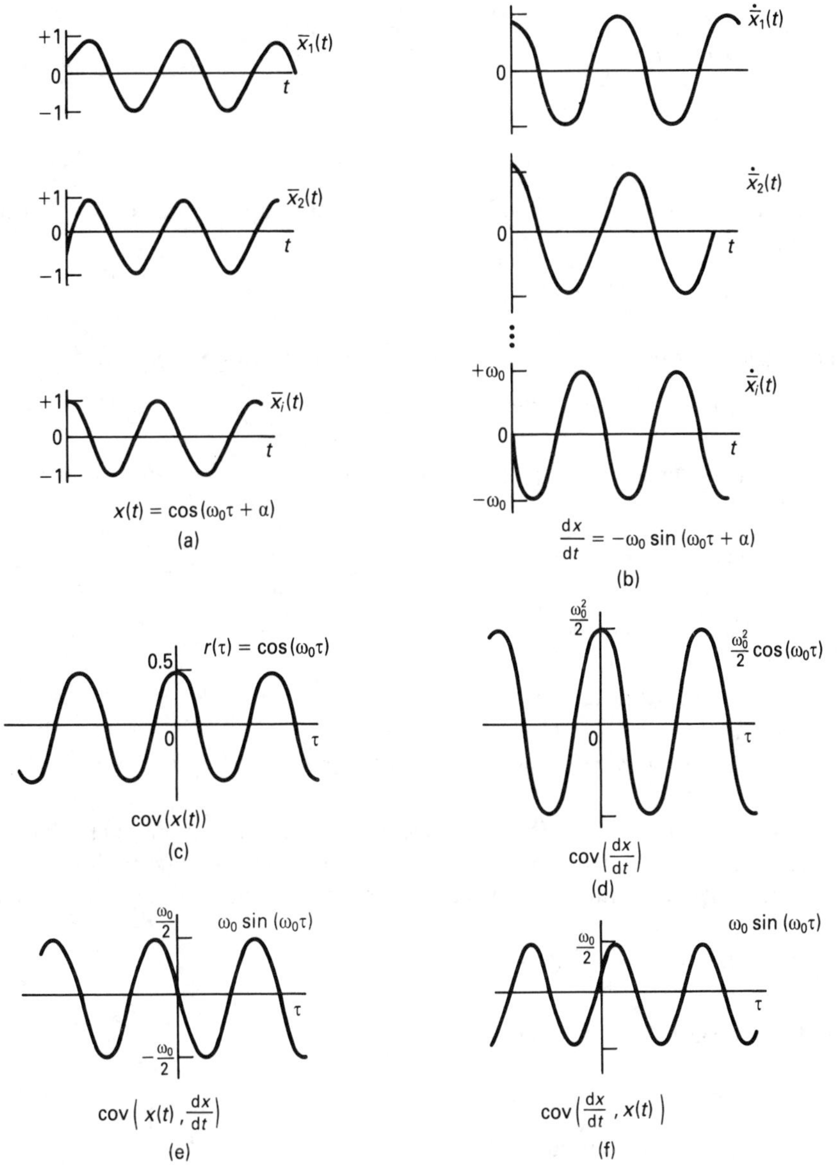

Figure 2.13 Differentiation of a stochastic process.

exist.

2. The mean of $x_{\mathrm{I}ab}$ is

$$\varepsilon(x_{\mathrm{I}ab}) = \int_a^b \varepsilon(x(t))\,\mathrm{d}t$$

3. The variance of $x_{\mathrm{I}ab}$ is

$$\begin{aligned}\operatorname{var}(x_{\mathrm{I}ab}) &= \varepsilon\left[\left(\int_a^b x(t)\,\mathrm{d}t - \varepsilon\int_a^b x(t)\,\mathrm{d}t\right)^2\right] \\ &= \int_a^b\int_a^b r(t,s)\,\mathrm{d}t\,\mathrm{d}s \qquad (2.28)\end{aligned}$$

Open integration is readily defined by taking the upper limit of integration to be variable

$$x_{\mathrm{I}}(t) = \int_{t_0}^t x(\lambda)\,\mathrm{d}\lambda \qquad (2.29)$$

This yields a stochastic process, $x_{\mathrm{I}}(t)$, with the following properties.

1. $x_{\mathrm{I}}(t)$ exists if and only if

$$\int_{t_0}^t \varepsilon(x(\lambda))\,\mathrm{d}\lambda \text{ and } \int_{t_0}^t\int_{t_0}^s r(\lambda,\mu)\,\mathrm{d}\lambda\,\mathrm{d}\mu$$

exist.

2. The mean of $x_{\mathrm{I}}(t)$ is

$$\varepsilon(x_{\mathrm{I}}(t)) = \int_{t_0}^t \varepsilon(x(\lambda))\,\mathrm{d}\lambda \qquad (2.30)$$

3. The covariance of $x_{\mathrm{I}}(t)$ is

$$r_{\mathrm{I}}(t,s) = \operatorname{cov}(x_{\mathrm{I}}(t)) = \int_{t_0}^t\int_{t_0}^s r(\lambda,\mu)\,\mathrm{d}\lambda\,\mathrm{d}\mu \qquad (2.31)$$

4. The integration of Eqn (2.29) can legitimately be interpreted in a mean square sense, and is consequently the inverse of mean square differentiation (Eqn (2.22)). The constant of integration, usually a random variable, must of course be remembered:

$$x(t) = \int_{t_0}^t \frac{\mathrm{d}x(\lambda)}{\mathrm{d}\lambda}\cdot\mathrm{d}\lambda + C \qquad (2.32)$$

where C is (typically) a random variable.

Example

Consider the process (Figure 2.10)

$$x(t) = \cos(\omega_0 t + \alpha), \quad \varepsilon(x(t)) = 0, \quad r(t,s) = \tfrac{1}{2}\cos(\omega_0(s-t))$$

ω_0 constant, α evenly distributed between $\pm\pi$. Then, for $t_0 = 0$ (Eqn (2.29)), the integral process is

$$x_I(t) = \int_{t_0}^{t} (\cos \omega_0 \lambda + \alpha)\, d\lambda = \frac{1}{\omega_0} \sin(\omega_0 t + \alpha)$$

The mean and covariance of $x_I(t)$ are (Eqns (2.30) and (2.31))

$$\varepsilon(x_I(t)) = \tfrac{1}{2} \int_{0}^{t} 0\, d\lambda = 0$$

$$r_I(t, s) = \int_{t_0}^{t} \int_{t_0}^{s} \cos \omega_0(\mu - \lambda)\, d\lambda\, d\mu$$

$$= \frac{1}{2\omega_0^2} \cos(\omega_0(s - t))$$

This is easily verified (Eqn (2.7)):

$$r_I(t, s) = \varepsilon\left(\frac{1}{\omega_0} \sin(\omega_0 t + \alpha) \cdot \frac{1}{\omega_0} \sin(\omega_0 s + \alpha)\right)$$

$$= \frac{1}{2\omega_0^2} \cos(\omega_0(s - t))$$

Finally (cf. Eqn (2.32)), the differential coefficient of $x_I(t)$ is

$$\frac{d'}{dt}(x_I(t)) = \frac{d}{dt}\left(\frac{1}{\omega_0} \sin(\omega_0 t + \alpha)\right)$$

$$= \cos(\omega_0 t + \alpha)$$

$$= x(t)$$

2.7 Some important processes

The stochastic processes which are recognized as fundamental to system analysis and modelling are considered in this section. Algorithms for generating approximations to some of these are described in Section 4.2.

2.7.1 Markov processes

A process $x(t)$ is 'Markov' if its statistics at time t_j depend at most on its value at any one previous time, $x(t_{j-1})$, and not on its longer term history. This is stated formally in terms of the conditional PDF describing $x(t_j)$:

$$p(\xi_j | x(t_{j-1}), x(t_{j-2}), \ldots, x(t_{j-k})) = p(\xi_j | x(t_{j-1})) \tag{2.33}$$

where ξ_j refers to $x(t_j)$.

Thus, for a Markov process, knowledge of $x(t_{j-1})$ is sufficient to specify the PDF of $x(t_j)$; the previous values, $x(t_{j-2}), x(t_{j-3}), \ldots, x(t_{j-k}), \ldots$, are irrelevant.

Since non-Markov processes are generally mathematically intractable, most processes are modelled and analyzed under the assumption that they are Markov, and this applies to all the processes considered to any depth in this book. Fortunately, many practical processes can usefully be modelled under this assumption.

The most important properties of Markov processes are as follows.

1. A Markov process, $x(t)$, can be specified 'totally' in terms of its (infinite dimensional) joint PDF (Eqn (2.1)) if an initial PDF $p(\xi_{j-k})$ and an expression for $p(\xi_j|x(t_{j-k})$ are known.

$$p(\xi_j, \xi_{j-1}, \ldots, \xi_{j-k}) = p(\xi_j|x(t_{j-1}))\cdot p(\xi_{j-1}|x(t_{j-2}))\ldots p(\xi_{j-k+1}|x(t_{j-k}))\cdot p(\xi_{j-k}) \tag{2.34}$$

2. If $x(t)$ is strictly stationary, the terms in Eqns (2.33) and (2.34) are time invariant, in which case $x(t)$ is specified totally by the second order PDF:

$$p(\xi_j, \xi_{j-1}) = p(\xi_j|x(t_{j-1}))\cdot p(\xi_{j-1})$$

This can be used with Eqn (2.34) to construct a joint PDF of any order.

3. A continuous time 'Markov chain' process has an ensemble of 'staircase' functions with discontinuities at regular intervals.

4. A non-stationary Markov process is 'homogeneous' if $p(\xi_j)$ is a function of $(j-1)$, i.e. of time, but $p(\xi_j|x(t_{j-1}))$ is not.

Examples

(a) All the examples considered so far in this chapter are Markov processes, since for every one of them:

$$p(\xi_j|x(t_{j-1}), x(t_{j-2}), \ldots, x(t_{j-k})) = p(\xi_j|x(t_{j-1}))$$

(b) A well-known experimental Markov process is the 'random walk', archetypically generated by tossing a coin at regular intervals, $t = nT$, and taking a step of length λ to the right if the result is a head, to the left if it is a tail. Sample functions, for which $x(t)$ represents the distance travelled from the starting point, are shown in Figure 2.14(a).

If $(t_j - t_{j-1}) < T$, $p(\xi_j|x(t_{j-1}))$ depends on t and is not perhaps of great interest. However, if

$$(t_j - t_{j-1}) \geqslant T$$

$$p(\xi_j|x(t_{j-1}), x(t_{j-2})\ldots) = p(\xi_j|x(t_{j-1}))$$

$$= 0.5[\delta(\xi_j - x(t_{j-1}) + \lambda) + \delta(\xi_j - x(t_{j-1}) - \lambda)]$$

Thus, for $(t_j - t_{j-1}) > T$, $x(t)$ is Markov. Furthermore, $x(t)$ is non-stationary, it is a Markov chain, and it is homogeneous. The PDF of $x(t)$ is easy to calculate. The number of heads, r, occurring in n tosses ($t = nT$), which defines the number of steps to the right, is given by the binomial distribution (Eqn (1.8)):

$$\text{Prob}(r \text{ heads}) = \binom{n}{r}(0.5)^n$$

Clearly,

$$x(nT) = (r - (n - r))\lambda = (2r - n)\lambda = m\lambda, \quad 0 \leqslant r \leqslant n$$

where

$$m = -n, -n+2, \ldots, -n-2, n$$

Thus,

$$\text{Prob}(x(nT) = m\lambda) = \binom{n}{\frac{n+m}{2}}(0.5)^n$$

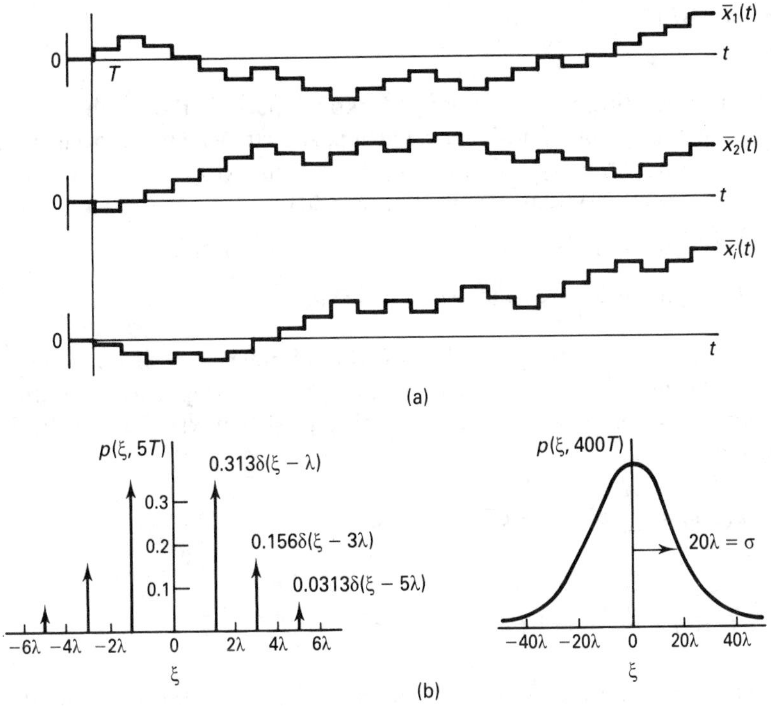

Figure 2.14 Random walk.

Equivalently (Figure 2.14(b)),

$$p(\xi, nT) = \sum_{m=-n}^{+n} \binom{n}{\frac{n+m}{2}} (0.5)^n \delta(\xi - m\lambda) \tag{2.35}$$

It is also instructive to regard $x(nT)$ as the sum of n independent, and therefore uncorrelated, steps:

$$x(nT) = e(T) + e(2T) + \cdots + e(nT)$$

where

$$e(iT) = \pm\lambda, \text{ and } \varepsilon(e(iT)) = 0, \ \varepsilon[(e(iT))^2] = \lambda^2$$

The mean and variance can now be calculated using Eqns (1.43) and (1.44):

$$\varepsilon(x(nT)) = n0 = 0$$

$$\sigma_x^2(nT) = \varepsilon[(x(nT))^2] = n\lambda^2$$

For n large, in accordance with the Central Limit Theorem (Section 1.13), $x(nT)$ becomes Gaussian:

$$p(\xi, nT) = \frac{1}{\lambda\sqrt{2\pi n}} \exp\left(-\frac{\xi^2}{2n\lambda^2}\right) \tag{2.36}$$

(c) An interesting non-stationary Markov process is the 'birth process', which describes the population of a biological group. A birth, which can occur at any time, increases the population by 1. Some sample functions, $x(0) = 1$, are shown in Figure 2.15(a). The probability that $x(t) = m$ at $(t + \Delta t)$ is

$$\text{Prob}(x(t + \Delta t) = m | x(t) = m) = 1 - q_m \Delta t, \quad m \geqslant 1$$

$$\text{Prob}(x(t + \Delta t) = m | x(t) = m - 1) = q_{m-1} \Delta t, \quad m > 1$$

where $q_m \Delta t$ is the probability of a birth within an interval Δt if $x(t) = m$.

If the probability of a birth in unit time is proportional to the existing population, $q_m = cm$, then it can be shown [ref. 3] that $x(t)$ has the PDF (Figure 2.15(b), $c = 0.1$)

$$p(\xi, t) = \sum_{i=1}^{m} \mathrm{e}^{-ct}(1 - \mathrm{e}^{-ct})^{\xi - 1} \cdot \delta(\xi - i)$$

Also,

$$\varepsilon(x(t)) = \mathrm{e}^{ct}, \quad \varepsilon(x^2(t)) = 2\mathrm{e}^{2ct} - \mathrm{e}^{ct}$$

Examination of these functions and Figure 2.15 shows $x(t)$ to be non-stationary, Markov, and non-homogeneous.

2.7.2 Gaussian processes

A process $x(t)$ is 'Gaussian' (or 'normal') if the joint PDF which specifies it totally is Gaussian (Eqn (1.34)):

$$p(\xi_1, \xi_2, \ldots, \xi_k; \quad t_1, t_2, \ldots, t_k) = [(2\pi)^{k/2} \det \mathbf{P}_k]^{-1} \exp\left[-\frac{1}{2}(\boldsymbol{\xi} - \mathbf{m}_k)^T \mathbf{P}_k^{-1} (\boldsymbol{\xi} - \mathbf{m}_k) \right] \tag{2.37}$$

where

$$\mathbf{m}_k = \varepsilon(x(t_1) \quad x(t_2) \quad \ldots \quad x(t_k))^T$$

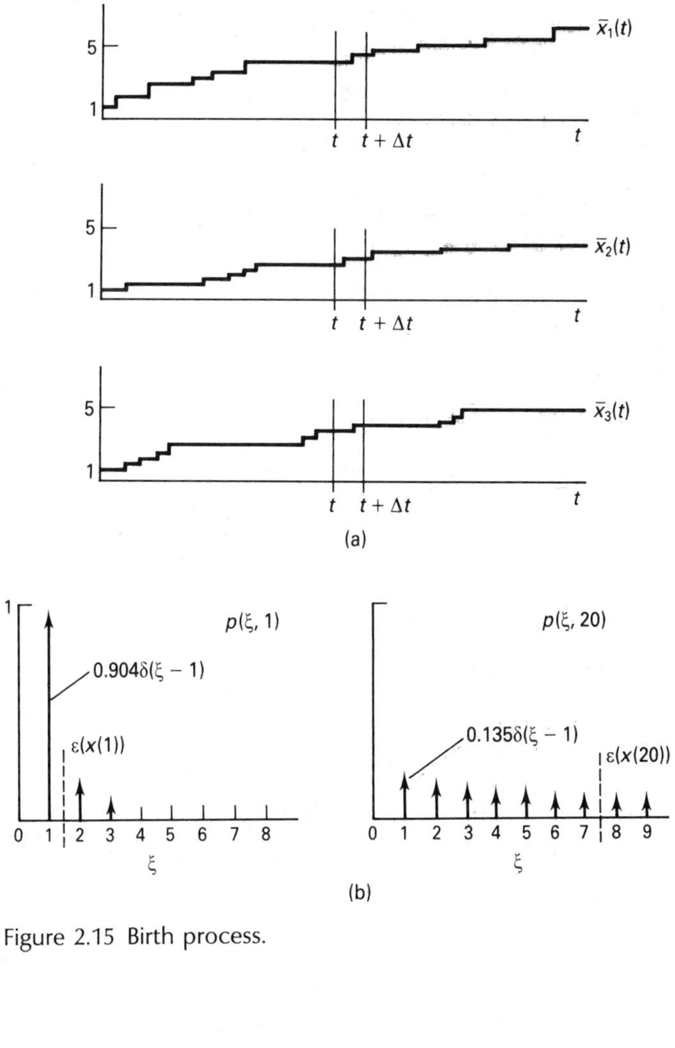

Figure 2.15 Birth process.

and

$$\mathbf{P}_k = \begin{pmatrix} r(t_1, t_1) & r(t_1, t_2) & \dots & r(t_1, t_k) \\ r(t_2, t_1) & r(t_2, t_2) & & r(t_2, t_k) \\ \vdots & \vdots & & \vdots \\ r(t_k, t_1) & r(t_k, t_2) & & r(t_k, t_k) \end{pmatrix}$$

It is easy to see that Eqn (2.37) can be evaluated if $\varepsilon(x(t))$ and $r(t, s)$ are known, and it follows that knowledge of these does specify a Gaussian process totally (Eqn (2.1)), in contrast to most other types of process.

If $x(t)$ is Gaussian, this naturally implies that its PDF $p(\xi, t)$ is Gaussian; the reverse is not necessarily true. Another interesting property, which is a consequence of the structure of Eqn (2.37), is that for Gaussian processes wide sense stationarity implies strict stationarity, again in contrast to most other types of process.

Many natural processes are Gaussian for reasons explained by the Central Limit Theorem (Section 1.13).

Example

The thickness, $x(t)$ (mm), of steel strip leaving a mill is found to be Gaussian with mean (mm) and covariance (mm^2):

$$\varepsilon(x(t)) = 5$$

$$r(t, s) = r(\tau) = 0.64e^{-|\tau|}, \quad \tau = s - t$$

$$r(0) = 0.64$$

$$\sigma_x = \sqrt{0.64} = 0.80$$

The standard normal form variate, q, is introduced (Section 1.9):

$$q = \frac{x - \varepsilon(x)}{\sigma_x}$$

Thus, from the SNF table (Appendix C, Figure C3.2):

$$\text{Prob}(x \leqslant 4) = \text{Prob}\left(q \leqslant -\frac{1}{0.8}\right) = 0.106$$

$$\text{Prob}(3 < x \leqslant 7) = \text{Prob}(-2.5 < q \leqslant +2.5) = 0.988$$

The joint PDF for any number of points can be constructed according to Eqn (2.37). Consider $t_1 = 1$, $t_2 = 3$; $\tau = t_2 - t = 2$. Then:

$$\mathbf{m}_2 = (5 \quad 5)^T$$

$$\mathbf{P}_2 = \begin{pmatrix} 0.640 & 0.087 \\ 0.087 & 0.640 \end{pmatrix}$$

Substitution in Eqn (2.37) yields $p(\xi_1, \xi_2; 1, 3)$. The calculation of probabilities from this is, of course, substantially more complicated than in the single-dimensional case.

2.7.3 White noise

White noise is defined as a wide sense stationary process with PSD (Eqn (2.16)) constant over all frequencies.

For such a process, $w(t)$, of 'strength' q the PSD is

$$S_w(\omega) = q, \quad -\infty \leqslant \omega \leqslant +\infty \tag{2.38}$$

The autocorrelation is given by the inverse Fourier transform (Eqn (2.17)):

$$\begin{aligned} \bar{r}_w(\tau) &= \mathscr{F}^{-1}(q) \\ &= q\,\delta\tau \end{aligned} \tag{2.39}$$

where $\delta\tau$ is a Dirac delta function (Appendix D, Section D1). This implies that $w(t)$ is 'infinitely' correlated with itself for $\tau = 0$ and otherwise uncorrelated with itself. Since $\omega(t)$ is stationary, this in turn implies that $\varepsilon(w(t)) = \varepsilon(w(t+\tau))$, and that

$$\bar{r}_w(\tau) = r_w(\tau) = q\,\delta\tau \tag{2.40}$$

Conversely,

$$\begin{aligned} S_w(\omega) &= \mathscr{F}(\bar{r}_w(\tau)) \\ &= \mathscr{F}(q\,\delta\tau) \\ &= q \end{aligned}$$

$S_w(\omega)$ and $\bar{r}_w(\tau)$ are illustrated in Figure 2.16(a, b) and a heuristically constructed

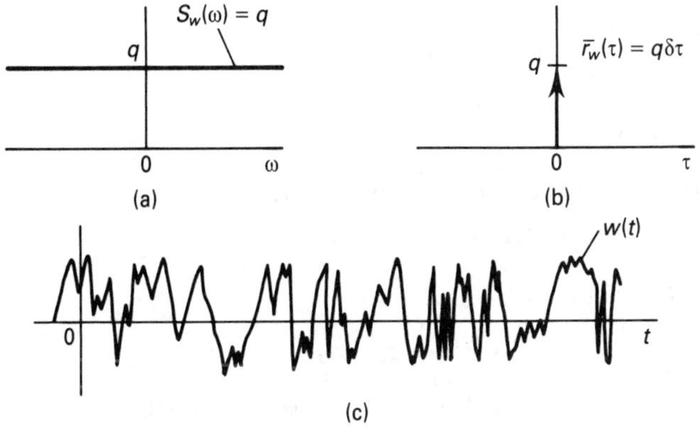

Figure 2.16 White noise PSD and autocorrelation.

sample function, in which the power per unit bandwidth is equal at all frequencies, or, equivalently, $x(t)$ is uncorrelated with $x(t + \tau)$, is shown in Figure 2.16(c).

The requirement that $S_w(\omega)$ is constant over all frequencies implies that white noise has infinite power – clearly a physical impossibility. Correspondingly, the variance of white noise, $\bar{r}_w(0)$, is infinite, implying that some sample functions have infinite amplitude – another physical impossibility. Figure 2.16(c), which is intended to convey the infinite bandwidth and zero autocorrelation of $w(t), \tau = 0$, is thus actually incorrect; true white noise is a mathematical fiction whose sample functions cannot be portrayed, but whose properties turn out to be useful in the analysis and modelling of stochastic systems. White noise may, or may not, be Gaussian; if it is Gaussian its wide sense stationarity implies strict stationarity (Section 2.7.2).

The concept of white noise can usefully be extended to a non-stationary process whose strength varies with time. This is 'extended white noise', $\bar{w}(t)$:

$$\varepsilon(\bar{w}(t)) = 0$$

$$r_{\bar{w}}(\tau) = \bar{r}_{\bar{w}}(t, s) = \bar{q}(t)\,\delta(s - t) \tag{2.41}$$

where $\bar{q}(t)$ is the strength of $\bar{w}(t)$ at time t. Since $\bar{w}(t)$ is not stationary, its PSD is meaningless. Like $w(t)$, $\bar{w}(t)$ is a mathematical fiction which proves useful in analysis and system modelling.

2.7.4 Processes with independent increments

A process, $x(t)$, has independent increments if

$$(x(t_j) - x(t_{j-1})), (x(t_{j-1}) - x(t_{j-2})), \ldots, (x(t_1) - x(t_0))$$

are independent (Section 1.14). (If these increments are merely uncorrelated, $x(t)$ is described as a process with statistically orthogonal increments.)

Using a slightly neater nomenclature, this implies that the increment $(x(s) - x(t))$ is independent of $x(t), s > t$. Also

$$\varepsilon(x(s) - x(t)) = 0$$

or

$$\varepsilon(x(s)) = \varepsilon(x(t))$$

The mean of $x(t)$ is thus constant (usually zero), and

$$\varepsilon[x(t)\cdot(x(s) - x(t))] = 0$$

Turning to the covariance of $x(t)$:

$$\begin{aligned} r(t, s) &= \varepsilon[(x(t) - \varepsilon(x(t)))(x(s) - \varepsilon(x(s)))] \\ &= \varepsilon[(x(t) - \varepsilon(x(t)))(x(s) - x(t) + x(t) - \varepsilon(x(s)))] \\ &= \varepsilon[(x(t) - \varepsilon(x(t)))^2] \end{aligned}$$

since

$$\varepsilon[(x(t) - \varepsilon(x(t)))(x(s) - x(t))] = 0$$

and

$$\varepsilon(x(s)) = \varepsilon(x(t))$$

Hence,

$$r(t, s) = r(t, t), \quad s \geqslant t$$

Similarly,

$$r(t, s) = r(s, s), \quad t \geqslant s$$

In summary, the covariance of $x(t)$ is the variance at the earlier of the two times involved:

$$r(t, s) = \begin{cases} r(t, t), & s \geqslant t \\ r(s, s), & t \geqslant s \end{cases} \tag{2.42}$$

A function of this kind is shown in Figure 2.17(a).

If $(x(s) - x(t))$ depends only on $\tau = (s - t)$, then $x(t)$ is a process with (strictly) stationary independent increments; a typical example is illustrated in Figure 2.17(b). The fact that the increments are stationary does not of course imply that $x(t)$ is stationary; indeed most such processes are non-stationary.

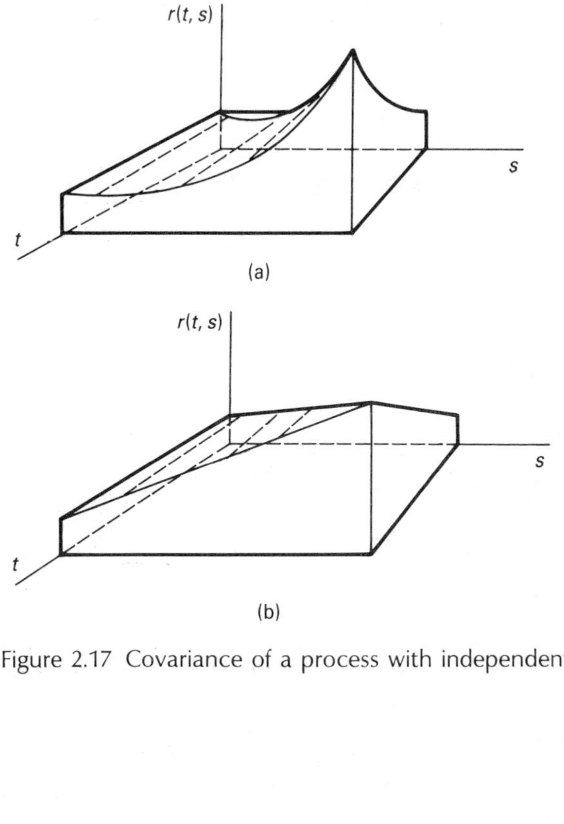

Figure 2.17 Covariance of a process with independent increments.

Example

The random walk of Section 2.7.1, Example(b), is an example of such a process, at least for $s - t \geqslant T$. Each increment $(x(s) - x(t))$, being determined by one or more coin tossings, is independent of any other non-overlapping increment, and these increments are obviously stationary, with zero means.

The sample functions of Figure 2.14 indicate that $x(t)$ is non-stationary and that $\varepsilon(x(t)) = 0$. Also, as shown previously, $\sigma_x^2(nT) = n\lambda^2$. Hence (Eqn (2.42)),

$$r(t, s) = r(nT, mT), \quad \begin{cases} nT \leqslant t < (n+1)T \\ mT \leqslant s < (m+1)T \end{cases}$$

$$= \begin{cases} n\lambda^2, & m \geqslant n \\ m\lambda^2, & n \geqslant m \end{cases}$$

This is illustrated in Figure 2.18 (cf. Figure 2.17(b)).

2.7.5 The Wiener process

The Wiener process (also known as Brownian motion), $\beta(t)$, which models the diffusion of a particle through a fluid, is defined as follows:

1. $\beta(t)$ is non-stationary with independent increments.
2. $\beta(t)$ is Gaussian.
3. The variance of any increment is proportional to the time over which it is measured:

$$\varepsilon(\beta(s) - \beta(t))^2 = q(s - t), \quad s \geqslant t \tag{2.43}$$

where q is the 'diffusion' of $\beta(t)$.

4. The initial value of $\beta(t)$ is traditionally taken as zero:

$$\beta(0) = 0$$

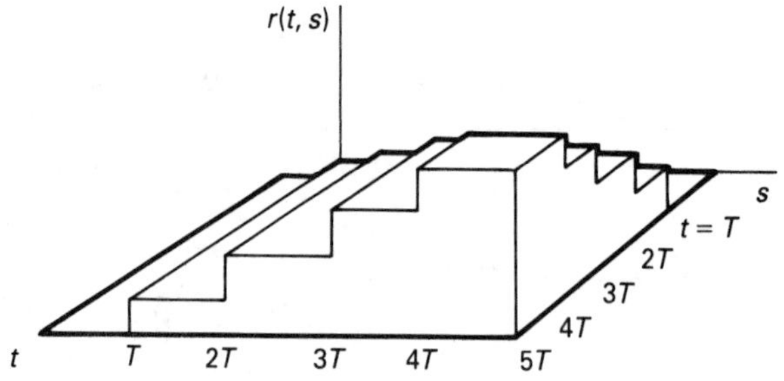

Figure 2.18 Covariance function.

It follows from (1) and (4) that

$$\varepsilon(\beta(t)) = 0 \tag{2.44}$$

and from (2) that

$$\varepsilon(\beta(s)|\beta(t)) = \beta(t), \quad s \geqslant t \tag{2.45}$$

It also follows from (1) and (4) that

$$r_\beta(t,t) = \varepsilon(\beta(t))^2 = qt$$

Thus (Eqn (2.42)),

$$r_\beta(t,s) = \begin{cases} r_\beta(t,t) = qt, & t \leqslant s \\ r_\beta(s,s) = qs, & s \leqslant t \end{cases} \tag{2.46}$$

Typical sample functions of $\beta(t)$ and $r_\beta(t,s)$ are shown in Figure 2.19(a) and (b) respectively.

The sample functions 'wander away' from $\beta(0) = 0$ as time progresses so that, although the mean $\varepsilon(\beta(\mathrm{t}))$ remains zero, the variance $r_\beta(t,t)$ increases (Eqn (2.46)).

The Wiener process turns out to be fundamental to the construction of models of stochastic dynamical systems, and in that context it is sometimes valuable to define an 'extended Wiener process', $\bar{\beta}(t)$, whose diffusion (Eqn (2.43)) can vary with time:

$$\varepsilon(\bar{\beta}(s) - \bar{\beta}(t))^2 = \int_t^s q(\lambda)\,\mathrm{d}\lambda, \quad s \geqslant t$$

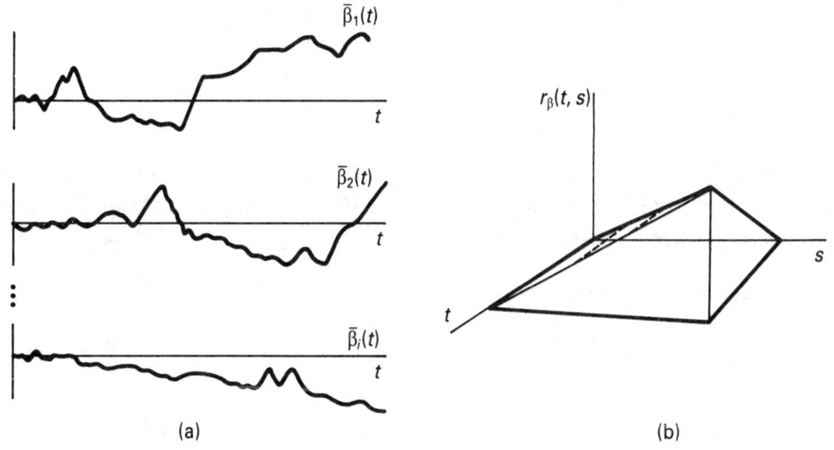

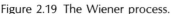

Figure 2.19 The Wiener process.

The covariance of $\bar{\beta}(t)$ is

$$
\left.\begin{aligned}
r_{\bar{\beta}}(t_1 s) = r_{\bar{\beta}}(t, t) &= \int_0^t \bar{q}(\lambda)\,\mathrm{d}\lambda, \quad s \geqslant t \\
r_{\bar{\beta}}(s, s) &= \int_0^s \bar{q}(\lambda)\,\mathrm{d}\lambda, \quad t \geqslant s
\end{aligned}\right. \tag{2.47}
$$

It is worth noting that $\beta(t)$ and $\bar{\beta}(t)$ both exhibit the 'quadratic variation property' that the square of the fundamental differential increment – as well as its expected value – is proportional to the corresponding time increment:

$$
(\mathrm{d}\bar{\beta}(t))^2 = \varepsilon(\mathrm{d}\bar{\beta}(t))^2 = \bar{q}(t)\,\mathrm{dt} \tag{2.48}
$$

This is of importance in dynamic system modelling theory.

Example

The random walk of Figure 2.14 is a Wiener process, at least for $(s - t) \geqslant T$. It is of interest to calculate its diffusion, q. Recall that

$$
\sigma_x^2(nT) = r(nT, nT) = n\lambda^2
$$

Equivalently,

$$
r(t, t) = \frac{t}{T}\lambda^2, \quad t = nT
$$

q is the diffusion at $t = nT$. Thus:

$$
q = \frac{\lambda^2}{T}, \quad t = nT
$$

2.7.6 White noise and the Wiener process

An examination of the means and covariances suggests that white noise is the (mean square) differential coefficient of the Wiener process and, conversely, that the Wiener process is the Riemann integral of white noise:

$$
\left.\begin{aligned}
\varepsilon(w(t)) = \varepsilon(\beta(t)) &= 0 \\
r_w(t, s) &= \frac{\partial^2 r_\beta(t, s)}{\partial t\,\partial s}
\end{aligned}\right\} \tag{2.49}
$$

$$
r_\beta(t, s) = \int_0^t \int_0^s r_w(\lambda, \mu)\,\mathrm{d}\lambda\,\mathrm{d}\mu \tag{2.50}
$$

where

$w(t)$ is white noise, strength q,
$\beta(t)$ is a Wiener process, diffusion q, and

$$r_w(t,s) = \mathrm{cov}(w(t)) = \begin{cases} q\delta(s-t), & s \geqslant t \\ q\delta(t-s), & t \geqslant s \end{cases}$$

$$r_\beta(t,s) = \mathrm{cov}(\beta(t)) = \begin{cases} qt, & s \geqslant t \\ qs, & t \geqslant s \end{cases}$$

Given these relationships, which are easily demonstrated, Eqns (2.24) and (2.31) suggest a differential–integral relationship between the Wiener process and white noise.

Close inspection, however, reveals that although Eqns (2.49) and (2.50) are certainly valid, white noise cannot be Riemann integrated, nor can the Wiener process be differentiated in the mean square sense. The difficulty is perhaps most easily seen by considering the variance of $w(t)$:

$$r_w(t,t) = r_w(0) = \delta(0) = \infty$$

This can be true only if the amplitudes of a significant number of sample functions of $w(t)$ are infinite at any instant. Such a sample function could also have infinite amplitude at the following instant – rendering it unintegrable: $w(t)$ is not integrable in the Riemann sense.

A detailed look at $\beta(t)$ also reveals a problem. From Eqn (2.22):

$$\frac{\mathrm{d}\beta(t)}{\mathrm{d}t} = \mathrm{LIM}\left[\frac{\bar{\beta}_i(t+\delta t) - \bar{\beta}_i(t)}{\delta t}\right]$$

Evaluation of this involves the expression

$$\lim_{\substack{i\to\infty \\ \delta t\to 0}} \left\{\varepsilon\left[\left(\frac{\bar{\beta}_i(t+\delta t) - \beta_i(t)}{\delta t}\right)^2\right]\right\} = \lim_{\delta t\to 0}\left(\frac{q\,\delta t}{(\delta t)^2}\right) = \lim_{\delta t\to 0}\left(\frac{q}{\delta t}\right) = \infty$$

No meaningful interpretation can thus be given to $\mathrm{d}\beta/\mathrm{d}t$ and so $\beta(t)$ is not differentiable in the mean square sense. Heuristically, more than 'a few' sample functions of $\beta(t)$ have jumps at any time, t, and $\mathrm{d}\beta/\mathrm{d}t$ does not converge in the sense required by Eqn (2.22).

In spite of these conclusions, it is often helpful to regard the differential coefficient of the Wiener process, diffusion q, to be white noise, strength q, and to regard the converse integral relationship as true. However, no mathematical arguments or deductions can be based on these fundamentally invalid relationships.

2.8 Stochastic integrals

It turns out that the differential and integral defined in Section 2.6 are only partly equal to the task of constructing consistent mathematical models of stochastic dynamical systems, and specialized stochastic integrals have to be defined to make up for their inadequacies.

The stochastic, or Wiener, integral and differential, used in the modelling of linear systems, are described first, and the more specialized Itô and Stratonovich integrals, used in non-linear models, are then examined in this section.

2.8.1 The stochastic or Wiener integral

The stochastic, or Wiener, integral of a deterministic signal, $a(t)$, which finds use in the modelling of linear dynamical systems, is defined as the stochastic process

$$I(t) = \int_{t_0}^{t} a(\lambda)\,\mathrm{d}\bar{\beta}(\lambda)$$

where

$a(t)$ is a deterministic signal,

$\bar{\beta}(t)$ is an extended Wiener process, diffusion $\bar{q}(t)$ (this covers the possibility that $\bar{q}(t)$ is constant).

Meaning is given to this integral by considering how the sample functions of $I(t)$ are formed. Considering a time increment $(t_{j+1} - t_j)$, the increment in $\bar{I}_i(t)$ at t_j is formed by multiplying $a(t_j)$ by $\bar{\bar{\beta}}_i(t_{j+1}) - \bar{\beta}(t_j)$. This is indicated in Figure 2.20 for N intervals over the period $[t_0, t]$.

$I(t)$ is formally defined as the limit in the mean square, in the sense of Eqns (2.20) and (2.21), of the summation of these product elements:

$$I(t) = \int_{t_0}^{t} a(\lambda)\,\mathrm{d}\bar{\beta}(\lambda)$$

$$= \mathrm{LIM}\left\{\lim_{N\to\infty}\left[\sum_{j=0}^{N-1} a_i(t_j)(\bar{\bar{\beta}}_i(t_{j+1}) - \bar{\bar{\beta}}_i(t_j))\right]\right\} \tag{2.51}$$

This definition allows 'a few' of the sample functions calculated in this way to fail to coincide with those of $I(t)$. Some properties of $I(t)$ are as follows:

1. $I(t)$ is an extended Wiener process of diffusion:

$$\bar{q}_{\mathrm{I}}(t) = a(t)\cdot\bar{q}(t) \tag{2.52}$$

2. The mean and convariance of $I(t)$ are

$$\varepsilon(I(t)) = 0 \tag{2.53}$$

$$r_{\mathrm{I}}(t,s) = \int_{t}^{s} (a(\lambda))^2\bar{q}(\lambda)\,\mathrm{d}\lambda \tag{2.54}$$

3. The evaluation of $I(t)$ follows the familiar rules of Riemann integration,

including integration by parts:

$$\int_{t_0}^{t} a(\lambda)\,\mathrm{d}\bar{\beta}(\lambda) = \int_{t_0}^{t_1} a(\lambda)\,\mathrm{d}\bar{\beta}(\lambda) + \int_{t_1}^{t} a(\lambda)\,\mathrm{d}\bar{\beta}(\lambda) \tag{2.55}$$

$$\int_{t_0}^{t} a(\lambda)\,\mathrm{d}\bar{\beta}(\lambda) = a(\lambda)\bar{\beta}(\lambda)|_{t_0}^{t} - \int_{t_0}^{t} \bar{\beta}(\lambda)\,\mathrm{d}(a(\lambda)) \tag{2.56}$$

Example

Consider the process

$$I(t) = \int_0^t \cos(\omega_0\lambda)\,\mathrm{d}\beta(\lambda)$$

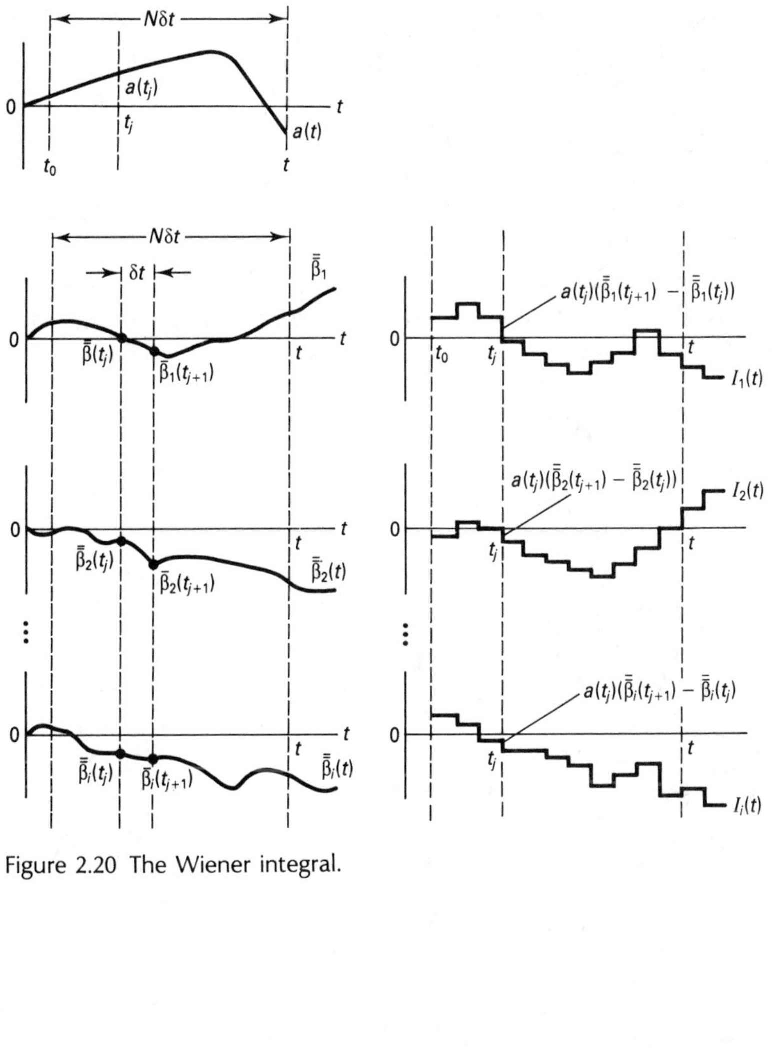

Figure 2.20 The Wiener integral.

where

ω_0 is constant,
$\beta(t)$ is a Wiener process, diffusion q.

From Eqns (2.53) and (2.54),

$$\varepsilon(I(t)) = 0$$

$$\begin{aligned} r_I(t,s) &= r_I(t,t), \quad s \geqslant t \\ &= \int_0^t \cos^2(\omega_0\lambda) q \, \mathrm{d}\lambda \\ &= \frac{q}{2}\left(t + \frac{1}{2\omega_0}\sin(2\omega_0 t)\right) \end{aligned}$$

2.8.2 The stochastic or Wiener differential

The stochastic differential is the inverse of the Wiener integral (Eqn (2.51)):

$$\mathrm{d}I(t) = I(t + \mathrm{d}t) - I(t) = a(t)\,\mathrm{d}\bar{\beta}(t) \tag{2.57}$$

where

$a(t)$ is a deterministic signal,
$\bar{\beta}(t)$ is an extended Wiener process, diffusion $\bar{q}(t)$,
$\mathrm{d}I(t)$ is the (infinitesimal) increment in $I(t)$ corresponding to the increment $\mathrm{d}\bar{\beta}(t)$ occurring over $\mathrm{d}t$.

No useful meaning can be ascribed to the differential coefficient $\mathrm{d}I(t)/\mathrm{d}t$ (which heuristically would be the deterministic process $a(t)$). However, $a(t)$ can be recovered from the covariance of $I(t)$ via the Riemann integral Eqn (2.54):

$$\begin{aligned} (a(t))^2\bar{q}(t) &= \frac{\mathrm{d}}{\mathrm{d}t}(r_I(t,s)) \\ &= \frac{\mathrm{d}}{\mathrm{d}t}(r_I(t,t)), \quad s \geqslant t \end{aligned}$$

Hence,

$$a(t) = \left[\frac{1}{\bar{q}(t)}\cdot\frac{\mathrm{d}}{\mathrm{d}t}(r_I(t,t))\right]^{1/2} \tag{2.58}$$

Example

Consider $I(t)$ of the last example:

$$r_I(t,t) = \frac{q}{2}\left[t + \frac{1}{2\omega_0}\sin(\omega_0 t)\right]$$

Hence,

$$a(t) = \left[\frac{1}{q}\cdot\frac{q}{2}(1 + \cos^2(\omega_0 t) - \sin^2(\omega_0 t))\right]^{1/2}$$
$$= \cos(\omega_0 t)$$

2.8.3 The Itô stochastic integral

The ideas of Section 2.8.2 can be extended to define the integration of a stochastic process, $x(t)$, with respect to an extended Wiener process, $\bar{\beta}(t)$. This finds application in the modelling of non-linear dynamical systems.

There are two accepted definitions with slightly different properties. The Itô integral is considered here, and the Stratonovich integral in Section 2.8.4.

The Itô integral of a stochastic process, $x(t)$, WRT the extended Wiener process, $\bar{\beta}(t)$, is the stochastic process

$$I_{\text{I}}(t) = \int_{t_0}^{t} x(\lambda)\,\mathrm{d}\bar{\beta}(\lambda)$$

where

$x(t)$ is a stochastic process,
$\bar{\beta}(t)$ is an extended Wiener process, diffusion $\bar{q}(t)$.

As with Eqn (2.51), the mean square definition is used to allow 'a few' sample functions to fail to converge:

$$I_{\text{I}}(t) = \int_{t_0}^{t} x(\lambda)\,\mathrm{d}\bar{\beta}(\lambda)$$
$$= \text{LIM}\left\{\lim_{N\to\infty}\left[\sum_{j=0}^{N-1} x_i(t_j)(\bar{\bar{\beta}}_i(t_{j+1}) - \bar{\bar{\beta}}_i(t_j))\right]\right\} \quad (2.59)$$

where, as before, the interval $[t_0, t]$ is divided into N subintervals. The construction of $I_{\text{I}}(t)$ is indicated in Figure 2.21. Defined in this way, $I_{\text{I}}(t)$ has the following properties:

1. $I_{\text{I}}(t)$ is a Markov, but not a Wiener, process:

   $$\varepsilon(I_{\text{I}}(t + \mathrm{d}t)) = I_{\text{I}}(t) \quad (2.60)$$

2. The mean is zero:

   $$\varepsilon(I_{\text{I}}(t)) = 0 \quad (2.61)$$

   Unfortunately, the covariance of $I_{\text{I}}(t)$ cannot be expressed easily, but its

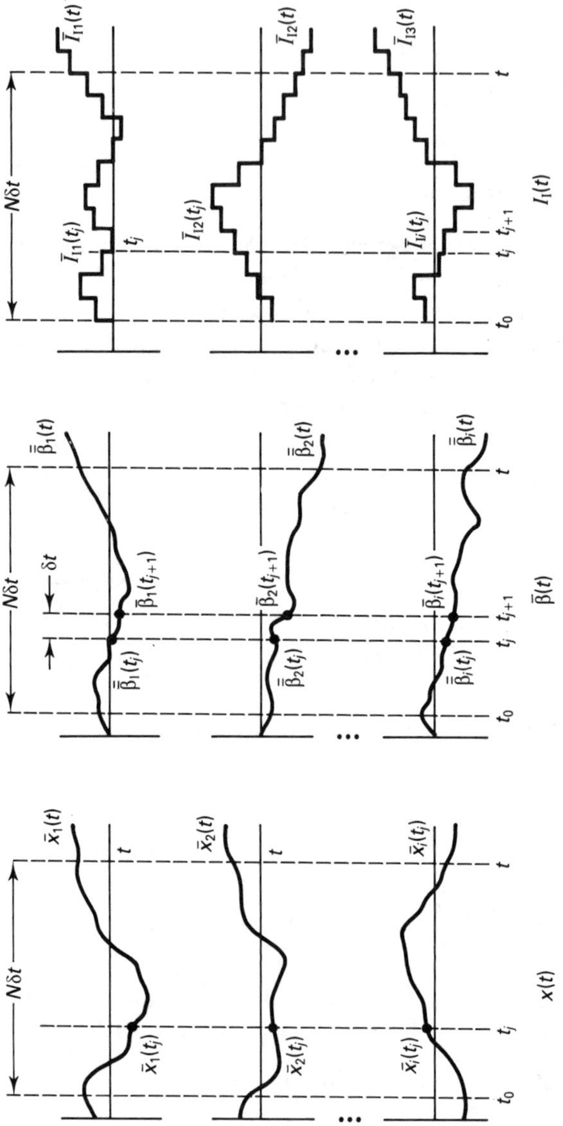

Figure 2.21 The Itô stochastic integral.

variance is

$$r_I(t,t) = \int_{t_0}^{t} \varepsilon(x(\lambda))^2 \bar{q}(\lambda)\,d\lambda$$

$$= \int_{t_0}^{t} \bar{r}(\lambda,\lambda)\bar{q}(\lambda)\,d\lambda \tag{2.62}$$

3. The Wiener integral can be regarded as a special case of the Itô integral, with $x(t)$ deterministic.

4. The differential inverse of the Itô integral is (cf. Eqn (2.57)):

$$dI_I(t) = I_I(t + dt) - I_I(t) = x(t)\,d\bar{\beta}(t) \tag{2.63}$$

As with the Wiener differential, no useful meaning attaches to

$$\frac{dI_I^{(t)}}{dt}$$

5. The calculation of $I_I(t)$, where this is analytically possible, does not follow the familiar rules of Riemann integration, but is dependent on the evaluation of Eqn (2.59) using any properties of $x(t)$, $\bar{\beta}(t)$ relevant to the case in hand [refs 2, 4].

Example

Consider the process

$$I_I(t) = \int_0^t 2\beta(\lambda)\,d\beta(\lambda)$$

where $\beta(t)$ is a Wiener process, diffusion q. From Eqn (2.59), with a slight abuse of nomenclature:

$$I_I(t) = \lim_{N\to\infty} \left\{ 2\sum_{j=0}^{N-1} [\beta(t_j)(\beta(t_{j+1}) - \beta(t_j))] \right\}$$

$$= \lim_{N\to\infty} \left[\sum_{j=0}^{N-1} (\beta^2(t_{j+1}) - \beta^2(t_j)) - \sum_{j=0}^{N-1} (\beta(t_{j+1}) - \beta(t_j))^2 \right]$$

Hence, recalling the quadratic variation property (Eqn (2.48)):

$$I_I(t) = \beta^2(t) - q(t)$$

This, of course, differs from Riemann integration, which would have yielded $\beta^2(t)$. The evaluation of this example depends on a special property of the integrand, $2\beta(t)$, which is naturally not shared by most integrands. The value of the Itô integral lies not in the analytic expressions it can be used to generate, but in the basis it provides for constructing models of non-linear dynamical systems.

2.8.4 The Stratonovich integral

Like the Itô integral, the Stratonovich integral is useful in the modelling of non-linear dynamical systems. Its integrand, however, is restricted to analytic functions of the extended Wiener process:

$$I_S(t) = \int_{t_0}^{t} f(\bar{\beta}(\lambda))\,\mathrm{d}\bar{\beta}(\lambda)$$

where $f(\bar{\beta}(t))$ is an analytic function of the extended Wiener process $\bar{\beta}(t)$, diffusion $\bar{q}(t)$.

$I_S(t)$ is defined in much the same way as the Itô integral, with the notable difference that the integral is evaluated not at t_j, t_{j+1}, but halfway through the time increments as shown in Figure 2.22 for the ith sample function, $I_{Si}(t)$:

$$I_S(t) = \int_{t_0}^{t} f(\bar{\beta}(\lambda))\,\mathrm{d}\bar{\beta}(\lambda)$$

$$= \mathrm{LIM}\left\{\lim_{N\to\infty}\left[\sum_{j=0}^{N-1} f\left(\bar{\bar{\beta}}_i\left(\frac{t_j + t_{j+1}}{2}\right)\right)(\bar{\bar{\beta}}_i(t_{j+1}) - \bar{\bar{\beta}}_i(t_j))\right]\right\} \qquad (2.64)$$

The Stratonovich integral has significantly more restricted application than

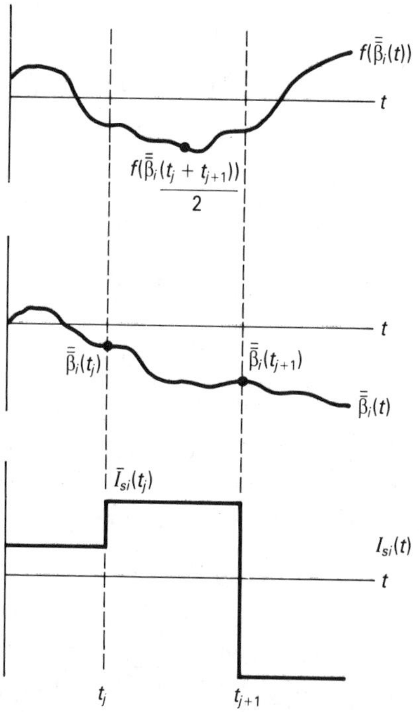

Figure 2.22 The Stratonovich stochastic integral.

the Itô integral, because of the restriction on its integrand, and also because $I_S(t)$ is not guaranteed to be Markov. It is not pursued further in this book, the interested reader being referred to refs 2 and 4.

2.9 Vector processes

The ideas of Sections 2.2–2.8 extend readily to vector processes.

Consider the vector stochastic process

$$\mathbf{x}(t) = (x_1(t) \quad x_2(t) \quad \ldots \quad x_m(t))^T$$

Each element $x_j(t), j = 1, 2, \ldots, m$, is a scalar stochastic process with its own ensemble, PDFs, mean, covariance and autocorrelation. Cross-covariance and cross-correlation functions describe relationships between these processes. These functions are represented neatly in the form of a mean vector and covariance matrix.

The mean vector (cf. Eqn (2.4)) is

$$\begin{aligned}\varepsilon(\mathbf{x}(t)) &= \mathbf{m}_x(t)\\ &= (\varepsilon(x_1(t)) \quad \varepsilon(x_2(t)) \quad \ldots \quad \varepsilon(x_m(t)))^T\end{aligned} \tag{2.65}$$

The covariance matrix (cf. Eqn (2.7)) is

$$\operatorname{cov}(\mathbf{x}(t)) = \mathbf{R}(t, s) = \varepsilon[(\mathbf{x}(t) - \varepsilon(\mathbf{x}(t)))(\mathbf{x}(s) - \varepsilon(\mathbf{x}(s)))^T)] \tag{2.66}$$

$\mathbf{R}(t, s)$ is a square matrix, the diagonal elements of which are the covariances of $x_j(t)$, the off-diagonal terms being the cross-covariances of $x_j(t), x_h(t)$:

$$r_{jj}(t, s) = \varepsilon[(x_j(t) - \varepsilon(x_j(t)))(x_j(s) - \varepsilon(x_j(s)))]$$

$$r_{jh}(t, s) = \varepsilon[(x_j(t) - \varepsilon(x_j(t)))(x_h(s) - \varepsilon(x_h(s)))]$$

where $r_{jh}(t, s)$ is the element of the jth row and hth column of $\mathbf{R}(t, s)$.

Examples

(a) Consider a vector of independent white noise processes,

$$\mathbf{w}(t) = (w_1(t) \quad w_2(t) \quad \ldots \quad w_m(t))^T, \text{ strengths } q_1, q_2, \ldots, q_m$$

Since $w_j(t)$ is independent of $w_h(t)$, the off-diagonal elements of the covariance matrix are 0:

$$\mathbf{R}_w(t, s) = \begin{pmatrix} q_1\delta\tau & 0 & 0 & \cdots & 0 \\ 0 & q_2\,\delta\tau & 0 & & 0 \\ 0 & 0 & q_3\,\delta\tau & & 0 \\ \vdots & & & & \vdots \\ 0 & 0 & 0 & \cdots & q_m\,\delta\tau \end{pmatrix}$$

where $\tau = (s - t)$.

The mean vector is zero:

$$\varepsilon(\mathbf{w}(t)) = \mathbf{0}$$

(b) Consider the vector process:

$$\mathbf{x}(t) = \begin{pmatrix} a + bt \\ a - bt \end{pmatrix}$$

where a, b are independent evenly distributed random numbers with zero means, and variances σ_a^2, σ_b^2.

$$\mathbf{R}(t, s) = \varepsilon\left[\begin{pmatrix} a + bt \\ a - bt \end{pmatrix}(a + bs \quad a - bs)\right]$$

$$= \begin{pmatrix} (\sigma_a^2 + \sigma_b^2 ts) & (\sigma_a^2 - \sigma_b^2 ts) \\ (\sigma_a^2 - \sigma_b^2 ts) & (\sigma_a^2 + \sigma_b^2 ts) \end{pmatrix}$$

Also,

$$\varepsilon(\mathbf{x}(t)) = \begin{pmatrix} \varepsilon(a + bt) \\ \varepsilon(a - bt) \end{pmatrix} = \mathbf{0}$$

2.10 Mathematical models of dynamical systems

The behaviour of deterministic continuous time dynamical systems is generally modelled by ordinary differential equations, usually cast in standard form as a set of first order 'state' equations and a further set of algebraic 'output' equations:

$$\dot{\mathbf{x}}(t) = \mathbf{f}(\mathbf{x}, \mathbf{u}, t) \tag{2.67}$$

$$\mathbf{y}(t) = \mathbf{q}(\mathbf{x}, \mathbf{u}, t) \tag{2.68}$$

where

$$\mathbf{x}(t) = (x_1(t) \quad x_2(t) \quad \dots \quad x_m(t))^T$$

$$\mathbf{u}(t) = (u_1(t) \quad u_2(t) \quad \dots \quad u_q(t))^T$$

$$\mathbf{y}(t) = (y_1(t) \quad y_2(t) \quad \dots \quad y_r(t))^T$$

the q system inputs are represented by the q-vector, $\mathbf{u}(t)$,
the m states by the m-vector, $\mathbf{x}(t)$,
the r outputs by the r-vector, $\mathbf{y}(t)$.

The model is shown in block form in Figure 2.23(a).

A useful subclass of systems is modelled by the 'linear time-variant' equations:

$$\dot{\mathbf{x}}(t) = \mathbf{A}(t)\mathbf{x}(t) + \mathbf{B}(t)\mathbf{u}(t) \tag{2.69}$$

$$\mathbf{y}(t) = \mathbf{C}(t)\mathbf{x}(t) + \mathbf{D}(t)\mathbf{u}(t) \tag{2.70}$$

where

$\mathbf{A}(t)$ is a time-dependent $(m \times m)$ matrix,
$\mathbf{B}(t)$ is a time-dependent $(m \times q)$ matrix,
$\mathbf{C}(t)$ is a time-dependent $(r \times m)$ matrix,
$\mathbf{D}(t)$ is a time-dependent $(r \times q)$ matrix.

In turn, a large and useful subclass of these is modelled by the 'linear time-invariant' (LTI) equations (Appendix A, Section A2):

$$\dot{\mathbf{x}}(t) = \mathbf{A}\mathbf{x}(t) + \mathbf{B}\mathbf{u}(t) \tag{2.71}$$

$$\mathbf{y}(t) = \mathbf{C}\mathbf{x}(t) + \mathbf{D}\mathbf{u}(t) \tag{2.72}$$

where **A**, **B**, **C**, **D** are constant matrices. The detailed properties of these models are described in many standard texts. The most important, including stability, controllability and observability, are outlined briefly in Appendix A, Section A2.

The question addressed here is how to extend these models to represent corruption by noise processes of the kind which occur in practice.

Two distinct types of noise must be considered: process noise, $\mathbf{w}(t)$, which amounts to an unknown – and unwanted – vector forcing function input to some part of the system, and measurement noise, $\mathbf{v}(t)$, which corrupts the system output before it can be measured. These are represented in Figure 2.23(b).

At first sight, it seems simple to extend Eqns (2.67) and (2.68) to accommodate $\mathbf{w}(t)$ and $\mathbf{v}(t)$, yielding

$$\dot{\mathbf{x}}(t) = \mathbf{f}(\mathbf{x}, \mathbf{u}, \mathbf{w}, t) \tag{2.73}$$

$$\mathbf{y}(t) = q(\mathbf{x}, \mathbf{u}, \mathbf{v}, t) \tag{2.74}$$

In fact, this model turns out to be invalid for two reasons. A brief examination of these gives a worthwhile insight into the problem, and leads to the structure of the models which are now accepted as standard.

Firstly, since $\mathbf{w}(t)$ is stochastic, Eqn (2.73) must be interpreted in a manner which accommodates this, for example so that $\dot{\mathbf{x}}(t)$ is defined by Eqn (2.22). Its solution, $\mathbf{x}(t)$, under any such interpretation, is also stochastic. However, if Eqn (2.73) is non-linear, it can be shown that $\mathbf{x}(t)$ is generally non-Markov, rendering it almost

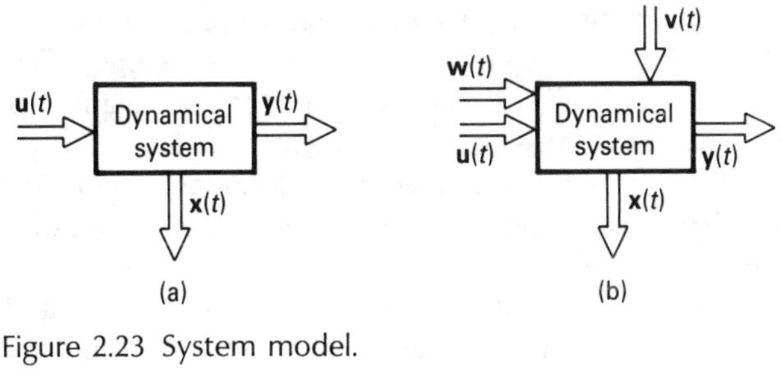

Figure 2.23 System model.

impossible to analyze, and not practically useful. The same argument applies to Eqn (2.74).

Secondly, and more seriously, for the equations to qualify as a general standard-form model of a causal dynamical system, it is necessary that $\mathbf{w}(t), \mathbf{w}(s)$ are independent, $s \neq t$, so that $\mathbf{x}(t)$ is defined only by the current state, input forcing and noise forcing functions; not by any previous values of these variables. This implies that $\mathbf{w}(t)$ must be white noise, and since the variance of white noise is infinite (Section 2.7.3), the variance of $\dot{\mathbf{x}}(t)$ must generally be infinite. This is unacceptable, since it means that the rate of change of a significant number of sample functions of $\mathbf{x}(t)$ is infinite at any time, after which these functions are undefined.

Equivalently, the solution of Eqn (2.73) generally involves integrating white noise WRT time, which (Section 2.7.6) is mathematically invalid.

It is thus clear that the model of Eqns (2.73) and (2.74) fails, and some alternative must be proposed. Since both of the above difficulties apply to non-linear models, but only the second to linear models, the former turn out to be significantly more complicated than the latter. Linear models are examined in this section, and non-linear models in Section 2.11.

For linear systems the attractive, but invalid, model corresponding to Eqns (2.73) and (2.74) is

$$\dot{\mathbf{x}}(t) = \mathbf{A}(t)\mathbf{x}(t) + \mathbf{B}(t)\mathbf{u}(t) + \mathbf{H}(t)\bar{\mathbf{w}}(t) \tag{2.75}$$

$$\mathbf{y}(t) = \mathbf{C}(t)x(t) + \mathbf{D}(t)\mathbf{u}(t) + \bar{\mathbf{v}}(t) \tag{2.76}$$

where $\bar{\mathbf{w}}(t)$ is an s-vector of extended white noise forcing functions, or 'process' noise, of 'strength' $\bar{\mathbf{Q}}(t)$ (Sections 2.7.3 and 2.9) such that

$$\varepsilon(\bar{\mathbf{w}}(t)\cdot\bar{\mathbf{w}}^T(t) = \bar{\mathbf{Q}}(t)\cdot\delta\tau, \quad \tau = (s - t)$$

$\bar{\mathbf{v}}(t)$ is an r-vector of extended white 'measurement' noise, of strength $\bar{\mathbf{R}}(t)$ such that

$$\varepsilon(\bar{\mathbf{v}}(t)\cdot\bar{\mathbf{v}}^T(t)) = \bar{\mathbf{R}}(t)\cdot\delta\tau, \quad \tau = (s - t)$$

and $\mathbf{H}(t)$ is the relevant process noise gain $(m \times s)$ matrix.

$\bar{\mathbf{w}}(t), \bar{\mathbf{v}}(t)$ are generally taken to be Gaussian, since this accords well with what occurs in practice, and their time-variant strengths automatically include the simpler, more usual, time-invariant cases. Figure 2.23(b) indicates the model structure. Since $\mathbf{x}(t), \mathbf{y}(t)$ are linear functions of $\mathbf{w}(t), \mathbf{v}(t)$, which are Gaussian, they too are Gaussian (Section 1.12), though generally they are not, of course, white. If the statistics of $\mathbf{w}(t), \mathbf{v}(t)$ are known, the model allows calculation of the means and covariances of $\mathbf{x}(t), \mathbf{y}(t)$ which, since they are Gaussian, specify them completely.

As with Eqn (2.73), the infinite variance of $\mathbf{w}(t)$ renders Eqn (2.75) invalid. (In contrast, Eqn (2.76), being algebraic, is valid, giving $\mathbf{y}(t)$ a white noise content and, correspondingly, elements with infinite variance.) A substitute for Eqn (2.75) must therefore be found before any mathematical manipulation can be undertaken.

The standard linear model, which retains the benefits of Eqn (2.75), but is mathematically valid, is written in differential form

$$\mathrm{d}\mathbf{x}(t) = \mathbf{A}(t)\mathbf{x}(t)\,\mathrm{d}t + \mathbf{B}(t)\mathbf{u}(t)\,\mathrm{d}t + \mathbf{H}(t)\,\mathrm{d}\bar{\boldsymbol{\beta}}(t) \tag{2.77}$$

where $\bar{\boldsymbol{\beta}}(t)$ is a vector of extended Wiener processes, diffusion $\bar{\mathbf{Q}}(t)$, and the differentials on the right-hand side imply stochastic Riemann integration (Section 2.6.2), straightforward deterministic Riemann integration, and Wiener integration (Section 2.8.1) respectively.

For heuristic purposes, Eqns (2.75) and (2.76) are usually regarded as meaningful, and models are virtually always synthesized in this form, but for any deductive or mathematical purposes, they must be interpreted as Eqns (2.77) and (2.76).

The formal – and invalid – solution of Eqn (2.75) is thus of no interest, but Eqn (2.77) yields the behaviour of $\mathbf{x}(t)$:

$$\mathbf{x}(t) = \mathbf{x}(t_0) + \int_{t_0}^{t} \mathbf{A}(\lambda)\mathbf{x}(\lambda)\,\mathrm{d}\lambda + \int_{t_0}^{t} \mathbf{B}(\lambda)\mathbf{u}(\lambda)\,\mathrm{d}\lambda + \int_{t_0}^{t} \mathbf{H}(\lambda)\,\mathrm{d}\bar{\boldsymbol{\beta}}(\lambda) \tag{2.78}$$

where the first and third integrals are as defined by Eqns (2.29) and (2.51) respectively, while the second is a straightforward Riemann integral. The solution of this (cf. Appendix A, Section A2.1) may be written

$$\mathbf{x}(t) = \varphi(t, t_0)\mathbf{x}(t_0) + \int_{t_0}^{t} \varphi(\lambda, t_0)\mathbf{B}(\lambda)\mathbf{u}(\lambda)\,\mathrm{d}\lambda + \int_{t_0}^{t} \varphi(\lambda, t_0)\mathbf{H}(\lambda)\,\mathrm{d}\bar{\boldsymbol{\beta}}(\lambda) \tag{2.79}$$

where $\varphi(t, t_0)$ is the 'state transition matrix' which satisfies the deterministic differential equation:

$$\frac{\mathrm{d}}{\mathrm{d}t}(\varphi(t, t_0)) = \mathbf{A}(t)\varphi(t, t_0) \tag{2.80}$$

A useful illustration of the procedures involved – and of the faulty consequences of using Eqn (2.75) directly – is provided by calculating the mean and covariance of $\mathbf{x}(t)$, given $\mathbf{u}(t)$, $\bar{\mathbf{Q}}(t)$, $\bar{\mathbf{R}}(t)$.

2.10.1 The mean of $\mathbf{x}(t)$

The behaviour of $\varepsilon(\mathbf{x}(t))$ is easily found from Eqn (2.77):

$$\varepsilon(\mathrm{d}\mathbf{x}(t)) = \varepsilon[\mathbf{A}(t)\mathbf{x}(t)\,\mathrm{d}t + \mathbf{B}(t)\mathbf{u}(t)\,\mathrm{d}t] + \mathbf{0}$$

This is a straightforward deterministic differential equation, which may be written, with substitution from Eqn (2.23):

$$\dot{\mathbf{m}}_x(t) = \frac{\mathrm{d}}{\mathrm{d}t}(\varepsilon(\mathbf{x}(t))) = \mathbf{A}(t)\varepsilon(\mathbf{x}(t)) + \mathbf{B}(t)\mathbf{u}(t) \tag{2.81}$$

Being identical in structure to the deterministic Eqn (2.69), this has the solution (Appendix A, Section A2.1)

$$\varepsilon(\mathbf{x}(t)) = \boldsymbol{\varphi}(t, t_0)\varepsilon(\mathbf{x}(t_0)) + \int_{t_0}^{t} \boldsymbol{\varphi}(\lambda, t_0)\mathbf{B}(\lambda)\mathbf{u}(\lambda)\,d\lambda \tag{2.82}$$

Slightly misleadingly, these same, correct conclusions could be reached by taking expectations of both sides of the invalid Eqn (2.75).

2.10.2 The covariance of $\mathbf{x}(t)$

First, consider $\mathbf{x}(s)$ in terms of $\mathbf{x}(t)$ (Eqn (2.79)):

$$\mathbf{x}(s) = \boldsymbol{\varphi}(s, t)\mathbf{x}(t) + \int_{t}^{s} \boldsymbol{\varphi}(\lambda, t)\mathbf{B}(\lambda)\mathbf{u}(\lambda)\,d\lambda + \int_{t}^{s} \boldsymbol{\varphi}(\lambda, t)\mathbf{H}(\lambda)\,d\bar{\boldsymbol{\beta}}(\lambda)$$

Hence,

$$\begin{aligned}\mathbf{R}(t, s) &= \varepsilon[(\mathbf{x}(t) - \varepsilon(\mathbf{x}(t)))(\mathbf{x}(s) - \varepsilon(\mathbf{x}(s)))^T] \\ &= \varepsilon[(\mathbf{x}(t) - \varepsilon(\mathbf{x}(t)))(\boldsymbol{\varphi}(s, t)x(t) - \boldsymbol{\varphi}(s, t)\varepsilon(\mathbf{x}(t)))^T]\end{aligned}$$

since, clearly,

$$\varepsilon[(\mathbf{x}(t) - \varepsilon(\mathbf{x}(t)))\cdot \int_{t}^{s} \boldsymbol{\varphi}(\lambda, t)\mathbf{B}(\lambda)\mathbf{u}(\lambda)\,d\lambda] = 0$$

and $\mathbf{x}(t)$ is uncorrelated with $\int_t^s \boldsymbol{\varphi}(\lambda, t)\mathbf{H}(\lambda)\,d\bar{\boldsymbol{\beta}}(\lambda)$. Thus

$$\begin{aligned}\mathbf{R}(t, s) &= \varepsilon[(\mathbf{x}(t) - \varepsilon(\mathbf{x}(t)))(\mathbf{x}(t) - \varepsilon(\mathbf{x}(t)))^T \boldsymbol{\varphi}^T(s, t)] \\ &= \mathbf{R}(t, t)\cdot \boldsymbol{\varphi}^T(s, t), \quad s \geqslant t\end{aligned} \tag{2.83}$$

It is equally easy to show that

$$\mathbf{R}(t, s) = \boldsymbol{\varphi}(t, s)\cdot \mathbf{R}(s, s), \quad s \geqslant t$$

It remains to develop an expression describing the behaviour of $\mathbf{R}(t, t)$:

$$\begin{aligned}\mathbf{R}(t, t) &= \varepsilon[(\mathbf{x}(t) - \varepsilon(\mathbf{x}(t)))(\mathbf{x}(t) - \varepsilon(\mathbf{x}(t)))^T] \\ &= \boldsymbol{\varphi}(t, t_0)\mathbf{R}(t, t_0)\boldsymbol{\varphi}^T(t, t_0) + \int_{t_0}^{t} \boldsymbol{\varphi}(\lambda, t_0)\mathbf{H}(\lambda)\bar{\mathbf{Q}}(\lambda)\mathbf{H}^T(\lambda)\boldsymbol{\varphi}^T(\lambda, t_0)\,d\lambda\end{aligned} \tag{2.84}$$

Both cross-multiplied terms have disappeared here since $\mathbf{x}(t_0)$ is uncorrelated with $d\bar{\boldsymbol{\beta}}(t), t \geqslant t_0$, and $\varepsilon(d\bar{\boldsymbol{\beta}}(t))^2 = \bar{\mathbf{Q}}(t)\cdot dt$ (Eqn (2.48)). This can be cast neatly as a differential equation in $\mathbf{R}(t, t)$ by differentiating both sides, a procedure which requires the use of Leibnitz's rule* to evaluate the differentiated integration term. Recalling Eqn (2.80) and differentiating Eqn (2.84):

* Leibnitz's rule:

$$\frac{d}{dt}\int_{a_1(t)}^{a_2(t)} f(t, \lambda)\,d\lambda = \int_{a_1(t)}^{a_2(t)} \frac{\partial f}{\partial t}\cdot d\lambda + f(t, a_2(t))\cdot \frac{da_2}{dt} - f(t, a_1(t))\cdot \frac{da_1}{dt}$$

$$\begin{aligned}\dot{\mathbf{R}}(t,t) &= \mathbf{A}(t)\boldsymbol{\varphi}(t,t_0)\mathbf{R}(t_0,t_0)\boldsymbol{\varphi}^T(t,t_0)\\ &\quad+\boldsymbol{\varphi}(t,t_0)\mathbf{R}(t_0,t_0)\boldsymbol{\varphi}^T(t,t_0)\mathbf{A}^T(t)+\mathbf{H}(t)\bar{\mathbf{Q}}(t)\mathbf{H}^T(t)\\ &\quad+\int_{t_0}^{t}\mathbf{A}(\lambda)\boldsymbol{\varphi}(\lambda,t_0)\mathbf{H}(\lambda)\bar{\mathbf{Q}}(\lambda)\mathbf{H}^T(\lambda)\boldsymbol{\varphi}^T(\lambda,t_0)\,\mathrm{d}\lambda\\ &\quad+\int_{t_0}^{t}\boldsymbol{\varphi}(\lambda,t_0)\mathbf{H}(\lambda)\bar{\mathbf{Q}}(\lambda)\mathbf{H}^T(\lambda)\boldsymbol{\varphi}^T(\lambda,t_0)\mathbf{A}^T(\lambda)\,\mathrm{d}\lambda\end{aligned}$$

Substituting from Eqn (2.84):

$$\dot{\mathbf{R}}(t,t)=\mathbf{A}(t)\mathbf{R}(t,t)+\mathbf{R}(t,t)\mathbf{A}^T(t)+\mathbf{H}(t)\bar{\mathbf{Q}}(t)\mathbf{H}^T(t) \tag{2.85}$$

This is soluble, at least by numerical methods, given the initial condition $\mathbf{R}(t_0, t_0)$.

Equations (2.83) and (2.85) together describe the behaviour of $\mathrm{cov}(\mathbf{x}(t))$, and Eqn (2.81) $\varepsilon(\mathbf{x}(t))$. Since $\mathbf{x}(t)$ is Gaussian, it follows (Section 2.7.2) that it is fully specified by these variables. Eqn (2.76) readily yields $\varepsilon(\mathbf{y}(t))$ and $\mathrm{cov}(\mathbf{y}(t))$ and, since $\mathbf{y}(t)$ is also Gaussian, this too is fully specified.

It is instructive to calculate the behaviour of $\mathbf{R}(t, t)$ using the incorrect model of Eqn (2.75). In this case, remembering that $\mathbf{u}(s)$ is uncorrelated with $\mathbf{x}(t), s \geqslant t$:

$$\begin{aligned}\dot{\mathbf{R}}(t,t) &= \frac{\mathrm{d}}{\mathrm{d}t}\{\varepsilon[(\mathbf{x}(t)-\varepsilon(\mathbf{x}(t)))(\mathbf{x}(t)-\varepsilon(\mathbf{x}(t)))^T]\}\\ &= \frac{\mathrm{d}}{\mathrm{d}t}\{\varepsilon(\mathbf{x}(t)\mathbf{x}^T(t))-\varepsilon(\mathbf{x}(t))\cdot\varepsilon(\mathbf{x}^T(t))\}\\ &= \varepsilon[\dot{\mathbf{x}}(t)\cdot\mathbf{x}^T(t)-\mathbf{x}(\mathbf{t})\cdot\dot{\mathbf{x}}^T(t)]\\ &= \mathbf{A}(t)\cdot\mathbf{R}(t,t)-\mathbf{R}(t,t)\cdot\mathbf{A}(t)\quad \text{(Eqn (2.75))}\end{aligned}$$

This is incorrect, since it does not depend on $\bar{\mathbf{Q}}(t)$, which it obviously should. The fault lies not in the calculation, but in the model Eqn (2.75).

Example

A 'quarter car' suspension system, a concept useful in analyzing the vertical dynamics of road vehicles, is modelled in simplified form by two masses, M_B (kg), representing the 'quarter car' body and M_A (kg), the axle, and two springs, one representing the damped suspension, stiffness K_S ($\mathrm{N\,m^{-1}}$) and the other tyre stiffness K_T ($\mathrm{N\,m^{-1}}$). The suspension damping factor is F_S ($\mathrm{N\,s\,m^{-1}}$). The arrangement is shown in Figure 2.24.

The behaviour of the system is modelled by the equations

$$\left.\begin{aligned}&M_B\ddot{z}_B+F_S(\dot{z}_B-\dot{z}_A)+K_S(z_B-z_A)=0\\ &M_A\ddot{z}_A+F_S(\dot{z}_A-\dot{z}_B)+K_S(z_A-z_B)+K_T(z_A-w)=0\\ &y=z_A-z_B\end{aligned}\right\} \tag{2.86}$$

where

z_B is the vertical displacement of M_B (m),
z_A is the vertical displacement of M_A (m),
$w(t)$ is the stochastic vertical road displacement (m),
$y(t)$ is the relative displacement of the two masses (m).

This model, which involves two differential operations, is easily recast in standard fourth order state equation form (Eqn (2.75)) by defining $\mathbf{x}(t)$ in a sensible way, and substituting accordingly. Thus, using the fairly typical values:

$$M_B = 250\,\text{kg};\quad M_A = 20\,\text{kg};\quad K_S = 20 \times 10^3\,\text{N m}^{-1};\quad F_S = 10^3\,\text{N s m}^{-1};$$

$$K_T = 150 \times 10^3\,\text{N m}^{-1}$$

and defining

$$\mathbf{x} = (z_B \quad \dot{z}_B \quad z_A \quad 0.02\dot{z}_A)^T$$

a state space model is

$$\dot{\mathbf{x}} = \begin{pmatrix} 0 & 1 & 0 & 0 \\ -80 & -4 & 80 & 200 \\ 0 & 0 & 0 & 50 \\ 20 & 1 & -170 & -50 \end{pmatrix} \mathbf{x} + \begin{pmatrix} 0 \\ 0 \\ 0 \\ 150 \end{pmatrix} \mathbf{w} \tag{2.87}$$

$$\mathbf{y} = (1 \quad 0 \quad -1 \quad 0)\mathbf{x}$$

Eqn (2.87), being of the form of Eqn (2.75), constitutes a heuristic model which can be realized (using band limited white noise to simulate $w(t)$) on a digital computer, to yield sample functions of $\mathbf{x}(t)$ corresponding to sample functions of $w(t)$. Thus, a computer realization of Eqn (2.87) yields the sample functions shown in

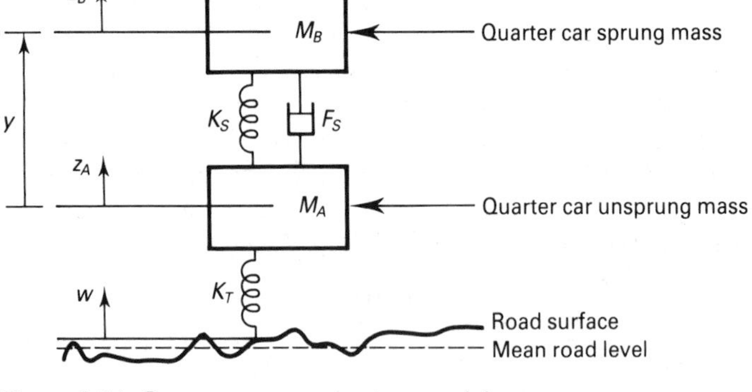

Figure 2.24 Quarter car suspension model.

Figure 2.25, which describe the quarter car behaviour when the vehicle strikes (a) a step and (b) a rough patch at $t = 0.5$ s (the road being smooth for $t < 0.5$ s), such that the strength of $w(t)$ is $6.25 \times 10^{-4}\,\text{m}^2$.

As might be expected, the displacement of the car body, $x_1(t)$, exhibits a substantially lower frequency behaviour than that of the axle, $x_3(t)$, which is not radically different from the road surface displacement, $w(t)$.

Any mathematical interpretation of this model must, of course, be undertaken using the form of Eqn (2.77), so that the mean and covariance of $\mathbf{x}(t)$ are given by Eqns (2.81), (2.83) and (2.85). Some results for the car hitting a rough patch at $t = 0.5$ s (cf. Figure 2.25(b)) are shown in Figure 2.26.

Figure 2.26 shows the variance functions settling down to constant values after some 1.0 s for the car body, and 0.25 s for the axle, and the covariance functions reflecting corresponding dynamics in terms of the time shift variable, τ. At this stage, of course, $\mathbf{x}(t)$ is virtually stationary.

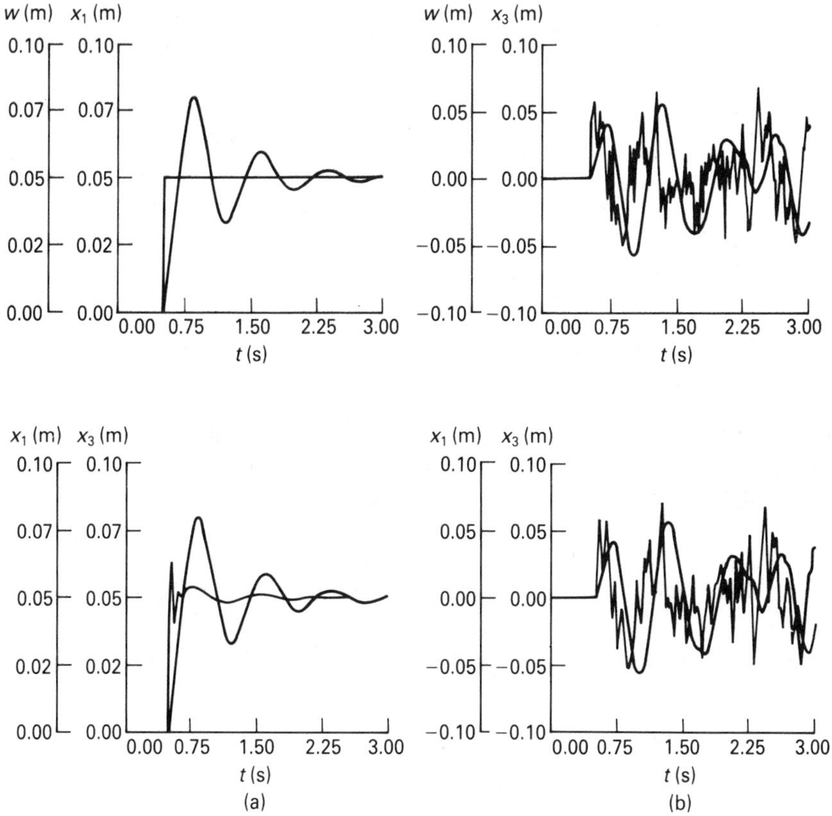

Figure 2.25 Model sample functions.

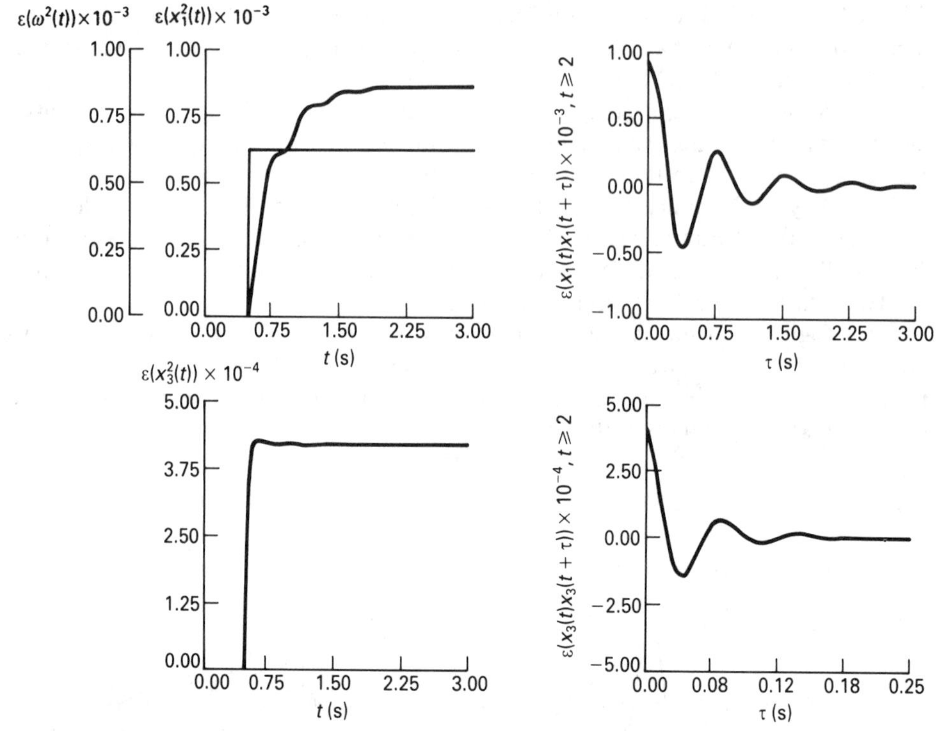

Figure 2.26 Covariance functions.

2.11 The modelling of non-white noise

The linear models of Section 2.10 can be extended quite readily to represent non-white process and measurement noise. This is done by incorporating linear filters with appropriate characteristics (Appendix B, Section B1) between white noise sources and the dynamical system, and augmenting the state space description of the system to include these. This arrangement is shown in Figure 2.27(a). The design of the filters is accomplished by simple frequency domain methods, as follows.

If a stochastic process (or indeed a deterministic signal), $f(t)$, of power spectral density $S_f(\omega)$ is fed to a linear time invariant system – in this case a filter – of Fourier transfer function $F(\mathrm{j}\omega)$, as shown in Figure 2.27(b), the PSD of the output, $n(t)$, is given by

$$S_n(\omega) = F(\mathrm{j}\omega)\cdot F(-\mathrm{j}\omega)\cdot S_f(\omega)$$

In particular, if $f(t)$ is chosen to be unit strength Gaussian white noise, the PSD of $n(t)$ is

$$S_n(\omega) = F(\mathrm{j}\omega)\cdot F(-\mathrm{j}\omega)\cdot 1 \tag{2.88}$$

Thus, if the PSD of $n(t)$ is specified, and if it can be 'spectrally factorized' into $F(j\omega).F(-j\omega)$, as required by Eqn (2.88), it should be possible to synthesize a filter with the appropriate transfer function. If spectral factorization is not possible, it is usually possible to find a transfer function – and so a filter – which approximately satisfies Eqn (2.88). A noise vector (Figure 2.27(a)) can of course be generated by an array of such filters, each excited by an independent white noise source.

Example

Consider the example of Section 2.10, in which a model of the behaviour of a quarter car suspension system is constructed, and the process noise, representing the road surface, is modelled by Gaussian white noise.

Clearly, a better representation of the road surface would be provided by noise with a PSD based on experimental data. Appropriate data for an 'average main' road* suggest a road surface with PSD as shown in Figure 2.28, where η is measured in radians per metre, and, for a vehicle velocity of λ m s^{-1}, this translates into an approximate spectral density

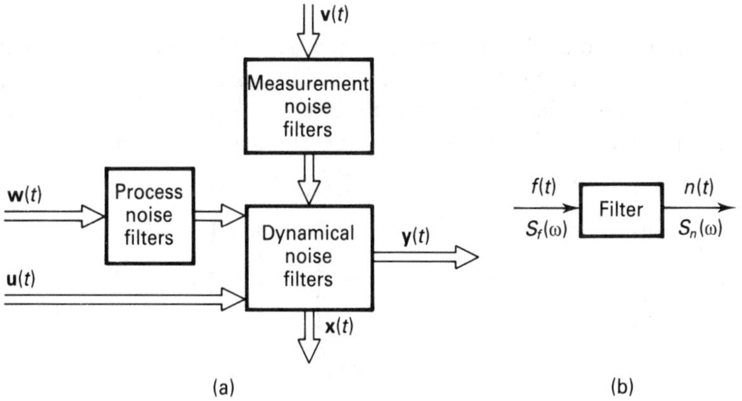

Figure 2.27 Augmented linear system.

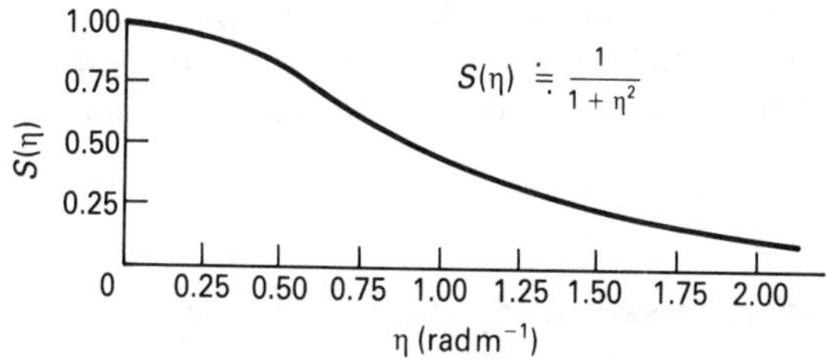

Figure 2.28 Road surface characteristics.

* La Barre, R. P., R. T. Forbes and S. Andrew, *The Measurement and Analysis of Road Roughness*, MIRA Report No. 1970/5, 1969.

$$S(\omega) = \frac{\lambda^2}{\lambda^2 + \omega^2} \tag{2.89}$$

This approximate expression for $S(\omega)$ is, of course, deliberately chosen so that it can be spectrally factorized (Eqn (2.88)):

$$F(j\omega)\cdot F(-j\omega) = \frac{\lambda}{(\lambda + j\omega)}\cdot\frac{\lambda}{(\lambda - j\omega)}$$

Clearly, there are two ways of selecting $F(j\omega)$ to satisfy this equation, and it is sensible to select that which yields a stable, non-minimum phase filter (i.e. one with poles and zeros in the open left-half s-plane). Thus, the filter transfer function is

$$F(j\omega) = \frac{\lambda}{\lambda + j\omega}$$

Correspondingly, its Laplace transfer function is (Appendix A, Section A1)

$$F(s) = \frac{\lambda}{\lambda + s}$$

It remains to model the filter in state equation form and to augment the system model accordingly. Standard state equation realization formulae (Appendix A, Section A2.4) are useful here, and, using the controllable realization of Eqn (A2.19) and $x_5(t)$ to represent the single filter state, the augmented model is

$$\dot{\mathbf{x}} = \begin{pmatrix} 0 & 1 & 0 & 0 & 0 \\ -80 & -4 & 80 & 200 & 0 \\ 0 & 0 & 0 & 50 & 0 \\ 20 & 1 & -170 & -50 & 150 \\ 0 & 0 & 0 & 0 & -\lambda \end{pmatrix}\mathbf{x} + \begin{pmatrix} 0 \\ 0 \\ 0 \\ 0 \\ \lambda \end{pmatrix} w$$

A computer realization of this for $\lambda = 20$ (m s^{-1}) over a flat surface, $0 \leqslant t < 0.5$, then an 'average' road, $t \geqslant 0.5$, yields the sample functions in Figure 2.29. The road noise is less violent than that shown in Figure 2.25, but the sprung mass displacement, $x_1(t)$, has larger amplitude – a resonant frequency of the system is being excited more violently. The sprung mass vertical velocity and acceleration are, however, reduced for the better road surface.

2.12 Non-linear mathematical models of dynamical systems

It would be attractive to extend the deterministic model of Eqns (2.67) and (2.68) to include white process and measurement noise in the most natural way, yielding the model of Eqns (2.73) and (2.74), but, as explained in Section 2.10 this is mathematically invalid. A somewhat more restricted, but still useful, model, using an additive noise term to generate the stochastic rate of change of state, and pure additive measurement noise, is written heuristically:

$$\dot{\mathbf{x}} = \mathbf{f}(\mathbf{x}, \mathbf{u}, t) + \mathbf{H}(\mathbf{x}, t) \cdot \bar{\mathbf{w}}(t) \tag{2.90}$$

$$\mathbf{y} = \mathbf{g}(\mathbf{x}, \mathbf{u}, t) + \bar{\mathbf{v}}(t) \tag{2.91}$$

where

$\mathbf{H}(\mathbf{x}, t)$ is an (m × s) matrix function of the stochastic state, $\mathbf{x}(t)$, and time, t,
$\bar{\mathbf{w}}(t)$, $\bar{\mathbf{v}}(t)$ are vectors of Gaussian extended white noise, strengths $\bar{\mathbf{Q}}(t)$, $\bar{\mathbf{R}}(t)$, respectively.

This model is mathematically invalid for the reason that Eqn (2.75) is invalid. Eqn (2.90) can, however, be recast in a valid form using the Itô stochastic integral (Eqn (2.59)):

$$\mathrm{d}(\mathbf{x}(t)) = \mathbf{f}(\mathbf{x}, \mathbf{u}, t) \cdot \mathrm{d}t + \mathbf{H}(\mathbf{x}, t)\, \mathrm{d}\bar{\boldsymbol{\beta}}(t) \tag{2.92}$$

where $\bar{\boldsymbol{\beta}}(t)$ is an s-vector of extended Wiener processes, diffusion $\bar{\mathbf{Q}}(t)$.

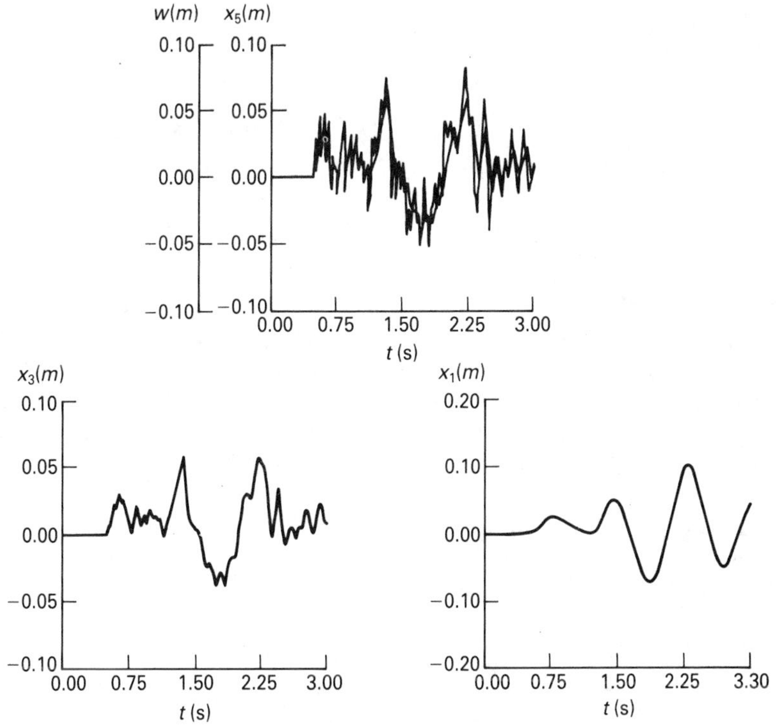

Figure 2.29 Model sample functions.

Eqn (2.92) guarantees that $\mathbf{x}(t)$ is Markov, owing to the underlying properties of the Itô integral. A similar model based on the Stratonovich integral (Eqn (2.64)) is also valid, though more restricted, and in this case $\mathbf{x}(t)$ is not guaranteed to be Markov. For these reasons, the Itô integral based model is now more popular.

In practice, non-linear models of stochastic dynamical systems are almost invariably cast in the form of Eqns (2.90) and (2.91) – and realized by a simulation algorithm using this form (with band limited white noise forcing functions). For any analytical purposes, however, Eqn (2.90) must be interpreted as Eqn (2.92). As in the linear case, the output equation model poses no particular problems.

Eqn (2.92) is equivalent to [ref. 5]

$$\mathbf{x}(t) = \mathbf{x}(t_0) + \int_{t_0}^{t} \mathbf{f}(\mathbf{x}, \mathbf{u}, \lambda)\, d\lambda + \int_{t_0}^{t} \mathbf{H}(\mathbf{x}, \lambda)\, d\bar{\boldsymbol{\beta}}(\lambda)$$

where the integrals are interpreted as Riemann (Eqn (2.27)) and Itô (Eqn (2.59)) respectively, and the variance, though not the covariance, of the latter is available from Eqn (2.62).

One important theoretical result which can be deduced from this [ref. 5] is the 'Fokker–Planck', or 'forward Kolmogorov', equation which describes the propagation of the joint PDF of $\mathbf{x}(t)$, given that at t_0 (Eqn (1.29)),

$$\begin{aligned}\frac{\partial}{\partial t}(p(\boldsymbol{\xi}, t)) = &-\sum_{i=1}^{m} \frac{\partial}{\partial \xi_i}(p(\xi_i, t) f_i(\boldsymbol{\xi}, \mathbf{u}, t)) \\ &+ \frac{1}{2}\sum_{i=1}^{m}\sum_{j=1}^{m} \frac{\partial^2}{\partial \xi_i\, \partial \xi_j}(p(\boldsymbol{\xi}, t)\mathbf{H}(\boldsymbol{\xi}, t)\bar{\mathbf{Q}}(t)\mathbf{H}^T(\boldsymbol{\xi}, t))\end{aligned} \tag{2.93}$$

where

$$\boldsymbol{\xi} = (\xi_1 \quad \xi_2 \quad \dots \quad \xi_m)^T$$

and $p(\boldsymbol{\xi}, t)$ is the joint PDF describing $\mathbf{x}(t)$; ξ_1 refers to x_1, ξ_2 to $x_2, \dots, \xi_m$ to x_m; $f_i(\boldsymbol{\xi}, \mathbf{u}, t)$ refers to the ith row of $\mathbf{f}(\mathbf{x}, \mathbf{u}, t)$ (Eqn (2.90)), $\mathbf{x}$ being replaced by $\boldsymbol{\xi}$.

An initial condition, $p(\boldsymbol{\xi}, 0)$, is, of course, required. The complexity of Eqn (2.93) renders it difficult to solve even by numerical methods [ref. 5], but where soluble it is useful, and the solution can be used to find the behaviour of the mean and variance – though not the covariance – of $\mathbf{x}(t)$:

$$\frac{d}{dt}[\varepsilon(\mathbf{x}(t))] = \varepsilon[\mathbf{f}(\mathbf{x}, \mathbf{u}, t)] \tag{2.94}$$

and

$$\begin{aligned}\dot{\mathbf{R}}(t, t) = &\ \varepsilon[\mathbf{f}(\mathbf{x}, \mathbf{u}, t)\cdot \mathbf{x}^T(t)] - \varepsilon[\mathbf{f}(\mathbf{x}, \mathbf{u}, t)]\cdot \varepsilon(\mathbf{x}^T(t)) \\ &+ \varepsilon[\mathbf{x}(t)\cdot \mathbf{f}^T(\mathbf{x}, \mathbf{u}, t)] - \varepsilon(\mathbf{x}(t))\cdot \varepsilon[\mathbf{f}^T(\mathbf{x}, \mathbf{u}, t)] \\ &+ \varepsilon[\mathbf{H}(\mathbf{x}, t)\cdot \bar{\mathbf{Q}}(t)\cdot \mathbf{H}^T(\mathbf{x}, t)]\end{aligned} \tag{2.95}$$

Obtaining results from these equations entails solving Eqn (2.93) and then evaluating Eqn (1.32). The great complexity of this renders these equations less useful than they appear at first sight.

For the special linear case, Eqns (2.94) and (2.95) simplify into Eqns (2.81) and (2.85), which are reasonably tractable.

Example

Consider a rod which is rotatable at one end round a horizontal axis and driven by a motor gearbox, the whole arrangement being mounted on a table subject to random vertical displacement. The system, which simulates, in simplified form, the vertical dynamics of the turret gun of a tank travelling over rough terrain, is shown in Figure 2.30.

The angular behaviour of the rod is modelled by the (heuristic) equation, derived using Lagrangian mechanics:

$$I\ddot{\theta} + f\dot{\theta} + Mgl\cos\theta = n - Ml\cos\theta \cdot w$$

where

θ is the angular position of the rod (rad),
u is the torque (N m) supplied by the motor,
w is the random vertical acceleration of the axis (m s^{-2}),
I is the moment of inertia of the rod about its axis (kg m^2),
f is the frictional torque (N m s rad^{-1}),
M is the mass of the rod (kg),
l is the distance from the axis to the centre of gravity (m),
g is the gravitational constant (m s^{-2}).

These equations are readily recast in heuristic state equation form (Eqns (2.93) and (2.94)), and for $I = 0.5$ kg m^2; $f = 0.1$ N m s rad^{-1}; $M = 1$ kg; $l = 0.5$ m; $g = 9.81$ m s^{-2}:

$$\dot{\mathbf{x}} = \begin{pmatrix} \dot{x}_1 \\ \dot{x}_2 \end{pmatrix} = \begin{pmatrix} x_2 \\ -9.81\cos x_1 - 0.2x_2 + u - w\cos x_1 \end{pmatrix} \tag{2.96}$$

$$y = x_1$$

where $x_1 = \theta$, $x_2 = \dot{\theta}$.

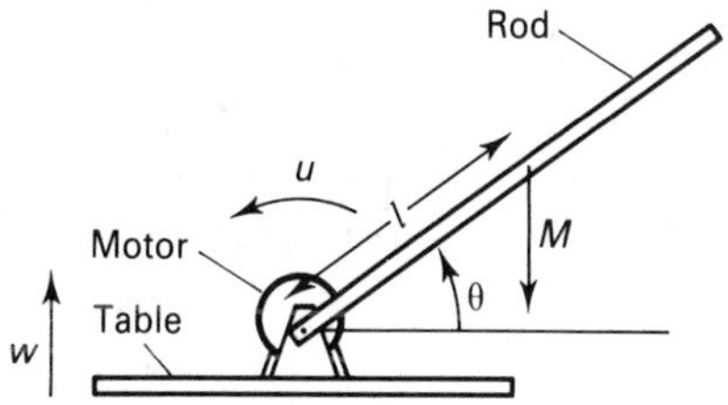

Figure 2.30 Experimental non-linear system.

The system is unstable for $\frac{\pi}{4} \leqslant x_1 \leqslant \frac{\pi}{2}$, making simulation rather unrevealing, but for $-\frac{\pi}{4} < x_1 < \frac{\pi}{4}$, it is stable for small perturbations, and its behaviour is illustrated in Figure 2.31 ($\mathbf{x}(0) = \begin{pmatrix} -\frac{\pi}{4} & 0 \end{pmatrix}^T$, $u = 9.81$ and w white noise, strength 0, $t < 0.5$, and 0.01, $t \geqslant 0.5$).

Theoretically, the PDF of $x(t)$ could be found by solving Eqn (2.93) numerically, and $x(t)$ and $\mathbf{R}(t, t)$ by solving Eqns (2.94) and (2.95), again numerically. However, as this involves the solution of partial differential equations, which is beyond the scope of this book, this is not pursued further here. This example is, however, taken further in Chapters 5 and 6.

EXERCISES

1 Consider the process:

$$x(t) = e^{-at}$$

where a is a random number evenly distributed between $+1$ and $+2$. Find $\varepsilon(x(t))$ and $\text{cov}(x(t))$.

2 Consider the process:

$$x(t) = \cos(\omega t)$$

where ω is a random number evenly distributed between $\Omega \pm \Delta$. Find $\varepsilon(x(t))$ and $\text{cov}(x(t))$.

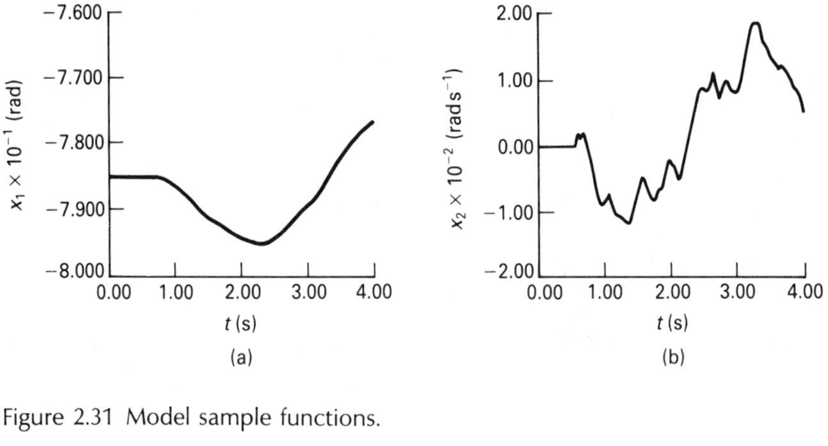

Figure 2.31 Model sample functions.

3 Consider the process:

$$x(t) = a\cos(\omega t) + b\sin(\omega t)$$

where a, b are independent random numbers with zero mean and variance σ^2. Find $\text{cov}(x(t))$.

4 Find $S(\omega)$ if:
(a) $\bar{r}(\tau) = e^{-a|\tau|}$,
(b) $\bar{r}(\tau) = e^{-a|\tau|} + \cos(b\tau)$,
(c) $\bar{r}(\tau) = 6 + 4\cos(b\tau)$,
(d) $\bar{r}(\tau) = e^{-a|\tau|}\cos(b\tau)$.

5 Find $\bar{r}(\tau)$ if:
(a)

$$S(\omega) = \begin{cases} 9 - \omega^2, & |\omega| \leqslant 3 \\ 0, & |\omega| > 3, \end{cases}$$

(b)

$$S(\omega) = \begin{cases} 1 - \dfrac{|\omega|}{2}, & |\omega| \leqslant 2 \\ 0, & |\omega| > 2. \end{cases}$$

6 Show that if

$$y = x(t+1) - x(t-1)$$

then:

$$\bar{r}_y(\tau) = 2\bar{r}_x(\tau) - \bar{r}_x(\tau+2) - \bar{r}_x(\tau-2)$$

and:

$$S_y(\omega) = 4S_x(\omega)\sin^2(\omega).$$

7 Consider the process $x(t)$ with $r(\tau) = e^{-|\tau|} - e^{-2|\tau|}$.
Find the greatest value of n such that $d^n x/dt^n$ exists.

8 Determine whether the processes are differentiable which have:
(a) $r(\tau) = e^{-a|\tau|}$,
(b) $r(\tau) = \dfrac{\sin(a\tau)}{\tau}$,
(c) $r(\tau) = \dfrac{2}{2+\tau^2}$.

9 A process $x(t)$ has covariance $r(\tau) = (1 + 2|\tau|)e^{-|\tau|}$. Show that $x(t)$ is differentiable and that $x(t)$, $\dot{x}(t)$ are independent.

10 Consider a Wiener process $\beta(t)$, diffusion q. Show that $\beta(t)$ is integrable and find the covariance of the integrated process.

11 Find the solution of the dynamical equation:

$$d\mathbf{x} = \begin{pmatrix} 0 & 1 \\ 0 & 0 \end{pmatrix} dt + \begin{pmatrix} 0 \\ 1 \end{pmatrix} d\beta, \quad \mathbf{x}(0) = \begin{pmatrix} 1 \\ 1 \end{pmatrix}$$

where

x represents system states,
β is a unit diffusion Wiener process.

12 A mass spring damper system subject to noise input is modelled by the heuristic equation:

$$\ddot{y} + 0.1\dot{y} + y = 2w, \quad y(0) = 1$$

where

y is the mass displacement,
w is unit strength white noise.

Cast this model in state equation form and develop equations describing the behaviour of $\varepsilon(y(t))$, $\mathrm{cov}(y(t))$, $\mathrm{cov}(\dot{y}(t))$.
Calculate the steady state covariances of $y(t)$ and $\dot{y}(t)$.

13 A system is modelled by the heuristic equation:

$$\dot{\mathbf{x}} = \begin{pmatrix} -2 & 0 \\ 0 & -3 \end{pmatrix}\mathbf{x} + \begin{pmatrix} 1 \\ 1 \end{pmatrix}u + \begin{pmatrix} 2 \\ 3 \end{pmatrix}w$$

where

x represents the system states,
u is a constant input,
w is unit strength white noise.

Calculate the steady state mean and covariance of $x_1(t)$ and $x_2(t)$.

14 Specify the transfer functions of filters which will process white noise, strength q, to generate processes with power spectral densities:

(a) $S(\omega) = \dfrac{3}{3 + \omega^2}$,

(b) $S(\omega) = \dfrac{4\omega^2}{(4 + \omega^2)(9 + \omega^2)}$.

15 The free dynamical behaviour of a pendulum of length 1 (m) is modelled by:

$$\ddot{\theta} + 0.1\dot{\theta} + \sin\theta = 0$$

where θ is the angular displacement (rad).
The pendulum bob is subject to a random horizontal wind force proportional to the horizontal velocity of the wind relative to the bob ($0.01\ \mathrm{N\,m^{-1}\,s}$), and the wind velocity is modelled by unit strength white noise.
(a) Construct a state space model in the form of Eqn (2.92) for the behaviour of this stochastic system, and express it in integral form. What is the variance of the process noise function?
(b) Assuming θ to be small (i.e. $\sin\theta \doteqdot \theta$), construct a linear state space model for the system and find the steady state covariances of $\theta, \dot{\theta}$.

REFERENCES

[1] Astrom, K. J., *Introduction to Stochastic Control Theory*, Academic Press, New York, 1970.
[2] Maybeck, P. S., *Stochastic Models, Estimation and Control*, Academic Press, New York, 1982.
[3] Papoulis, A., *Probability, Random Variables and Stochastic Processes*, McGraw-Hill, New York, 1984.
[4] Helstrom, C. W., *Probability and Stochastic Processes for Engineers*, Macmillan, New York, 1991.
[5] Press, W. H., Flannery, B. P., Teukolsky, S. A. and Vetterling, W. T., *Numerical Recipes: The art of scientific computing*, CUP, Cambridge, 1986.

3 Discrete time stochastic processes and dynamical systems

3.1 Introduction

This chapter deals with the properties of discrete time stochastic processes and with the construction and analysis of standard-form mathematical models for linear dynamical systems. The derivation of these from models synthesized in the continuous time domain (Chapter 2) is also considered, but the non-linear case, being incapable of generalized treatment, is not explored.

The sequence of topics follows that established in Chapter 2 and, to avoid repetition, frequent reference to that chapter is made. Sections 3.2–3.4 contain definitions, Section 3.5 contains frequency domain considerations, and Section 3.6 contains a description of some important stochastic processes. Sections 3.7–3.10 describe the construction and analysis of models of dynamical systems.

3.2 Discrete time stochastic processes

The descriptions of Section 2.2 apply almost directly to discrete time processes with the continuous independent variable t replaced by the discrete variable $n, n = -\infty, \ldots, -1, 0, 1, 2, 3, \ldots, +\infty$. Thus the stochastic process $x(n)$ represents any discrete time sample function $\bar{x}_i(n)$ from its ensemble, $i = 1, 2, \ldots, \infty$, as illustrated in Figure 3.1.

Formally, $x(n)$ is a time-dependent random variable in accordance with the definition of Section 1.5, and its amplitude is described by a probability density function which is, in principle, a function of time, $p(\xi, n)$.

The dynamic behaviour of $x(n)$ is wholly characterized by the infinite dimensional joint PDF:

$$p(\xi_1, \xi_2, \ldots, \xi_n, \ldots; \quad 1, 2, \ldots, n, \ldots) \tag{3.1}$$

where ξ_1 refers to $n = 1$, ξ_2 to $n = 2$, and so on. As with continuous time processes, the complexity of this function renders it of little practical value, and $x(n)$ is usually described by its mean, covariance, and autocorrelation functions. These are defined in Section 3.3. The terms 'strictly stationary', 'wide sense stationary' and 'weakly stationary' apply to discrete time just as they do to continuous time processes (Section 2.2), with the obvious difference that times and time difference measurements are now discrete. Thus, for example, $x(n)$ is wide sense stationary if

$$\varepsilon(x(n)) = \varepsilon(x(n+k)) = m_x$$

$$\varepsilon(x(n)\cdot x(n+k)) = r_x(k)$$

The term 'ergodic' also applies to discrete time processes, carrying the same meaning as for the continuous time case.

Example

Consider tossing a die once every second and sampling the result after each toss, i.e. the process of Figure 2.3 sampled at intervals of 1 second, to yield $x(n)$. Two sample functions and $p(\xi, n)$ are shown in Figure 3.2.

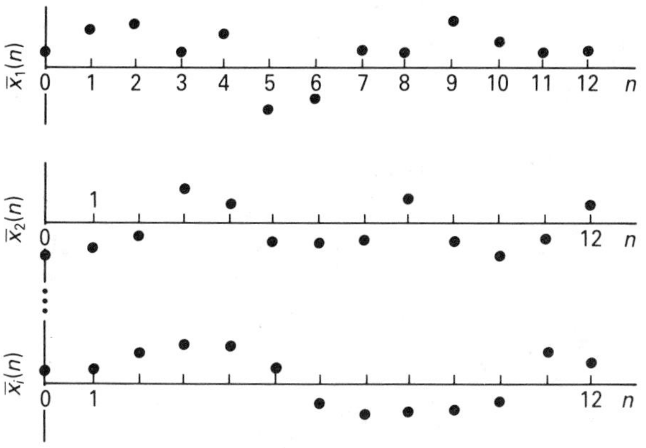

Figure 3.1 Ensemble of sample functions.

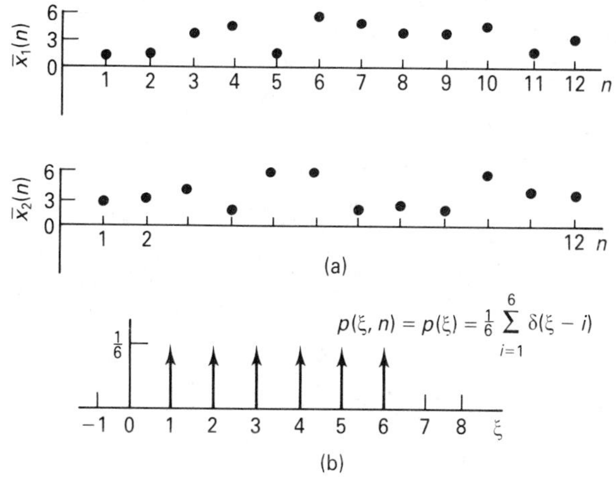

Figure 3.2 Discrete ensemble and PDF.

3.3 Ensemble properties of discrete time stochastic processes (cf. Section 2.3)

The unmanageable complexity of the joint PDF of Eqn (3.1), which fully describes the behaviour of a stochastic process, leads to the use of mean, variance, correlation and covariance functions, which provide useful, if rather coarse, descriptions.

3.3.1 The mean and variance (cf. Sections 1.8 and 2.3.1)

The mean is taken across the ensemble at time n:

$$\varepsilon(x(n)) = m_x(n)$$
$$= \int_{-\infty}^{+\infty} \xi p(\xi, n)\,\mathrm{d}\xi \tag{3.2}$$

The variance or 'spread' of $x(n)$ is

$$\sigma_x^2(n) = \varepsilon(x(n) - \varepsilon(x(n))^2)$$
$$= \int_{-\infty}^{+\infty} (\xi - \varepsilon(x(n)))^2 p(\xi, n)\,\mathrm{d}\xi \tag{3.3}$$

For $x(n)$ stationary (in any sense), the mean and variance are constant:

$$m_x(n) = m_x$$
$$\sigma_x^2(n) = \sigma_x$$

3.3.2 Autocorrelation and covariance (cf. Sections 1.14 and 2.3.2)

These functions are measures of the rate of change, or dynamics, of a process $x(n)$, and also of its amplitude.

The autocorrelation is

$$\bar{r}(n, l) = \varepsilon(x(n) \cdot x(l))$$
$$= \int_{-\infty}^{+\infty}\int_{-\infty}^{+\infty} \xi_n \xi_l p(\xi_n, \xi_l; n, l)\,\mathrm{d}\xi_n\,\mathrm{d}\xi_l \tag{3.4}$$

where $p(\xi_n, \xi_l; n, l)$ is the joint PDF describing $x(n)$, $x(l)$.

The covariance is essentially the same function, centralized about the mean values $\varepsilon(x(n))$, $\varepsilon(x(l))$:

$$r(n, l) = \mathrm{cov}(x(n))$$
$$= \varepsilon[(x(n) - \varepsilon(x(n)))(x(l) - \varepsilon(x(l)))] \tag{3.5}$$

As in the continuous time case (Eqn (2.8)),

$$r(n, l) = \bar{r}(n, l) - \varepsilon(x(n)) \cdot \varepsilon(x(l)) \tag{3.6}$$

If $x(n)$ is wide sense stationary, $\bar{r}(n, l)$ and $r(n, l)$ are even functions of $k = (l - n)$:

$$\bar{r}(n, l) = \bar{r}(l - n) = \bar{r}(k) = \bar{r}(-k) \tag{3.7}$$

$$r(n, l) = r(l - n) = r(k) = r(-k) \tag{3.8}$$

Examples

(a) Consider the mean and covariance of the process of the preceding example (Figure 3.2):

$$m_x(n) = m_x = \frac{1}{6}(1 + 2 + \cdots + 6) = 3.5$$

$$\sigma_x^2(n) = \sigma_x^2 = \frac{1}{6}(1^2 + 2^2 + \cdots + 6^2) - (3.5)^2 = 2.917$$

$$r(n, l) = r(k) = 2.917\delta(k)$$

where

$\delta(k)$ is a Kronecker delta function (Appendix D, Section D2),
$r(n, l) = r(k)$ is illustrated in Figure 3.3(a, b).

(b) Consider the discrete time process illustrated in Figure 3.4(a):

$$x(n) = a + bn$$

where a, b are independent random variables with PDFs, as shown in Figure 3.4(b, c).

As in the equivalent continuous time process (Section 2.3.2, Example (c)), the PDF of $x(n)$, $p_x(\gamma, n)$ is evaluated using Eqn (1.42) (cf. Figure 2.4). This is illustrated in Figure 3.4(d).

$\varepsilon(x(n))$ and $r(n, l)$ are readily calculated:

$$\varepsilon(x(n)) = \varepsilon(a + bn)$$
$$= \varepsilon(a) + n\varepsilon(b) = 0$$

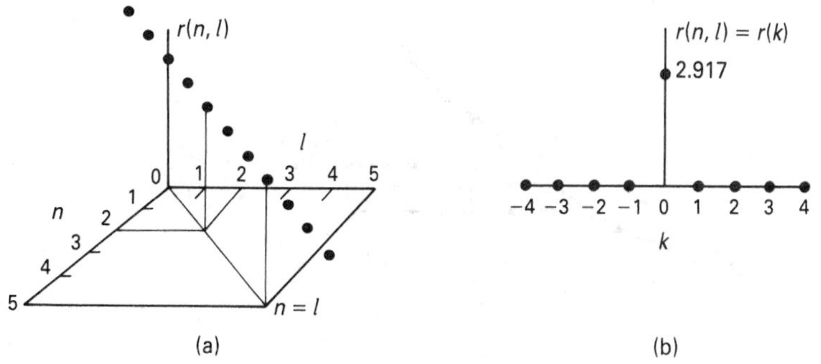

Figure 3.3 Covariance functions.

$$\bar{r}(n, l) = \varepsilon((a + bn)(a + bl))$$
$$= \frac{4}{3}(1 + nl)$$

Also (Eqn (3.6)),

$$r(n, l) = \frac{4}{3}(1 + nl)$$

This is illustrated in Figure 3.4(e) (cf. Figure 2.7(b)).

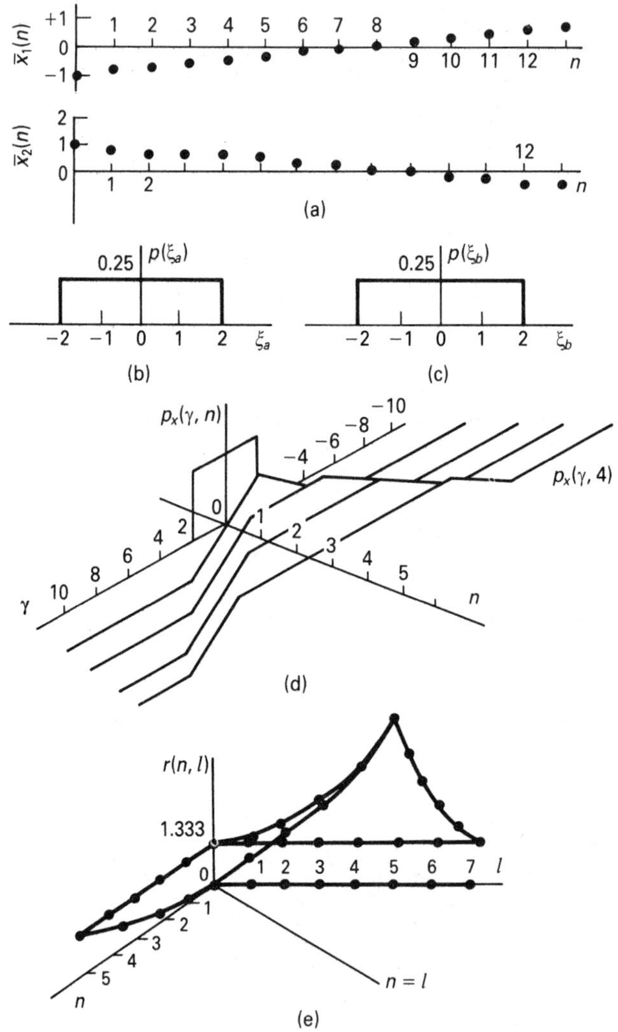

Figure 3.4 Stochastic process.

Finally, the variance of $x(n)$, represented on the diagonal $n = l$ in Figure 3.4(e), is

$$r(n, n) = \frac{4}{3}(1 + n^2)$$

3.3.3 Cross-correlation and cross-covariance (cf. Section 2.3.3)

A measure of the relationship between two processes $x(n)$, $y(n)$ is given by the cross-correlation function:

$$\bar{r}_{xy}(n, l) = \varepsilon(x(n) \cdot x(l))$$
$$= \int_{-\infty}^{+\infty} \int_{-\infty}^{+\infty} \xi_n \xi_l p(\xi_n, \xi_l; n, l) \, d\xi_n \, d\xi_l \tag{3.9}$$

where $p(\xi_n, \xi_l; n, l)$ is the joint PDF describing $x(n)$, $y(l)$, much as indicated in Figure 2.8(b).

The cross-covariance of $x(n)$, $y(l)$ is defined:

$$r_{xy}(n, l) = \varepsilon(x(n) - \varepsilon(x(n)))(x(l) - \varepsilon(x(l))) \tag{3.10}$$

$\bar{r}_{xy}(n, l)$, $r_{xy}(n, l)$ are related:

$$r_{xy}(n, l) = \bar{r}_{xy}(n, l) - \varepsilon(x(n)) \cdot \varepsilon(y(l)) \tag{3.11}$$

Moreover, if $x(n)$, $y(n)$ are wide sense stationary:

$$\bar{r}_{xy}(n, l) = \bar{r}_{xy}(l - n) = \bar{r}_{xy}(k) = \bar{r}_{yx}(-k) = \bar{r}_{xy}(n - l)$$

where $k = (l - n)$.

Similarly,

$$r_{xy}(n, l) = r_{xy}(l - n) = r_{xy}(k) = r_{yx}(-k) = r_{yx}(n - l)$$

Example

Consider the discrete time equivalent of the example of Figure 2.9, where two dice are thrown simultaneously, the results sampled immediately after each throw, and $x(n)$ defined as the result on one die, $y(n)$ the sum of the two results. The PDFs of $x(n)$, $y(n)$ are shown in Figure 2.9(a, b).

The means, cross-correlation and cross-covariance functions are easily calculated:

$$\varepsilon(x(n)) = \frac{1}{6}(1 + 2 + \cdots + 6) = 3.5$$

$$\varepsilon(y(l)) = \frac{1}{36}(2 + 3 \times 2 + 4 \times 3 + \cdots + 12) = 7.0$$

$$\bar{r}_{xy}(k) = \begin{cases} \frac{1}{36}(1 \times (2+3+\cdots+7) + 2(3+4+\cdots+8) + \cdots + 6(7+8+\cdots+12)) = 27.417, & k=0 \\ 3.5 \times 7.0 = 24.500, & k \neq 0 \end{cases}$$

$$r_{xy}(k) = \begin{cases} 27.417 - 3.5 \times 7.0 = 2.917, & k=0 \\ 24.500 - 3.5 \times 7.0 = 0, & k \neq 0 \end{cases}$$

3.4 Time-averaged properties of discrete time stochastic processes (cf. Section 2.4)

The time-averaged properties of ergodic stochastic processes, which correspond to the ensemble properties of Section 3.3, are defined as follows.

3.4.1 Time-averaged mean

Considering $\bar{x}_i(n)$ the ith sample function of $x(n)$ (Figure 3.1):

$$m_n = \lim_{N\to\infty}\left(\frac{1}{2N+1}\sum_{n=-N}^{+N}\bar{x}_i(n)\right)$$

Since $x(n)$ must be ergodic for m_n to be meaningful,

$$m_n = \lim_{N\to\infty}\left(\frac{1}{2N+1}\sum_{n=-N}^{+N}x(n)\right) \tag{3.12}$$

The time-averaged and ensemble means (Eqn (3.2)) are equal, $x(n)$ ergodic:

$$m_n = m_x$$

3.4.2 Time-averaged autocorrelation

Considering the ith sample function $\bar{x}_i(n)$ of an ergodic process, $x(n)$, the autocorrelation of $\bar{x}_i(n)$ is

$$\bar{r}_n(k) = \lim_{N\to\infty}\left(\frac{1}{2N+1}\sum_{n=-N}^{+N}\bar{x}_i(n)\bar{x}_i(n+k)\right)$$

Since $x(n)$ is ergodic,

$$\bar{r}_n(k) = \lim_{N\to\infty}\left(\frac{1}{2N+1}\sum_{n=-N}^{+N}x(n)x(n+k)\right) \tag{3.13}$$

Naturally, $\bar{r}_n(k)$ equals the ensemble correlation (Eqn (3.4)), $x(n)$ ergodic:

$$\bar{r}_n(k) = \bar{r}(k)$$

3.4.3 Time-averaged cross-correlation

The time-averaged cross-correlation between sample functions $\bar{x}_i(n)$, $\bar{y}_j(n)$, from two ergodic processes $x(n)$, $y(n)$ is defined:

$$\bar{r}_{nxy}(k) = \lim_{N\to\infty}\left(\frac{1}{2N+1}\sum_{n=-N}^{+N}\bar{x}_i(n)\bar{y}_j(n+k)\right)$$

Since $x(n)$, $y(n)$ are ergodic:

$$\bar{r}_{xy}(k) = \lim_{N\to\infty}\left(\frac{1}{2N+1}\sum_{n=-N}^{+N}x(n)y(n+k)\right)$$

Naturally, the ensemble cross-correlation Eqn (3.9) is equal to the time-averaged function, $x(n)$, $y(n)$ ergodic:

$$\bar{r}_{nxy}(k) = \bar{r}_{xy}(k)$$

Example

Consider the example of the preceding subsection concerning two dice being tossed repeatedly, and the results sampled $(x(n))$ and summed $(y(n))$.

By considering sample functions and using the methods of Section 1.5, it is not difficult to show that

$$m_x = 3.5$$

$$m_y = 7.0$$

$$\bar{r}_{nxy}(k) = \begin{cases} 27.417, k = 0 \\ 24.500, k \neq 0 \end{cases}$$

3.5 Frequency domain considerations (cf. Section 2.5)

As with continuous time processes the Fourier transform (Appendix A, Section A1.3(b)) of a deterministic signal allows it to be examined in terms of its amplitude and phase content, expressed as deterministic functions of frequency. Less usefully, the Fourier transform of a stochastic process, $x(n)$, is a complex stochastic function of frequency, β (radians per sample) (Appendix A, Eqn (A1.26)):

$$\mathscr{F}(x(n)) = \sum_{n=-\infty}^{+\infty} x(n)\mathrm{e}^{-\mathrm{j}\beta n}$$

$$= \bar{X}(\mathrm{e}^{\mathrm{j}\beta})$$

In contrast to the integration in the continuous time case, no special definition need be developed for the summation operator. $\bar{X}(\mathrm{e}^{\mathrm{j}\beta})$ is a complex stochastic function of the continuous independent variable β, of interest over the interval $-\pi < \beta \leqslant +\pi$ according the Nyquist sampling criterion [ref. 1].

The inverse relationship, transforming the continuous frequency stochastic function $\bar{X}(e^{j\beta})$ to the discrete time one, $x(n)$, is

$$x(n) = \mathscr{F}^{-1}(\bar{X}(e^{j\beta})) = \frac{1}{2\pi}\int_{-\pi}^{+\pi} \bar{X}(e^{j\beta})e^{j\beta n}\,d\beta$$

A practical measure of the frequency content of $x(n)$ is provided by its power spectral density (PSD) or power per unit bandwidth (typically watts per radian per sample). As in the continuous time case, the PSD and autocorrelation function form a Fourier transform pair. This is the Wiener–Khinchine theorem in the discrete time domain (cf. Eqns (2.16) and (2.17)):

$$\bar{S}(\beta) = \mathscr{F}(\bar{r}(k)) = \sum_{k=-\infty}^{+\infty} \bar{r}(k)e^{-j\beta k}, \quad -\pi < \beta \leqslant +\pi \tag{3.14}$$

Conversely:

$$\bar{r}(k) = \mathscr{F}^{-1}(\bar{S}(\beta)) = \frac{1}{2\pi}\int_{-\pi}^{+\pi} \bar{S}(\beta)e^{j\beta k}\,d\beta \tag{3.15}$$

Since $\bar{r}(k)$ is real and even, it follows that $\bar{S}(\beta)$ is even and real (Appendix A, Section A1.3).

The cross-spectral density, $\bar{S}_{xy}(e^{j\beta k})$, of two ergodic discrete time processes $x(n)$, $y(n)$ is correspondingly related to their cross-correlation:

$$\bar{S}_{xy}(e^{j\beta}) = \mathscr{F}(\bar{r}_{xy}(k)) = \sum_{k=-\infty}^{+\infty} \bar{r}_{xy}(k)e^{-j\beta k} \tag{3.16}$$

$$\bar{r}_{xy}(k) = \mathscr{F}^{-1}(\bar{S}_{xy}(e^{j\beta})) = \frac{1}{2\pi}\int_{-\pi}^{+\pi} \bar{S}_{xy}(e^{j\beta})e^{+j\beta k}\,d\beta \tag{3.17}$$

Since $\bar{r}_{xy}(k)$ is real, but not usually even, $\bar{S}_{xy}(e^{j\beta})$ usually has an even real and odd imaginary part (Appendix A, Section A1.3).

Examples

As in the continuous time case, practical examples normally involve computer-generated results, analytical examples being useful mainly for illustrative purposes.

Consider the process

$$x(n) = \cos(\beta_0 n + \alpha)$$

where β_0 is constant and α is evenly distributed between $\pm\pi$. A sample function is shown in Figure 3.5(a). The Fourier transform of $x(n)$, which is not perhaps of great interest, is a complex stochastic variable with an ensemble similar to that in Figure 2.11(b).

The mean and autocorrelation of $x(n)$ are easily found (Eqns (3.2) and (3.4)):

$$\varepsilon(x(n)) = \varepsilon(\cos(\beta_0 n + \alpha)) = 0$$

$$\begin{aligned}\bar{r}(n, l) &= \varepsilon(\cos(\beta_0 n + \alpha)\cos(\beta_0 l + \alpha)) \\ &= \frac{1}{2}\cos(\beta_0 k) \\ &= \bar{r}(k)\end{aligned}$$

where $k = (l - n)$. This is illustrated in Figure 3.5(b).

The Fourier transform of this (Eqn (3.14)) gives, after some manipulation,

$$\begin{aligned}\bar{S}(\beta) &= \sum_{k=-\infty}^{+\infty} \tfrac{1}{2}\cos(\beta_0 k)\, \mathrm{e}^{-j\beta k} \\ &= \frac{\pi}{2}\delta(\beta - \beta_0) + \frac{\pi}{2}\delta(\beta + \beta_0)\end{aligned}$$

$\bar{r}(k)$ is even and real, so $\bar{S}(\beta)$ is real and even, as shown in Figure 3.5(c).

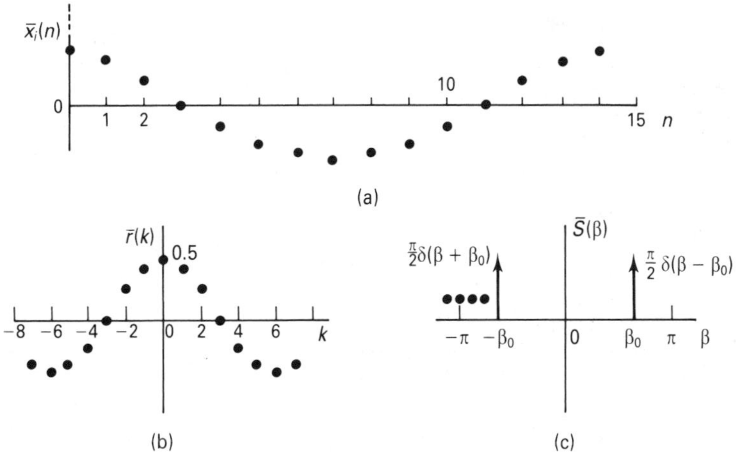

Figure 3.5 Autocorrelation and PSD.

3.6 Some important processes (cf. Section 2.7)

The discrete time equivalents of the continuous time stochastic processes of Section 2.7 are considered in this section. Since many of their properties are common to both cases, the treatment here is brief.

3.6.1 Markov processes (cf. Section 2.7.1)

A discrete time Markov process, $x(n)$, has the same essential property as its continuous time counterpart, namely that the conditional PDF describing $x(l)$, given $x(n)$, does not depend on $x(n-1)$, $x(n-2)$, $l>$n:

$$p(\xi_l | x(n), x(n-1), \ldots) = p(\xi_l | x(n)) \tag{3.18}$$

The properties of continuous time Markov processes apply virtually directly to discrete time processes, and are not repeated here.

Example

Consider a discrete time random walk, $x(n)$, generated by tossing a coin at times $n = 0, 1, 2, \ldots$ and specifying a step, size λ, to the right for a head and to the left for a tail. The process is non-stationary and Markov. A sample function, $\bar{x}_i(n)$, is given in Figure 3.6(a) (cf. Figure 2.14).

As explained in Section 2.7.1, the amplitude of $x(n)$ has the binomial PDF

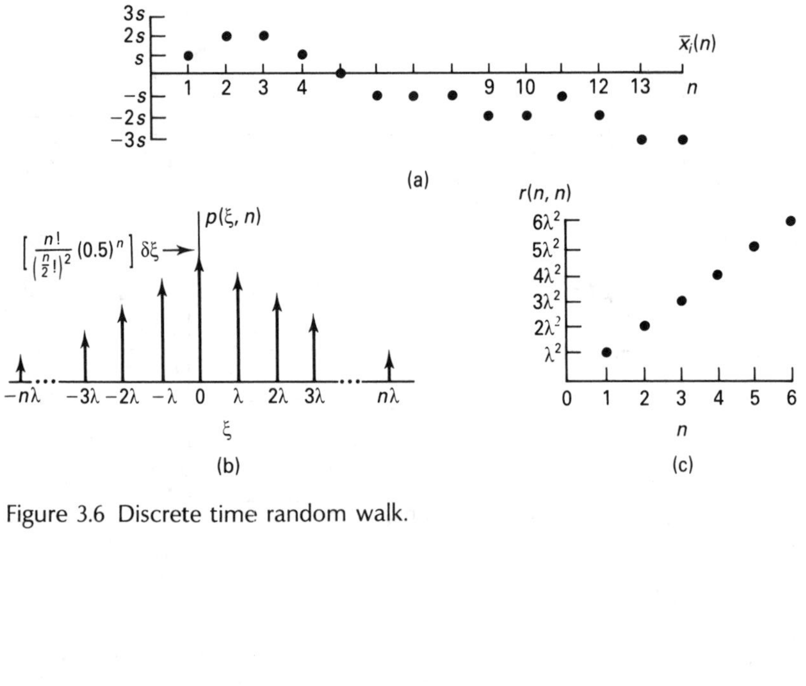

Figure 3.6 Discrete time random walk.

(Eqn (2.35)):

$$p(\xi, n) = \sum_{m=-n}^{+n} \begin{pmatrix} n \\ \frac{n+m}{2} \end{pmatrix} (0.5)^n \delta(\xi - m\lambda)$$

Recalling the mean and variance:

$$\varepsilon(x(n)) = 0$$

$$\sigma_x^2(n) = r(n, n) = n\lambda^2$$

These functions are illustrated in Figure 3.6(b, c).

3.6.2 Gaussian processes (cf. Section 2.7.2)

A discrete time process is Gaussian if the joint PDF which specifies it wholly is Gaussian (cf. Eqn (2.37)). The properties outlined for the continuous time Gaussian process apply equally to the discrete time case, perhaps the most important being that the mean and covariance are sufficient to specify the process fully.

Example

A discrete time Gaussian process is generated by sampling a continuous time Gaussian process via an ADC (analog to digital converter). Consider the thickness, $x(n)$, of steel strip leaving a mill, monitored by a transducer and sampled by an ADC, as illustrated in Figure 3.7(a).

Suppose the mean and covariance of the thickness (in continuous time) are (Section 2.7.2, Example):

$$\varepsilon(x(t)) = 5 \text{ (mm)}; \quad r(\tau) = 0.64e^{-|\tau|} \text{ (mm}^2)$$

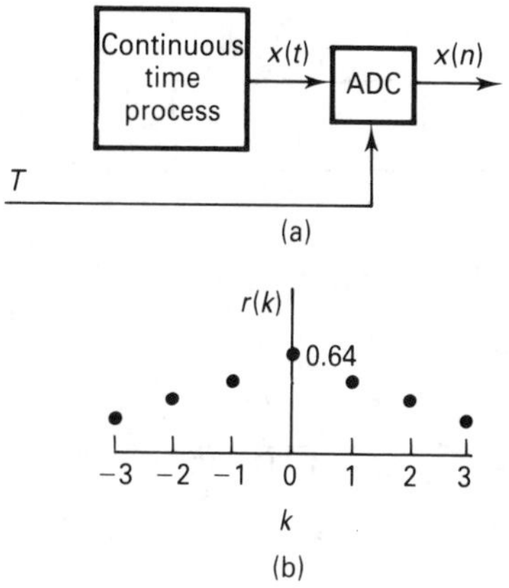

Figure 3.7 Discrete time Gaussian process.

If the ADC is clocked at intervals $T = 0.1$ (s), then,

$$\varepsilon(x(n)) = 5; \quad r(k) = 0.64e^{-0.1|k|}$$

$r(k)$ is illustrated in Figure 3.7(b).

Probabilities can be calculated using the SNF table as in the continuous time case.

3.6.3 White noise (cf. Section 2.7.3)

Discrete time white noise is defined as a wide sense stationary process whose power spectral density (Eqn (3.14)) is constant over all the frequencies of interest. For such a process, $w(n)$, strength q:

$$\bar{S}_w(\beta) = \begin{cases} q, & -\pi \leqslant \beta < +\pi \\ 0, & \text{elsewhere} \end{cases} \tag{3.19}$$

The autocorrelation is given by the inverse Fourier transform of the PSD (Eqn (3.15)):

$$\bar{r}_w(k) = \frac{1}{2\pi}\int_{-\pi}^{+\pi} q\, e^{j\beta k}\, d\beta = q\,\delta k \tag{3.20}$$

where δk is a Kronecker delta function (Appendix D, Section D2). This implies that $w(n)$ is correlated with itself, $k = 0$, and uncorrelated with itself, $k \neq 0$. Hence,

$$\varepsilon(w(n)) = \varepsilon(w(n+k)) = 0$$

$$r_w(k) = \bar{r}_w(k) = q\,\delta k$$

Thus, the variance of $w(n)$ is $r_w(0) = q$

$r_w(k)$, $\bar{S}_w(\beta)$ are illustrated in Figure 3.8(a, b), and a typical sample function, $\bar{w}_i(n)$, is shown in Figure 3.8(c).

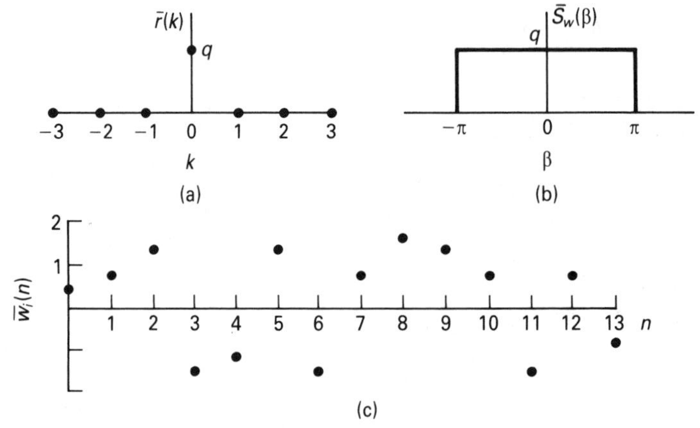

Figure 3.8 White noise PSD and autocorrelation.

In contrast to continuous time white noise, $w(n)$ is meaningful in practice and good approximations to it are fairly easily generated (Section 4.2). This is because it is band limited and consequently has finite power (and finite variance, q). The concept of white noise can usefully be extended to a non-stationary process whose strength varies with time. This is 'extended white noise', $\bar{w}(n)$:

$$\left.\begin{aligned} \varepsilon(\bar{w}(n)) &= 0 \\ r_{\bar{w}}(n,l) = \bar{r}_{\bar{w}}(n,l) &= \bar{q}(n)\,\delta(l-n) \end{aligned}\right\} \tag{3.21}$$

Being non-stationary, $\bar{w}(n)$ has no meaningful PSD.

3.6.4 Processes with independent increments (cf. Section 2.7.4)

A process, $x(n)$, has independent increments if

$$(x(n)-x(n-1)), (x(n-1)-x(n-2)), \ldots, (x(1)-x(0))$$

are independent. (If they are merely uncorrelated, $x(n)$ is a process with orthogonal increments.)

As with an equivalent continuous time process, this definition implies that $(x(l)-x(n))$ is independent of $x(n)$, $l > n$, and that the covariance of $x(n)$ equals the variance at the earlier time (cf. Eqn (2.42):

$$r(n,l) = \begin{cases} r(n,n), & n \leqslant l \\ r(l,l), & l \leqslant n \end{cases} \tag{3.22}$$

If $(x(l)-x(n))$ depends only on $k=(l-n)$, $x(n)$ is said to have stationary increments. This does not of course imply that $x(n)$ is stationary (in any sense) – usually, in fact, such processes are non-stationary.

Example

The random walk of Figure 3.6 is a good example of such a process. The increments are determined by coin tosses, and non-overlapping increments are therefore independent. They are also strictly (and therefore wide sense) stationary, while $x(n)$ is clearly non-stationary.

Recalling that $\varepsilon(x(n)) = 0$, it follows that (Eqn (3.22))

$$r(n,l) = \begin{cases} n\lambda^2, & l \geqslant n \\ l\lambda^2, & n \geqslant l \end{cases}$$

This is illustrated in Figure 3.9.

3.6.5 The discrete time Wiener process (cf. Section 2.7.5)

The discrete time Wiener process (Brownian motion), $\beta(n)$, is a process with independent increments defined in a manner analogous to the continuous time case.

$\beta(n)$ has the following properties:

1. $\beta(n)$ is non-stationary with stationary independent increments.
2. $\beta(n)$ is Gaussian.
3. The variance of an increment is proportional to the time over which it is measured:

$$\varepsilon(\beta(l) - \beta(n))^2 = q(l - n), l \geqslant n \tag{3.23}$$

where q is the diffusion of $\beta(n)$.

4. $\beta(0) = 0$.

It follows from (1), (3) and (4) that

$$r_\beta(n, n) = \varepsilon(\beta(n))^2 = qn$$

and that (cf. Eqn (2.46))

$$r_\beta(n, l) = \begin{cases} r_\beta(n, n) = q^n, & l \geqslant n \\ r_\beta(l, l) = q^l, & n \geqslant l \end{cases} \tag{3.24}$$

Typical sample functions, $\bar{\beta}_i(n)$ and $r_\beta(n, l)$, are shown in Figure 3.10(a, b).

The sample functions wander away from $\beta(0) = 0$, so that although $\varepsilon(\beta(n)) = 0$, the variance, $\varepsilon(\beta(n))^2$, increases with time. Unlike its continuous time counterpart, the discrete time Wiener process has no special significance in the construction of standard-form models of dynamical systems.

As with white noise, the concept of the Wiener process can be 'extended' to

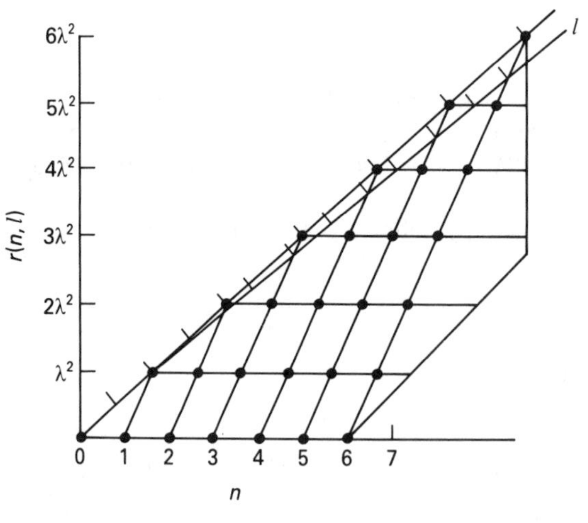

Figure 3.9 Covariance function.

a process whose diffusion is a function of time. In this case (cf. Eqns (2.47) and (2.49)),

$$\begin{aligned} r(n, l) &= \varepsilon(\beta(l) - \beta(n))^2 \\ &= \sum_{j=n}^{l} \bar{q}(j) \\ &= \begin{cases} \sum\limits_{j=0}^{n} \bar{q}(j), & l \geqslant n \\ \sum\limits_{j=0}^{l} \bar{q}(j), & n \geqslant l \end{cases} \end{aligned} \tag{3.25}$$

where

$$\varepsilon(\bar{\beta}(j+1) - \bar{\beta}(j))^2 = \bar{q}_j$$

A function of this type is shown in Figure 3.11.

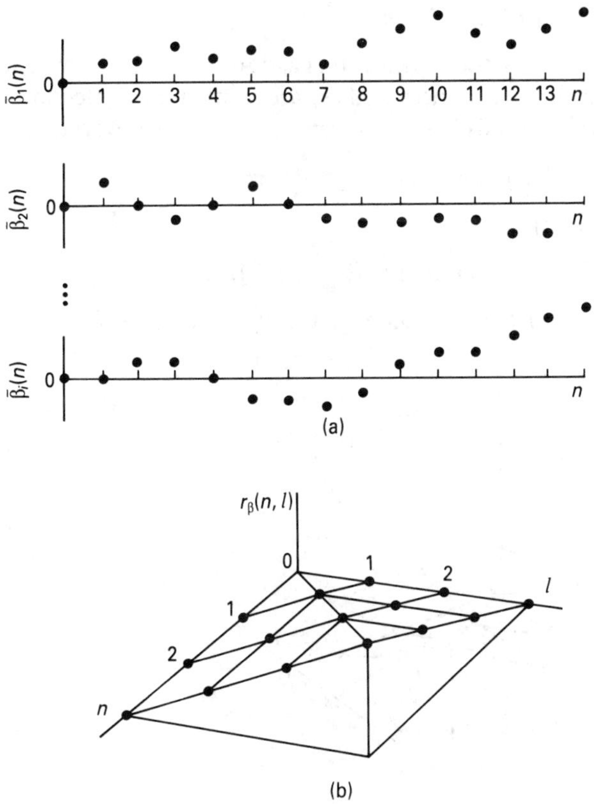

Figure 3.10 Discrete time Wiener process.

3.6.6 White noise and the Wiener process (cf. Section 2.7.6)

Since both processes are readily realizable, their relationship is simple: the Wiener process is generated by summing white noise over time:

$$\beta(n) = \sum_{j=0}^{n} w(j) \tag{3.26}$$

If the strength (or variance) of $w(n)$ is q, the diffusion of $\beta(n)$ is q. Eqn (3.26) also applies to extended white noise and the corresponding extended Wiener process:

$$\bar{\beta}(n) = \sum_{j=0}^{n} \bar{w}(j)$$

3.7 Vector processes (cf. Section 2.9)

The definitions of Sections 3.2–3.6 are readily extended to deal with vector stochastic processes. Consider the vector process

$$\mathbf{x}(n) = (x_1(n) \quad x_2(n) \quad \ldots \quad x_m(n))^T$$

Each element $x_i(n)$ represents a discrete time stochastic process with its own descriptive parameters. The means and covariances of these are neatly represented in vector and matrix format (cf. Eqns (2.65) and (2.66)) by the mean vector and covariance matrix:

$$\varepsilon(\mathbf{x}(n)) = (\varepsilon(x_1(n)) \quad \varepsilon(x_2(n)) \quad \ldots \quad \varepsilon(x_m(n)))^T \tag{3.27}$$

$$\begin{aligned} \text{cov}(\mathbf{x}(n)) &= \mathbf{R}(n, l) \\ &= \varepsilon[(\mathbf{x}(n) - \varepsilon(\mathbf{x}(n)))(\mathbf{x}(l) - \varepsilon(\mathbf{x}(l)))^T] \end{aligned} \tag{3.28}$$

The covariance matrix $\mathbf{R}(n, l)$ is a square matrix, the diagonal elements of which

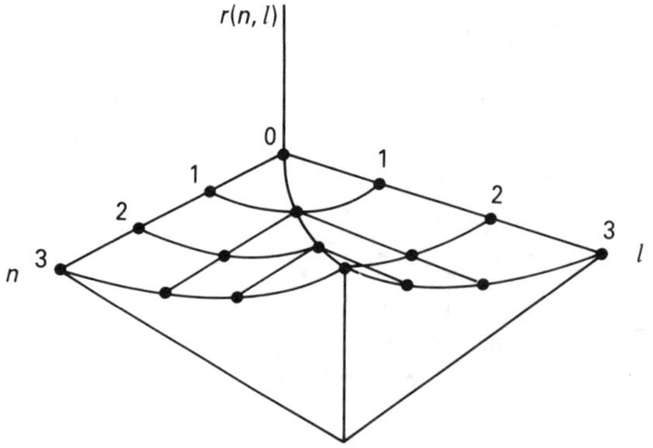

Figure 3.11 Covariance of extended Wiener process.

represent the covariances of $x_j(n), j = 1, 2, \ldots, m$, and the off-diagonal elements the cross-covariances:

$$\left.\begin{aligned} r_{jj}(n, l) &= \varepsilon[(x_j(n) - \varepsilon(x_j(n)))(x_j(l) - \varepsilon(x_j(l)))^T] \\ r_{jh}(n, l) &= \varepsilon[(x_j(n) - \varepsilon(x_j(n)))(x_h(l) - \varepsilon(x_h(l)))^T] \end{aligned}\right\} \tag{3.29}$$

Examples

(a) Consider the vector process $\mathbf{x}(n)$ whose elements, $x_j(n)$, are white noise, strength (or variance) q_j, each element independent of the others. Then

$$\varepsilon(\mathbf{x}(n)) = 0$$

$$\mathbf{R}(n, l) = \mathbf{R}(k)$$

$$= \begin{pmatrix} q_1\,\delta k & 0 & 0 & \ldots & 0 \\ 0 & q_2\,\delta k & 0 & & 0 \\ \vdots & & & & \vdots \\ 0 & 0 & 0 & \cdots & q_m\,\delta k \end{pmatrix}$$

where δk is a Kronecker delta function.

(b) Consider $\mathbf{x}(n) = (x_1(n) \quad x_2(n))^T$, white noise, in which $\varepsilon(x_1(n))^2 = \varepsilon(x_2(n))^2 = q$ and $x_1(n)$ is derived from $x_2(n)$ by

$$x_1(n+1) = x_2(n)$$

The cross-correlations are

$$\varepsilon(x_1(n)x_2(n+k)) = q\,\delta(k-1)$$

$$\varepsilon(x_2(n)x_1(n+k)) = q\,\delta(k+1)$$

where δk is a Kronecker delta function.

The mean and covariance of $\mathbf{x}(n)$ are

$$\varepsilon(\mathbf{x}(n)) = \mathbf{0}$$

$$\mathbf{R}(n, l) = \mathbf{R}(k)$$

$$= \begin{pmatrix} q\,\delta(k) & q\,\delta(k-1) \\ q\,\delta(k+1) & q\,\delta(k) \end{pmatrix}$$

3.8 Discrete time system models (cf. Section 2.10)

The behaviour of discrete time dynamical systems is modelled by difference equations, usually cast in standard form as a set of first order difference equations together with a further set of algebraic output equations (cf. Eqns (2.67) and (2.68)):

$$\mathbf{x}(n+1) = \mathbf{f}(\mathbf{x}(n), \mathbf{u}(n), n) \tag{3.30}$$

$$\mathbf{y}(n) = \mathbf{q}(\mathbf{x}(n), \mathbf{u}(n), n) \tag{3.31}$$

where

$$\mathbf{x}(n) = (x_1(n) \quad x_2(n) \quad \ldots \quad x_m(n))^T$$

$$\mathbf{u}(n) = (u_1(n) \quad u_2(n) \quad \ldots \quad u_q(n))^T$$

$$\mathbf{y}(n) = (y_1(n) \quad y_2(n) \quad \ldots \quad y_r(n))^T$$

The inputs, outputs and states are represented by the q-vector $\mathbf{u}(n)$, r-vector $\mathbf{y}(n)$, and m-vector $\mathbf{x}(n)$ respectively. The model is shown in block form in Figure 3.12(a).

It has to be said that the model of Eqn (3.30) cannot usually be synthesized in practice, since its realization depends on synthesizing a continuous time model in the form of Eqn (2.67), and then solving this in analytical form to yield Eqn (3.30) – which generally cannot be done. Approximate discrete time models, based, for example, on Runge–Kutta algorithms (Section 4.4) can of course be developed from Eqn (2.67), but, being approximate, they are irrelevant to considerations on the format generally appropriate to dynamical models. No difficulty attaches to the algebraic Eqn (3.31), which is simply a copy, in discrete time, of Eqn (2.68).

Eqns (3.30) and (3.31) therefore constitute on ideal rather than a practical model for the general case, but a large and useful subset of systems can be modelled – theoretically and practically – by the linear time-variant difference (cf. Eqns (2.69) and (2.70)):

$$\mathbf{x}(n+1) = \varphi(n)\mathbf{x}(n) + \psi(n)\mathbf{u}(n) \tag{3.32}$$

$$\mathbf{y}(n) = \mathbf{C}_d(\mathbf{n})\mathbf{x}(n) + \mathbf{D}_d(n)\mathbf{u}(n) \tag{3.33}$$

where

$\varphi(n)$ is a time-dependent $(m \times m)$, 'state transition matrix',
$\psi(n)$ is a time-dependent $(m \times q)$ matrix,
$\mathbf{C}_d(n)$ is a time-dependent $(r \times m)$ matrix,
$\mathbf{D}_d(n)$ is a time-dependent $(r \times q)$ matrix.

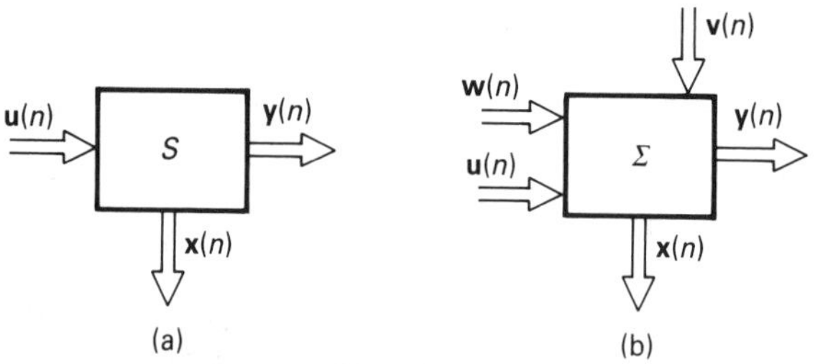

Figure 3.12 Discrete system model.

As in the general case, the model of Eqn (3.32) is usually derived from a model synthesized in the continuous time domain. Since the solution of Eqn (2.69) is known, this is a straightforward practical proposition (Appendix A, Eqns (A3.18)–(A3.21)). Eqn (3.33), being algebraic, is effectively a copy of the primary model Eqn (2.70).

In turn, a large and useful subclass of linear systems can be modelled by the linear time-invariant (LTI) equations (cf. Eqns (2.71) and (2.72)):

$$\mathbf{x}(n) = \varphi\mathbf{x}(n) + \psi\mathbf{u}(n) \tag{3.34}$$

$$\mathbf{y}(n) = \mathbf{C}_d\mathbf{x}(n) + \mathbf{D}_d\mathbf{u}(n) \tag{3.35}$$

where φ, ψ, $\mathbf{C}_d$ and $\mathbf{D}_d$ are constant matrices of appropriate dimension. The detailed properties of all these models are not explored here; the most important are outlined in Appendix A, Section A3.

Model extension to account for noise corruption is far more easily accomplished than for continuous time models, essentially because discrete time white noise is a valid practical process with unexceptional mathematical properties.

For non-linear systems, for example, Eqns (3.30 and 3.31) are readily extended:

$$\mathbf{x}(n+1) = \mathbf{f}(\mathbf{x}(n), \mathbf{u}(n), \mathbf{w}(n), n)$$

$$\mathbf{y}(n) = \mathbf{g}(\mathbf{x}(n), \mathbf{u}(n), \mathbf{v}(n), n)$$

where $\mathbf{w}(n)$ represents process noise, which corrupts the state behaviour of the system; $\mathbf{v}(n)$ represents measurement noise which corrupts the output. The model is shown in block form in Figure 3.12(b). For generality, $\mathbf{w}(n)$, must be vectors of white (or extended) white noise (Section 2.10).

This model is unfortunately less useful than it might appear, because its solution, $\mathbf{x}(n)$, is in general non-Markov, and so is unquantifiable without great complication. However, since such a model is almost impossible to synthesize in any case, as explained above, this is not important in practice – non-linear discrete time models are synthesized in approximate form, using Runge–Kutta or similar methods (Section 4.4). The linear models of Eqns (3.32), (3.33), (3.34) and (3.35), however, extended to account for noise, do yield Markov solutions, and are of use in the design of discrete time estimation and control systems (Chapters 5 and 6).

The linear-time variant model (Eqns (3.32) and (3.33)) becomes

$$\mathbf{x}(n+1) = \varphi(n)\mathbf{x}(n) + \psi(n)\mathbf{u}(n) + \mathbf{\Gamma}(n)\bar{\mathbf{w}}(n) \tag{3.36}$$

$$\mathbf{y}(n) = \mathbf{C}_d(n)\mathbf{x}(n) + \mathbf{D}_d(n)\mathbf{u}(n) + \bar{\mathbf{v}}(n) \tag{3.37}$$

where

$\bar{\mathbf{w}}(n)$ is an s-vector of Gaussian extended white noise,
$\bar{\mathbf{v}}(n)$ is an r-vector of Gaussian extended white noise,
$\mathbf{\Gamma}(n)$ is an $(m \times s)$ time-dependent gain matrix,

and

$$\text{cov}(\bar{\mathbf{w}}(n)) = \bar{\mathbf{Q}}_d(n)\,\delta k, \quad \bar{\mathbf{Q}}_d(n), \text{ an } (m \times m) \text{ matrix}$$

$$\text{cov}(\bar{\mathbf{v}}(n)) = \bar{\mathbf{R}}_d(n)\,\delta k, \quad \bar{\mathbf{R}}_d(n), \text{ an } (r \times r) \text{ matrix}$$

The linear time-invariant model (Eqns (3.34) and (3.35)) is

$$\mathbf{x}(n+1) = \varphi\mathbf{x}(n) + \psi\mathbf{u}(n) + \mathbf{\Gamma}\mathbf{w}(n) \tag{3.38}$$

$$\mathbf{y}(n) = \mathbf{C}_d\mathbf{x}(n) + \mathbf{D}_d\mathbf{u}(n) + \mathbf{v}(n) \tag{3.39}$$

where

φ, ψ, $\mathbf{C}_d$, $\mathbf{D}_d$ are constant matrices of appropriate dimension, and $\mathbf{w}(n)$, $\mathbf{v}(n)$ are (unextended) white noise vectors, strengths $\mathbf{Q}_d$, $\mathbf{R}_d$ respectively.

Both these linear models are mathematically valid since the white noise processes are unattended by the difficulties found in continuous time. The solution of Eqn (3.36) is easily formulated (cf. Eqn (2.79)):

$$\mathbf{x}(n+1) = \left[\prod_{j=n_0}^{n} \varphi(j)\right] \cdot \mathbf{x}(n_0) + \sum_{j=n_0}^{n}\left(\prod_{k=j+1}^{n} \varphi(k)\right) \cdot \psi(j)\mathbf{u}(j)$$

$$+ \sum_{j=n_0}^{n}\left(\prod_{k=j+1}^{n} \varphi(k)\right) \cdot \mathbf{\Gamma}(j)\bar{\mathbf{w}}(j) \tag{3.40}$$

In the LTI case (Eqn (3.38)), this simplifies slightly, since

$$\prod_{j=n_0}^{n} \varphi(j) = \varphi^{n-n_0+1} \text{ and } \prod_{k=j+1}^{n} \varphi(k) = \varphi^{n-j}$$

It is quite easy to formulate equations describing the propagation of $\varepsilon(\mathbf{x}(n))$ and $\text{cov}(\mathbf{x}(u))$:

$$\varepsilon(\mathbf{x}(n+1)) = \varphi(n)\varepsilon(\mathbf{x}(n)) + \psi(n)\mathbf{u}(n) \tag{3.41}$$

As in the continuous time case, since the terms involving $\varepsilon(\mathbf{x}(n)) - \varepsilon(\mathbf{x}(n))$ and $\mathbf{u}(j)$, $w(j)$, $j \geqslant n$ are zero,

$$\begin{aligned}\mathbf{R}(n, l) &= \text{cov}(\mathbf{x}(n)) \\ &= \varepsilon[(\mathbf{x}(n) - \varepsilon(\mathbf{x}(n)))(\mathbf{x}(l) - \varepsilon(\mathbf{x}(l)))] \\ &= \varphi(l, n) \cdot \mathbf{R}(n, n)\end{aligned} \tag{3.42}$$

where

$$\varphi(l, n) = \prod_{j=n}^{j=l-1} \varphi(j)$$

The behaviour of $\mathbf{R}(n, n)$ is described by

$$\mathbf{R}(n+1, n+1) = \varphi(n)\mathbf{R}(n, n)\varphi^T(n) + \mathbf{\Gamma}(n)\bar{\mathbf{Q}}_d(n)\mathbf{\Gamma}^T(n) \tag{3.43}$$

where $\mathbf{R}(n_0, n_0)$, the initial condition, is known.

In many cases, of course, $\bar{\mathbf{Q}}_d(n) = \mathbf{Q}_d$ (i.e. the process noise is stationary). And in the LTI case: $\boldsymbol{\varphi}(n) = \boldsymbol{\varphi}$; $\boldsymbol{\psi}(n) = \boldsymbol{\psi}$; $\boldsymbol{\Gamma}(n) = \boldsymbol{\Gamma}$. In the LTI case, if the system is stable (Appendix A, Section A3) and $\mathbf{u}(n)$ is constant, $\mathbf{x}(n)$ becomes stationary as $n \to \infty$, while $\mathbf{R}(n, n)$ converges to a value, $\mathbf{R}(\infty, \infty)$, which satisfies

$$\mathbf{R}(\infty, \infty) = \boldsymbol{\varphi}\mathbf{R}(\infty, \infty)\boldsymbol{\varphi}^T + \boldsymbol{\Gamma}\mathbf{Q}_d\boldsymbol{\Gamma}^T \tag{3.44}$$

Example

Consider a system modelled by:

$$\mathbf{x}(n+1) = \begin{pmatrix} 1.0 & 1.0 \\ -0.5 & 0.0 \end{pmatrix}\mathbf{x}(n) + \begin{pmatrix} -1.0 \\ 0.5 \end{pmatrix}u(n) + \begin{pmatrix} -1.0 \\ 0.5 \end{pmatrix}w(n)$$

$$y(n) = (1.0 \quad 0.0)\mathbf{x}(n) + v(n)$$

where $w(n)$, $v(n)$ are (1×1) vectors of white noise with strengths

$$\mathbf{Q}_d = 0.1, \mathbf{R}_d = 0.01$$

and initial conditions

$$\mathbf{x}(0) = \begin{pmatrix} 5 \\ 0 \end{pmatrix}, \varepsilon(\mathbf{x}(0)) = \begin{pmatrix} 5 \\ 0 \end{pmatrix}, \mathbf{R}(0,0) = \begin{pmatrix} 0 \\ 0 \end{pmatrix}$$

Graphs illustrating the behaviour of the system for $u(n) = 0$, $w(n) = 0$, $w(n-1) = 0$, $0 \leqslant n < 2$ are shown in Figure 3.13: $\mathbf{R}(n, n)$ reaches steady state at about $n = 10$, and $\mathbf{R}(k) = \mathbf{R}(l - n)$ becomes zero at about $k = 10$.

3.9 Sampled data systems

Continuous time systems are readily converted to discrete time, or sampled data systems, via simultaneously clocked DACs (digital to analog converters) and ADCs (analog to digital converters), as illustrated in Figure 3.14.

If the system S is modelled by the linear Eqns (2.75) and (2.76), and the characteristics of $\bar{\mathbf{w}}(t)$, $\bar{\mathbf{v}}(t)$, and the sampling period, T, are known, it is easy to formulate a model, Σ (Eqns (3.36) and (3.37)), in which $\bar{\mathbf{w}}(n)$, $\bar{\mathbf{v}}(n)$ are such that their effects on $\mathbf{x}(n)$, $\mathbf{y}(n)$ are the same as those of $\mathbf{w}(t, \mathbf{v}(t)$, 'seen' through virtual DACs.

Eqns (2.79) and (2.80) provide the necessary relationship between $\mathbf{x}(n)$ (i.e. $\mathbf{x}(nT)$) and $\mathbf{x}(n+1)$ (i.e. $\mathbf{x}((n+1)T)$):

$$\mathbf{x}(n+1) = \boldsymbol{\varphi}(n)\mathbf{x}(n) + \int_{nT}^{(n+1)T} \boldsymbol{\varphi}(\lambda, nT)\mathbf{B}(\lambda)\mathbf{u}(\lambda)\,\mathrm{d}\lambda + \int_{nT}^{(n+1)T} \boldsymbol{\varphi}(\lambda, nT)\mathbf{H}(\lambda)\,\mathrm{d}\bar{\boldsymbol{\beta}}(\lambda) \tag{3.45}$$

where $\boldsymbol{\varphi}(t, nT)$ is the solution of

$$\left.\begin{aligned} &\frac{\mathrm{d}}{\mathrm{d}t}(\boldsymbol{\varphi}(t, nT) = \mathbf{A}(t)\cdot\boldsymbol{\varphi}(t, nT) \\ &\boldsymbol{\varphi}(nT, nT) = \mathbf{I} \end{aligned}\right\} \tag{3.46}$$

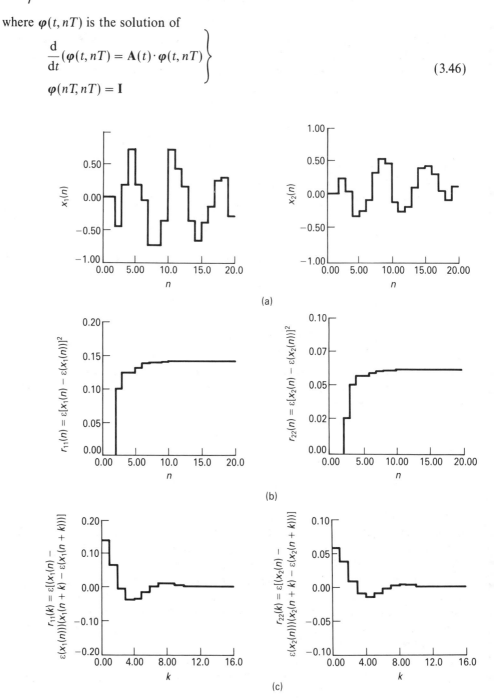

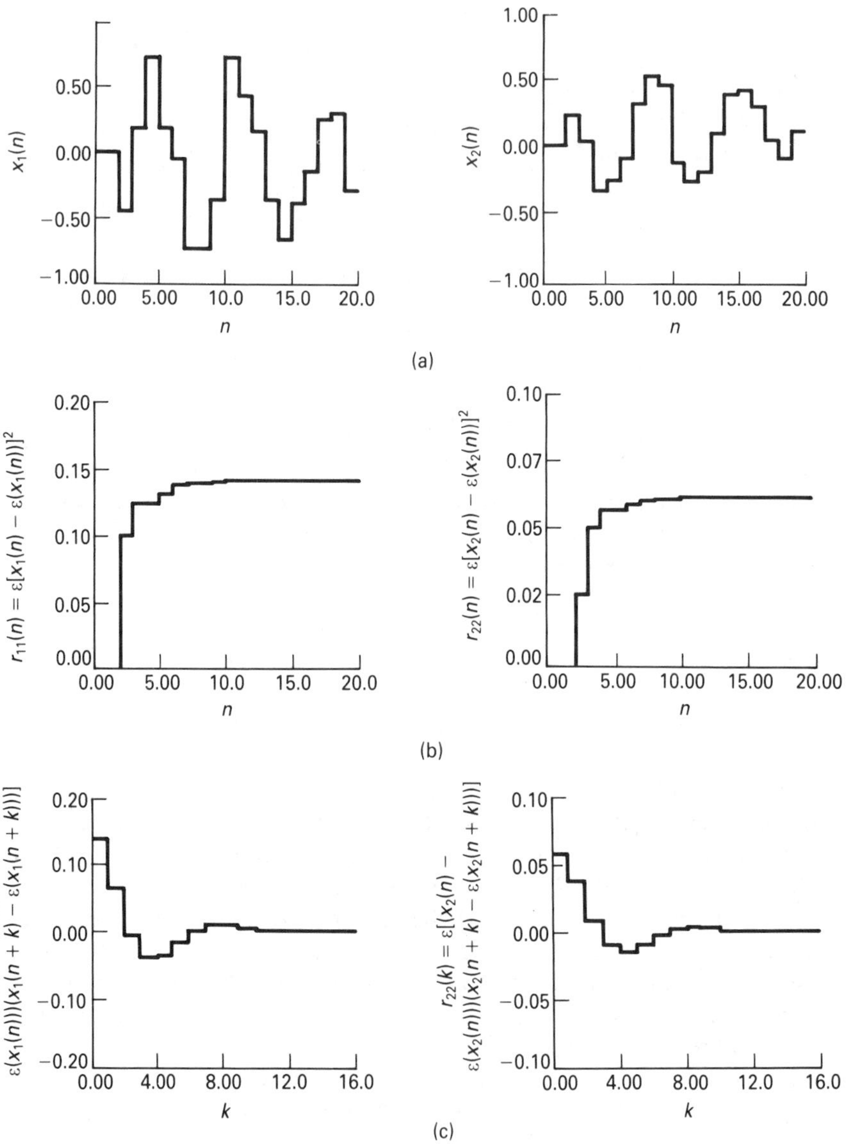

Figure 3.13 Discrete time system behaviour.

and

$$\varphi(n) = \varphi((n+1)T, nT)$$

Eqn (3.45) is the required state equation model (cf. Eqn (3.36)) in which $\varphi(n)$ is defined by Eqn (3.46):

$$\psi(n) = \int_{nT}^{(n+1)T} \varphi(\lambda, nT)\mathbf{B}(\lambda)\,\mathrm{d}\lambda \tag{3.47}$$

$$\Gamma(n) = \int_{nT}^{(n+1)T} \varphi(\lambda, nT)\mathbf{H}(\lambda)\,\mathrm{d}\lambda \tag{3.48}$$

It can be shown [ref. 2] that $\bar{\mathbf{w}}(n)$ is discrete time extended white noise with strength (matrix):

$$\bar{\mathbf{Q}}_d(n) = \int_{nT}^{(n+1)T} \varphi(\lambda, nT)\bar{\mathbf{Q}}(\lambda)\varphi^T(\lambda, nT)\,\mathrm{d}\lambda \tag{3.49}$$

where $\mathbf{Q}(t)$ is the strength (matrix) of $\bar{\mathbf{w}}(t)$, or diffusion of $\bar{\boldsymbol{\beta}}(t)$. The output equation (Eqn (3.37)), being algebraic, is identical in form to the continuous time equation (Eqn (2.76)):

$$\mathbf{C}_d(n) = \mathbf{C}(nT); \mathbf{D}_d(n) = \mathbf{D}(nT) \tag{3.50}$$

$\bar{\mathbf{v}}(n)$ is discrete time extended white noise with strength (matrix):

$$\bar{\mathbf{R}}_d(n) = \bar{\mathbf{R}}(nT) \tag{3.51}$$

where $\bar{\mathbf{R}}(nT)$ is the strength (matrix) of $\bar{\mathbf{v}}(t)$, $t = nT$.

To summarize, if a linear model of the continuous time system, S, is available, and the sampling time, T, is specified, a linear discrete time model of Σ can be

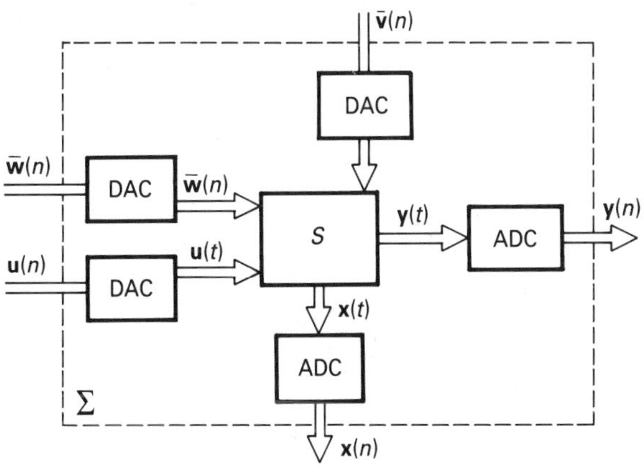

Figure 3.14 Sampled data system.

constructed using Eqns (3.45)–(3.51). These relationships simplify slightly for LTI systems, in which case the model equations are (Eqns (3.38) and (3.39))

$$\mathbf{x}(n+1) = \varphi\mathbf{x}(n) + \psi\mathbf{u}(n) + \mathbf{\Gamma}\mathbf{w}(n)$$

$$\mathbf{y}(n) = \mathbf{C}_d\mathbf{x}(n) + \mathbf{D}_d\mathbf{u}(n) + \mathbf{v}(n)$$

Recalling the constant parameters of S, namely $\mathbf{A}$, $\mathbf{B}$, $\mathbf{C}$, $\mathbf{D}$, $\mathbf{Q}$, $\mathbf{R}$:

$$\varphi = e^{\mathbf{A}T}$$

$$= \mathbf{I} + \mathbf{A}T + \frac{1}{2!}\mathbf{A}^2T^2 + \frac{1}{3!}\mathbf{A}^3T^3 + \cdots \tag{3.52}$$

$$\psi = \int_0^T e^{\mathbf{A}\lambda}\mathbf{B}\,d\lambda$$

$$= \mathbf{IB}T + \frac{1}{2!}\mathbf{AB}T^2 + \frac{1}{3!}\mathbf{A}^2\mathbf{B}T^3 + \cdots \tag{3.53}$$

$$\mathbf{\Gamma} = \int_0^T e^{\mathbf{A}\lambda}\mathbf{H}\,d\lambda$$

$$= \mathbf{IH}T + \frac{1}{2!}\mathbf{AH}T^2 + \frac{1}{3!}\mathbf{A}^2\mathbf{H}T^3 + \cdots \tag{3.54}$$

$\mathbf{w}(n)$, $\mathbf{v}(n)$ are (unextended) white Gaussian noise vectors of strengths

$$\mathbf{Q}_d = \int_0^T e^{\mathbf{A}\lambda}\cdot\mathbf{Q}\cdot(e^{\mathbf{A}\lambda})^T\,d\lambda \tag{3.55}$$

$$\mathbf{R}_d = \mathbf{R} \tag{3.56}$$

If the sampling period, T, is short compared with the system dynamics, these formulae can themselves be simplified:

$$\varphi \doteqdot \mathbf{I} + \mathbf{A}T \tag{3.57}$$

$$\psi \doteqdot \mathbf{B}T \tag{3.58}$$

$$\mathbf{\Gamma} \doteqdot \mathbf{H}T \tag{3.59}$$

$$\mathbf{Q}_d \doteqdot \mathbf{Q}T \tag{3.60}$$

$$\mathbf{R}_d = \mathbf{R} \tag{3.61}$$

Example

Consider the system of Figure 2.24 buffered by DACs and ADCs, as indicated in Figure 3.14, the sampling period being 0.050 s. In this case (Eqn (2.88),

$$\mathbf{A}=\begin{pmatrix}0 & 1 & 0 & 0\\ -80 & -4 & 80 & 200\\ 0 & 0 & 0 & 50\\ 20 & 1 & -170 & -50\end{pmatrix},\ \mathbf{B}=\begin{pmatrix}0\\0\\0\\0\end{pmatrix},\ \mathbf{H}=\begin{pmatrix}0\\0\\0\\150\end{pmatrix}$$

Applying Eqns (3.52)–(3.54) and Eqn (3.50) yields the model for the sampled data system:

$$\mathbf{x}(n+1)=\begin{pmatrix}0.9337 & 0.0456 & -0.1442 & 0.0438\\ -2.7686 & 0.7953 & -3.8042 & -0.2864\\ 0.1168 & 0.0109 & -0.2029 & -0.1445\\ -0.0753 & -0.0014 & 0.5089 & -0.0145\end{pmatrix}\mathbf{x}(n)+\begin{pmatrix}0.2104\\ 6.5728\\ 1.0862\\ -0.4334\end{pmatrix}w(n) \tag{3.62}$$

$$\mathbf{y}(n)=(1\quad 0\quad -1\quad 0)\mathbf{x}(n) \tag{3.63}$$

The strength of the road surface noise, $w(n)$, corresponding to that of $0.025\ \mathrm{m^2\,s^{-4}}$ in Figure 2.26, is given by Eqn (3.55), or approximately by Eqn (3.60):

$$\mathbf{Q}_d = 0.025 \times 0.05 = 1.25 \times 10^{-4}$$

The propagations of $\varepsilon(\mathbf{x}(n))$ and $\mathrm{cov}(\mathbf{x}(n))$ are described by Eqns (3.41), (3.42) and (3.43). Computer realizations of $r_{11}(n,n)$, $r_{33}(n,n)$ and $r_{11}(n+k)$ are shown in Figure 3.15(c–f) respectively (seen via a DAC for clarity). They agree well with the graphs of Figures 2.25 and 2.26.

3.10 The modelling of non-white noise

As in the continuous time models of Section 2.11, the linear dynamical models of Sections 3.8 and 3.9 can be augmented quite readily to represent non-white process and measurement noise. This is done by incorporating linear digital filters (Appendix B, Section B2) between white noise sources and the dynamical system, as shown in Figure 3.16(a).

If a stochastic process (or, for that matter, a deterministic signal), $f(n)$, power spectral density $\bar{S}_f(\beta)$ is fed to a linear time-invariant system – in this case a digital filter – as shown in Figure 3.16(b), the PSD of the output, $r(n)$, is given by

$$\bar{S}_r(\beta) = \bar{F}(\mathrm{e}^{+\mathrm{j}\beta})\bar{F}(\mathrm{e}^{-\mathrm{j}\beta})\bar{S}_f(\beta) \tag{3.64}$$

In particular, if $f(n)$ is chosen to be unit strength Gaussian white noise, which has unity PSD, then

$$\bar{S}_r(\beta) = \bar{F}(\mathrm{e}^{+\mathrm{j}\beta})\bar{F}(\mathrm{e}^{-\mathrm{j}\beta}) \tag{3.65}$$

Thus, if the PSD of $r(n)$ is specified so that it can be spectrally factorized according to Eqn (3.65), a filter algorithm with the required properties can be designed. This is not usually easy to achieve and a more practical approach is simply to conduct the exercise in the continuous time domain, discretize the filter and augment the discrete time state equations accordingly.

Example

Consider the example of Section 3.9 in which a discrete time model of the behaviour of a quarter car suspension system is constructed (Eqns (3.62) and (3.63)). The road surface noise, $w(n)$, is taken as white; the sampling period is 0.050 s.

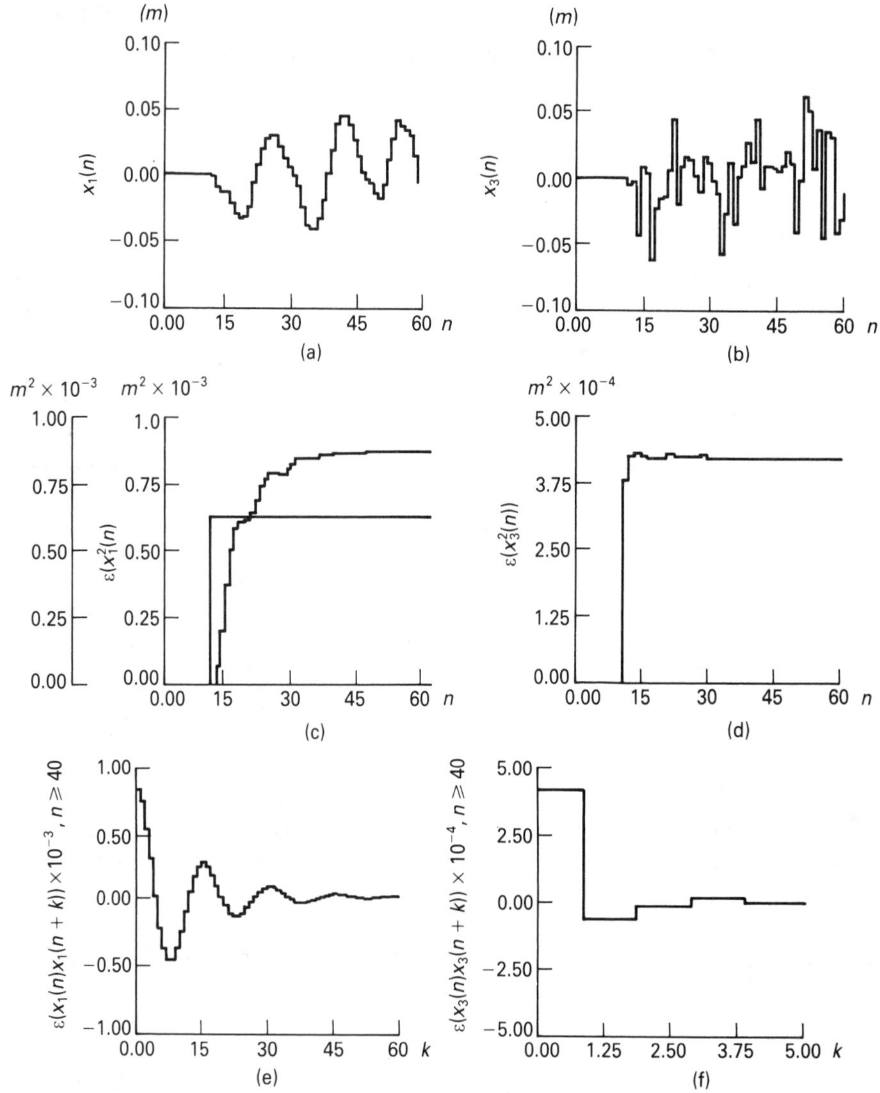

Figure 3.15 Sampled data system behaviour.

A more accurate model of road surface noise is provided by white Gaussian noise passed through a filter (Section 2.11) with transfer function:

$$\bar{F}(z) = \frac{z(1 - e^{-0.05\lambda})}{z - e^{-0.05\lambda}}$$

This is readily discretized (Appendix A, Eqn (A1.19)) to yield the digital filter transfer function

$$\bar{F}(z) = \frac{z(1 - e^{-0.05\lambda})}{z - e^{-0.05\lambda}}$$

Equivalently, in the time domain (Appendix A, Eqns (A1.15) and (A1.17)),

$$x_5(n+1) = e^{-0.05\lambda}x_5(n) + (1 - e^{-0.05\lambda})w(n)$$

where $x_5(n)$ is the filter output and $w(n)$ the white noise filter input.

Eqn (3.62) is readily augmented to accommodate this:

$$\mathbf{x}(n+1) = \begin{pmatrix} 0.938 & 0.046 & -0.144 & 0.044 & 0.210 \\ -2.769 & 0.795 & -3.804 & -0.286 & 6.573 \\ 0.117 & 0.011 & -0.203 & -0.145 & 1.086 \\ -0.075 & -0.001 & 0.509 & -0.015 & -0.433 \\ 0 & 0 & 0 & 0 & e^{-0.05\lambda} \end{pmatrix} \mathbf{x}(n) + \begin{pmatrix} 0 \\ 0 \\ 0 \\ 0 \\ 1 - e^{-0.05\lambda} \end{pmatrix} w(n) \tag{3.66}$$

$$\mathbf{y}(n) = (1 \quad 0 \quad -1 \quad 0 \quad 0)\mathbf{x}(n) \tag{3.67}$$

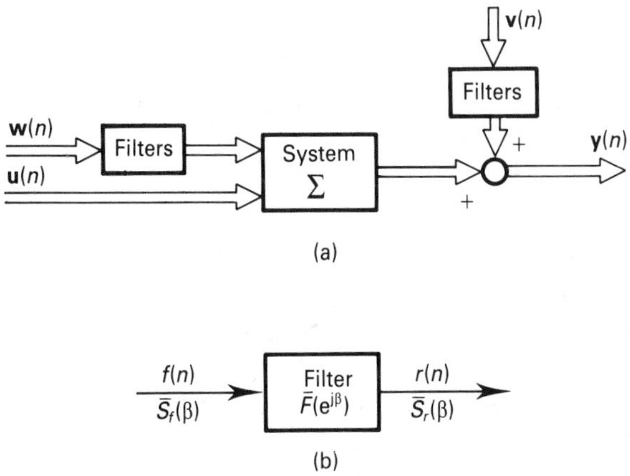

Figure 3.16 Linear model augmented to represent non-white noise.

The behaviour of this system is naturally less violent than that illustrated in Figure 3.15, to a degree dependent on the value of λ.

The same result is obtained by constructing augmented state equations in the continuous time domain, and discretizing the result using Eqns (3.45)–(3.49). Thus, Eqns (3.66) and (3.67) are also obtained by discretizing the augmented equations of Section 2.11, $T = 0.050$.

Another possibility, which suits computer-aided design methods, is to use Fourier transform algorithms to develop finite impulse response digital filters with appropriate characteristics. This approach is mentioned in Section 4.2.7.

EXERCISES

1 Consider the process:

$$x(n) = e^{-an}$$

where a is a random number evenly distributed between 0 and 2. Find $\varepsilon(x(n))$ and $\text{cov}(x(n))$.

2 Consider the process:

$$x(n) = \cos(an)$$

where a is a random number evenly distributed between $\pm\,\Delta$. Find $\varepsilon(x(n))$ and $\text{cov}(x(n))$.

3 Consider the process:

$$x(n) = a\cos(\beta n) + b\sin(\beta n)$$

where a, b are independent random numbers with zero mean and variance σ^2, and β is constant. Find $\text{cov}(x(n))$.

4 Develop expressions for $S(\beta)$ if:

(a) $\bar{r}(k) = 2e^{-|k|}$, $k = -1, 0, 1$,

(b) $\bar{r}(k) = \cos\left(\dfrac{\pi_k}{8}\right)$, $-4 \leqslant k \leqslant +4$.

5 Specify filter algorithms which can be used to generate noise with the following PSDs, assuming unit strength white noise filter inputs:

(a) $S(\beta) = \dfrac{9(1 + \cos\beta)}{5 + 3\cos\beta}$

(b) $S(\beta) = \dfrac{1.25 + \cos\beta}{(1.36 - 1.2\cos\beta)(2 - 2\cos\beta)}$

6 Consider a random walk defined in the following way. A coin is tossed, and if it shows a head no step is taken, while if it shows a tail another coin is tossed and a step, size λ, is taken to the right if this shows a head, to the left if a tail. Sketch some sample functions of this and develop a PDF describing the excursion, $x(n)$, of the progress from 0 after n sets of coin tosses. Calculate $\varepsilon(x(n))$, $\sigma_x^2(n)$.

7 A discrete time dynamical system is described by:

$$\mathbf{x}(n+1) = \begin{pmatrix} 0.80 & 0.05 \\ -4.0 & 0.90 \end{pmatrix} \mathbf{x}(n) + \begin{pmatrix} 0.10 \\ 4.00 \end{pmatrix} w(n), \quad \mathbf{x}(0) = \begin{pmatrix} 1 \\ 0 \end{pmatrix}$$

where

$\mathbf{x}(n)$ represents the system states,
$w(n)$ is unit strength white noise.

Develop expressions for the behaviour of $\varepsilon(\mathbf{x}(n))$ and $\mathrm{cov}(\mathbf{x}(n))$ and evaluate the steady state covariance of $\mathbf{x}(n)$.

8 A continuous time system is modelled by the heuristic equations:

$$\dot{\mathbf{x}} = \begin{pmatrix} -2 & 0 \\ 0 & -3 \end{pmatrix} \mathbf{x} + \begin{pmatrix} 1 \\ 1 \end{pmatrix} u + \begin{pmatrix} 2 \\ 3 \end{pmatrix} w, \quad \mathbf{x}(0) = \begin{pmatrix} 1 \\ 4 \end{pmatrix}$$

where

$\mathbf{x}$ represents the states,
u is a deterministic input,
w is unit strength white noise.

Construct a model for the equivalent sampled data system assuming a sampling period $T = 0.1$, and develop expressions for $\varepsilon(\mathbf{x}(n))$, $\mathrm{cov}(\mathbf{x}(n))$, $n \to \infty$.

REFERENCES

[1] Rabiner, L. R. and Gold, C., *Theory and Application of Digital Signal Processing*, Prentice Hall, Englewood Cliffs, NJ, 1975.
[2] Maybeck, P. S., *Stochastic Models, Estimation and Control*, Academic Press, New York, 1979.

4 Computer methods for stochastic processes

4.1 Introduction

A number of computer algorithms and allied techniques which are useful in the simulation and analysis of stochastic system behaviour are described in this chapter. Section 4.2 deals with generating sample functions of useful stochastic processes, Section 4.3 with the evaluation of Fourier transforms, and Section 4.4 with the solution of model equations using Runge–Kutta algorithms. All these methods can be elaborated upon significantly; appropriate references are given.

4.2 Algorithms for generating stochastic processes

Useful, simple algorithms, some realizable in hardware, for generating approximations to (band-limited) white noise, Wiener processes, and various coloured noise processes are described in this section.

4.2.1 White noise with uniform PDF

Discrete time white noise, with a uniform or 'flat' PDF is traditionally generated by the 'congruential multiplication' formula:

$$w(n+1) = (\alpha w(n) + \eta)\text{modulo } \gamma^p \tag{4.1}$$

where

$w(n)$ is the white noise process (Section 3.6.3),
p is the word length,
α, η are constants,
γ is the numerical radix used in the calculation.

If α, η are properly selected, $w(n)$ is distributed (not particularly evenly) between 0 and $(\gamma^p - 1)$, the sequence length being γ^p.

One recommended scheme [ref. 1] which allows the multiplication in Eqn (4.1) to be replaced by shifting and adding suggests

$$\alpha = \gamma^{(p/2+1)} + 3, \quad \eta = 0 \tag{4.2}$$

Thus, for a word of six decimal digits:

$$\alpha = 10^4 + 3 = 10\,003, \quad \eta = 0$$

$w(n)$ is (not very) evenly distributed between 0 and 999 999. For a word of sixteen binary digits:

$$\alpha = 2^9 + 3 = 515, \quad \eta = 0$$

$w(n)$ is (not very) evenly distributed between 0 and 65 535. Other (not very) even distributions are readily obtained by scaling and adding appropriate constants to each element as it is generated.

Significantly better results are obtained by 'shuffling' the numbers, as follows. One hundred numbers (typically) are generated, using Eqn (4.1), and stored in a table. An address, $A(n)$, is chosen and its content is output and replaced with the next number generated by Eqn (4.1), $w(n+1)$. The least significant two decimal digits of $w(n+1)$ define $A(n+1)$, whose content is then output and replaced with $w(n+2)$, and so on.

The congruential multiplication formula and its reshuffled variant are slow, owing to the number of multiplications involved; and for applications where speed is important, an almost equally simple algorithm [ref. 2] is available. In this case, k (typically $k = 50$) random numbers $w(n-i)$, $-R \leqslant w(n-i) \leqslant +R$ are chosen from a table of evenly distributed random numbers (Appendix C, Figure C3.1) and stored in a buffer, where they are designated $w(n-1), w(n-2), \ldots, w(n-k)$, as indicated in Figure 4.1.

A new random number, $w(n)$, is generated by an algorithm which guarantees the retention of a uniform PDF (in spite of its addition of two random numbers (Section 1.13)):

$$v(n) = w(n-1) + w(n-k)$$

$$w(n) = \begin{cases} v(n), & -R \leqslant v(n) \leqslant +R \\ v(n) + R, & v(n) < -R \\ v(n) - R, & v(n) > +R \end{cases} \tag{4.3}$$

where $w(n)$ is the random number output. The buffer is now updated to accommodate $w(n)$ and lose $w(n-k)$.

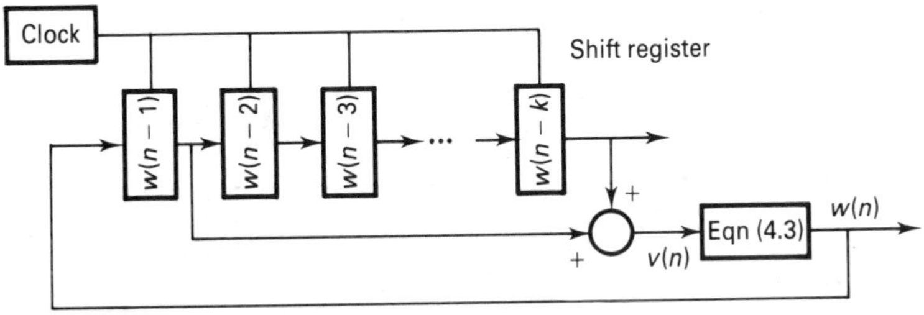

Figure 4.1 Random number generator.

4.2.2 Pseudo-random binary processes

A pseudo-random binary process, or signal (PRBS), which is white noise with only two amplitudes, can be generated by an algorithm of the kind shown in Figure 4.2(a), which is also easily realized in hardware.

The kth- and mth-stage outputs of a binary shift register, length k, are added, modulo 2, and inverted using an exclusive NOR operation. The result, $w(n)$, is output and fed back to the first stage. The maximum possible unrepeated sequence length, λ^{k-1}, is achieved if k, m are chosen so that

$$\lambda^k + \lambda^{k-m} + 1 = 0 \tag{4.4}$$

has no real roots [ref. 1]. This choice also guarantees that the shift register pattern never collapses to $11\ldots1$ – which must also be avoided at switch-on, typically by resetting to $00\ldots0$. Some possible values of k, m, are given in Figure 4.2(b).

4.2.3 Gaussian white noise

The random numbers generated by the methods above can be used to generate Gaussian white noise simply by adding them together in sufficiently large groups:

$$w(n) = \frac{1}{N}\sum_{i=0}^{N-1} w_f(Nn + i) \tag{4.5}$$

where

$w(n)$ is Gaussian white noise,
$w_f(n)$ is white noise of uniform PDF,
N is the number of numbers in each group (typically, $N > 10$).

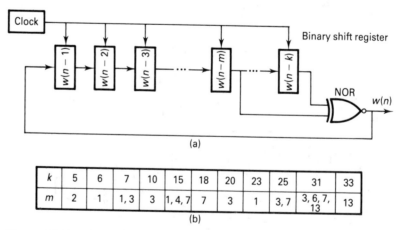

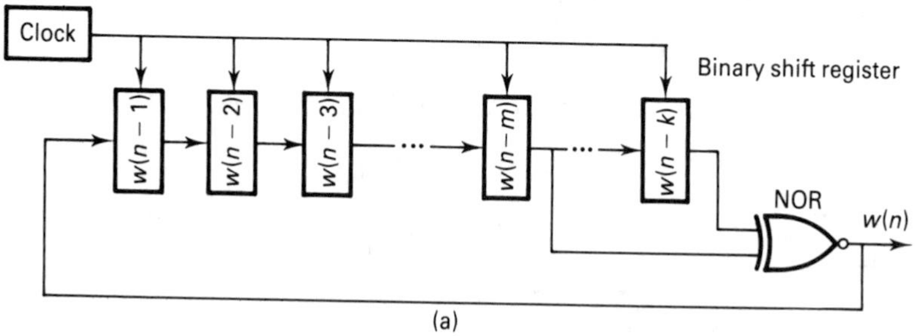

(a)

k	5	6	7	10	15	18	20	23	25	31	33
m	2	1	1, 3	3	1, 4, 7	7	3	1	3, 7	3, 6, 7, 13	13

(b)

Figure 4.2 Pseudo-random binary process generator.

As explained by the Central Limit Theorem (Section 1.13), the PDF of $w(n)$ becomes Gaussian with increasing accuracy as N increases.

Another scheme which is easily realized in hardware, and which generates continuous time Gaussian band-limited 'white' noise, is illustrated in Figure 4.3. A PRBS generator (Figure 4.2) and low pass analog filter (Appendix B, Section B1) are used.

A k-stage PRBS generator (with bias adjustment to ensure zero mean) feeds a low-pass analog filter. Over a time interval corresponding to the shift register period (kT, where T is the clock period), the PRBS pattern has a binomial distribution. If the filter cut-off frequency is $1/kT$ rad s^{-1}, its output is Gaussian, to an accuracy which increases with k. A variable gain amplifier allows adjustment of the output variance.

4.2.4 Correlated white noise processes

It is easy to generate correlated white noise processes, $w_1(n), w_2(n)$, uniformly distributed or Gaussian, using the white noise, $w(n)$, generated by one of the algorithms above.

Suppose the cross-correlation of $w_1(n), w_2(n)$ is specified:

$$r_{w_1w_2} = \varepsilon(w_1(n)w_2(n+k)) = q_{12}\,\delta(k)$$

Two schemes for generating $w_1(n)$, $w_2(n)$ from $w(n)$ are as follows:

1. Let

$$w_1(n) = w(n)$$

$$w_2(n) = \frac{q_{12}}{w_1(n)}$$

2. Assign $(N+1)/2$ elements of each group of N elements of $w(n)$ to $w_1(n)$, N odd, and $(N-1)/2$ elements to $w_2(n)$. The remaining element in the group of $(N+1)/2$ elements of $w_2(n)$ is:

$$w_2(n) = \frac{Nq_{12}}{w_1(n)}$$

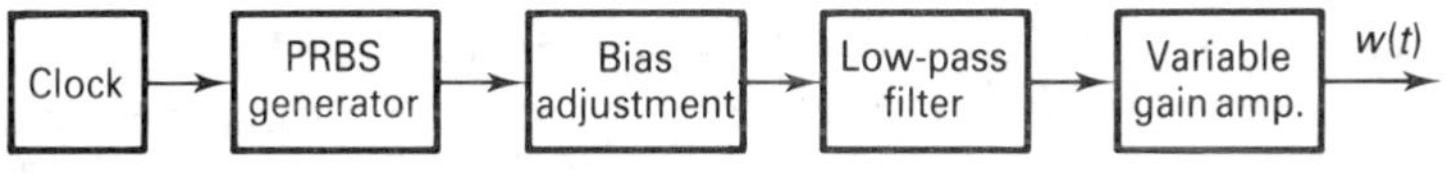

Figure 4.3 Gaussian (band-limited) white noise generator.

4.2.5 White noise with specified PDF

Theoretically, white noise with any PDF can be generated using uniformly distributed white noise as a basis by exploiting the relationships (Section 1.12) between the PDFs of two dependent variates.

Consider two variates, x, y with PDFs $p(\xi)$, $p(\chi)$, such that

$$p(\xi) = \begin{cases} 0, & \xi < 0 \\ 1, & 0 \leqslant \xi < 1 \\ 0, & \xi \geqslant 1 \end{cases} \tag{4.6}$$

If $p(\chi)$ is specified, it is in principle possible to find the function $y = f(x)$ which satisfies this specification. Given Eqn (4.6), the inverse function is

$$x = f^{-1}(y) = \int_{-\infty}^{y} p(\chi)\,\mathrm{d}\chi \tag{4.7}$$

If Eqn (4.7) can be inverted to give $y = f(x)$, then it is straightforward to generate white noise $y(n)$, with the specified PDF, $p(\chi)$, by generating uniformly distributed white noise, $x(n)$, and then calculating $y(n) = f(x(n))$ from Eqn (4.7).

Example

Consider

$$p(\chi) = \begin{cases} 0, \chi < 0 \\ \dfrac{1}{\sigma}\,\mathrm{e}^{-\chi/\sigma}, \chi \geqslant 0 \end{cases}$$

From Eqn (4.7):

$$x = f^{-1}(y) = \int_{0}^{y} \frac{1}{\sigma}\,\mathrm{e}^{-\chi/\sigma}\,\mathrm{d}\chi$$

$$= 1 - \mathrm{e}^{-y/\sigma}$$

This function happens to be analytically invertible, allowing the direct calculation of $y(n)$:

$$y(n) = -\sigma \ln(1 - x) \tag{4.8}$$

If $f^{-1}(y)$ cannot be expressed as a closed function, x – regarded as an area under the curve $p(\chi)$ (Eqn (4.7)) – can usually be calculated numerically. Then x is specified in the form of a look-up table of areas, each corresponding to a value of y, and this is used to generate $y(n)$, knowing $x(n)$. Gaussian white noise can be generated by this method using the table of Appendix C, Figure C3.2.

4.2.6 The Wiener process

A discrete time Wiener process is readily generated by summing Gaussian white noise:

$$\beta(n) = \sqrt{q} \sum_{k=0}^{n} w(k) \tag{4.9}$$

where

$\beta(n)$ is a Wiener process, diffusion q,
$w(n)$ is unit strength Gaussian white noise.

4.2.7 Coloured noise

As described in Sections 2.11 and 3.10, coloured noise can be generated by applying white noise to a linear filter whose transfer function is chosen according to the PSD of the noise to be generated. To synthesize such a filter analytically, however, that PSD must be capable of spectral factorization, and this represents a significant constraint.

Considerably more complicated design methods exist which permit the computer-aided design of finite impulse response (FIR) digital filters for this purpose [refs. 3, 4]. A PSD is specified at N frequencies (typically 32 or 64), and these are used to generate a $2N$ order FIR filter, essentially by taking the inverse Fourier transform of their square roots. Very similar methods can be used to design sets of filters to generate two or more processes with specified cross-correlation or cross-spectral density. These specialized methods are, however, beyond the scope of this book.

4.3 Algorithms for analyzing stochastic processes

Some simple algorithms for calculating probability density functions, means, variances, Fourier transforms and inverse Fourier transforms are outlined in this section.

4.3.1 Probability density functions, means, variances

The probability density function of a set of random numbers is readily found by selecting R equal-amplitude ranges, counting the numbers which lie in each range, and evaluating:

$$p(\xi_r) = \frac{N_r}{N(\xi_{r+1} - \xi_r)} \tag{4.10}$$

where

N_r elements lie in the range $\xi_r < x \leqslant \xi_{r+1}$,

$r = 0, 1, 2, \ldots, R$,

$N = \sum_{r=0}^{R-1} N_r$.

The mean and variance of x are calculated (Eqns (1.10) and (1.13)):

$$\varepsilon(x) = \frac{1}{N} \sum_{r=0}^{R-1} N_r \xi_r \tag{4.11}$$

$$\begin{aligned} \sigma_x^2 &= \varepsilon(x - \varepsilon(x))^2 \\ &= \frac{1}{N} \sum_{r=0}^{R-1} N_r (\xi_r)^2 - (\varepsilon(x))^2 \end{aligned} \tag{4.12}$$

4.3.2 Fourier transforms

The evaluation of Fourier transforms – and their inverses – by computer algorithm is important in the analysis of stochastic processes and systems. Efficient fast Fourier transform (FFT) algorithms are nowadays popular and one of these is described here. Elaborations, many of which deal with the effects of using finite numbers of samples, are explored in ref. 3.

The evaluation of the Fourier transform of a process $x(n)$ is considered by way of example, but the method applies directly to the calculation of correlation functions with the appropriate change of independent variable (Section 4.3.4).

The formula to be evaluated is (cf. Appendix 1, Eqns (A1.20) and (A1.28))

$$\begin{aligned} \mathscr{F}(x(n)) &= \mathscr{F}(x(nT)) = X(\mathrm{j}k\,\Delta\omega) \\ &= X(k) \\ &= \sum_{n=0}^{N-1} x(n)\,\mathrm{e}^{-\mathrm{j}(2\pi/N)kn} \end{aligned} \tag{4.13}$$

where

$x(n)$, $0 \leqslant n \leqslant (N-1)$, is a set of N numbers sampled at intervals T from the process $x(t)$, $0 \leqslant t \leqslant (N-1)T$,

$X(k)$, $0 \leqslant k \leqslant (N-1)$, is a set of N numbers describing the Fourier transform $X(\mathrm{j}\omega)$ at intervals $\Delta\omega$, $0 \leqslant \omega \leqslant (N-1)\,\Delta\omega$.

Thus, there are N time increments over which $x(t)$ is sampled, and N frequency increments over which $X(\mathrm{j}\omega)$ is evaluated; $X(\mathrm{j}\omega)$ is, in general, complex. (The FFT algorithm actually makes the assumption that $x(n)$ is cyclic with period $n = N$, and this results in the number of frequencies, N, namely $0, \Delta\omega, 2\Delta\omega, \ldots, (N-1)\,\Delta\omega$, at which $X(\mathrm{j}\omega)$ is evaluated.)

The inverse transformation is (cf. Appendix 1, Eqns (A1.21) and (A1.29)):

$$x(n) = x(nT) = \frac{1}{2\pi} \sum_{k=0}^{N-1} X(k)\,\mathrm{e}^{\mathrm{j}(2\pi/N)kn} \tag{4.14}$$

The straightforward evaluation of Eqns (4.13) and (4.14) involves an inordinate amount of calculation if, as is typical, $N > 1000$. FFT algorithms substantially reduce this, provided N is a power of 2 (typically, $N = 1024$).

Considering the Fourier transform (Eqn (4.13)), the sequence $(x(n))$ is divided into its even and odd numbered terms, $(x(2n))$, $(x(2n+1))$, $n = 0, 1, 2, \ldots, (N/2) - 1$ respectively:

$$X(k) = \sum_{n=0}^{(N/2)-1} x(2n)\,\mathrm{e}^{-\mathrm{j}(2\pi/N)k\cdot 2n} + \sum_{n=0}^{(N/2)-1} x(2n+1)\,\mathrm{e}^{-\mathrm{j}(2\pi/N)(2n+1)k} \tag{4.15}$$

$$= X_1(k) + W^k X_2(k), \qquad k = 0, 1, 2, \ldots, \left(\frac{N}{2} - 1\right)$$

where

$W = \mathrm{e}^{-\mathrm{j}(2\pi/N)}$,
$N = 2^\gamma$, γ an integer, and
$X_1(k)$, $X_2(k)$ are the $(N/2)$-point Fourier transforms of the 'even' and 'odd' sequences respectively.

Since Eqn (4.15) specifies $X(k)$ only for the $N/2$ points $k = 0, 1, \ldots, N/2 - 1$, it is necessary to define it for the remaining $N/2$ points, $k = N/2$, $N/2 + 1$, …, $(N - 1)$. It is not difficult to show that, for these points,

$$X(k) = X_1\left(k - \frac{N}{2}\right) + \mathrm{e}^{-\mathrm{j}(2\pi/N)(k-(N/2))} X_2(k)$$

$$= X_1\left(k - \frac{N}{2}\right) - W^k X_2\left(k - \frac{N}{2}\right), \quad \frac{N}{2} \leqslant k \leqslant N - 1$$

since

$$\mathrm{e}^{-\mathrm{j}\pi} = -1$$

The complete Fourier transform is thus, with a slight abuse of nomenclature:

$$X(k) = \begin{cases} X_1(k) + W^k X_2(k), & 0 \leqslant k \leqslant \dfrac{N}{2} - 1 \\[2ex] X_1(k) - W^k X_2(k), & \dfrac{N}{2} \leqslant k \leqslant N - 1 \end{cases} \tag{4.16}$$

The component parts of the right-hand side of Eqn (4.16) can be split in the same way:

$$X_1(k) = \begin{cases} X_{11}(k) + W^{2k}X_{12}(k), & 0 \leqslant k \leqslant \frac{N}{4} - 1 \\ X_{11}(k) - W^{2k}X_{12}(k), & \frac{N}{4} \leqslant k \leqslant \frac{N}{2} - 1 \end{cases}$$

$$X_2(k) = \begin{cases} X_{21}(k) + W^{2k}X_{22}(k), & 0 \leqslant k \leqslant \frac{N}{4} - 1 \\ X_{21}(k) - W^{2k}X_{22}(k), & \frac{N}{4} \leqslant k \leqslant \frac{N}{2} - 1 \end{cases}$$

The components of $X_{11}(k)$, $X_{12}(k)$, $X_{21}(k)$, $X_{22}(k)$ can in turn be split:

$$X_{11}(k) = \begin{cases} X_{111}(k) + W^{4k}X_{112}(k), & 0 \leqslant k \leqslant \frac{N}{8} - 1 \\ X_{111}(k) - W^{4k}X_{112}(k), & \frac{N}{8} \leqslant k \leqslant \frac{N}{4} - 1 \end{cases}$$

This splitting process can be continued until the right-hand side of Eqn (4.16) has been split into terms such as

$$X_{111\ldots1}(k) = \begin{cases} x(0) + W^{(N/2)k} \cdot x\left(\frac{N}{2}\right), & k = 0 \\ x(0) - W^{(N/2)k} \cdot x\left(\frac{N}{2}\right), & k = 1 \end{cases}$$

A glance at the above sequence shows that the terms appearing in the final two-term transforms are simply terms of $(x(n))$ arranged in a complicated order. Actually they are shuffled according to a 'reverse binary' algorithm, in which, if the initial order is represented in binary, the reshuffled order is in reverse binary. The process, which is quite difficult to follow, is clarified by a manageable example, $N = 8$.

Example

Consider the evaluation of an eight-point Fourier transform, $\mathscr{F}(x(n)) = X(k)$, by the above algorithm. The data shuffling recipe and calculation sequence are shown in Figure 4.4.

In Figure 4.4, the first stage of calculations generates

$X_{11}(0) = x(0) + x(4)$, since $e^{-j(2\pi/2)\cdot 0} = 1$

$X_{11}(1) = x(0) + x(4)$, since $e^{-j(2\pi/2)\cdot 1} = -1$

$\vdots$

$X_{22}(1) = x(3) + x(7)$, since $e^{-j(2\pi/2)\cdot 1} = -1$

The second stage generates

$X_1(0) = X_{11}(0) + X_{12}(0)$, since $e^{-j(2\pi/4)\cdot 0} = 1$

$X_1(1) = X_{11}(1) - jX_{12}(1)$, since $e^{-j(2\pi/4)\cdot 1} = -j$

$\vdots$

$X_2(3) = X_{21}(1) - jX_{22}(1)$, since $e^{-j(2\pi/4)\cdot 3} = +j$

The third, final, stage generates

$X(0) = X_1(0) + X_2(0)$, since $e^{-j(2\pi/8)\cdot 0} = 1$

$X(1) = X_1(1) + (\sqrt{2} - j\sqrt{2})X_2(1)$, since $e^{-j(2\pi/8)\cdot 1} = \sqrt{2} - j\sqrt{2}$

$\vdots$

$X(7) = X_1(3) + (\sqrt{2} + j\sqrt{2})X_2(3)$, since $e^{-j(2\pi/8)\cdot 7} = \sqrt{2} + j\sqrt{2}$

n (decimal)	*n* (binary)	*n* (reverse binary)	*n* (shuffled decimal)
0	000	000	0
1	001	100	4
2	010	010	2
3	011	110	6
4	100	001	1
5	101	101	5
6	110	011	3
7	111	111	7

(a) Shuffling process, $N = 8$

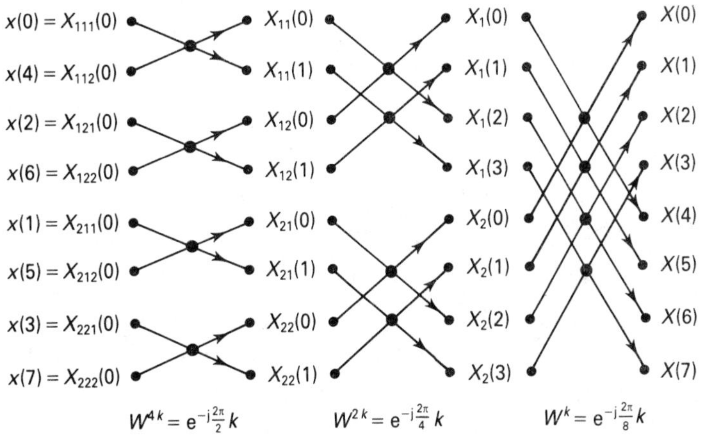

(b) FFT sequence, $N = 8$

Figure 4.4 FFT procedure, $N = 8$.

This sequence can be extended without undue complication to accommodate any reasonable power of 2. Figure 4.4(b) shows why the procedure is sometimes referred to as a 'butterfly' algorithm.

A valuable feature of the algorithm is that each variable is used only twice, after which its storage location may be overwritten by a new variable. Popular elaborations of the basic idea are due to Cooley, Lewis and Welch [in ref. 4] and Tukey [in ref. 4].

4.3.3 Inverse Fourier transforms

The natural corollary to Eqn (4.13) is Eqn (4.14), which generates the inverse Fourier transform as a set of N values, $x(n)$, $n = 0, 1, \ldots, (N-1)$, given a set of N values, $X(k)$, $k = 0, 1, \ldots, (N-1)$.

With the differences that there is division by 2π, a positive imaginary power of e, and that $X(k)$ is generally complex, this has the same structure as Eqn (4.16) and can therefore be evaluated in the same way – using the identical computer algorithm (Figure 4.4) – with obvious adjustments to accommodate the differences.

4.3.4 Correlation functions and power spectral densities

FFTs are useful in evaluating cross- and autocorrelation functions given ergodic process sample function data, and PSDs given correlation functions.

Consider the cross-correlation of two processes (Section 3.4.3), presented as two sets of numbers $x(n)$, $y(n)$, where N is a power of 2 to suit the FFT procedure:

$$\bar{r}_{xy}(k) = \frac{1}{2N}\sum_{n=0}^{N-1} x(n)y(n+k), \quad k = 0, 1, 2, \ldots, (N-1) \tag{4.17}$$

As before, direct evaluation of this is enormously lengthy for any reasonable N (typically $N = 1024$); the use of FFTs greatly reduces the load.

It can be shown [refs. 3, 4] that the cross-spectral density of $x(n)$, $y(n)$ can be written (cf. Eqn (4.13)):

$$S_{xy}(\mathrm{j}k\,\Delta\omega) = S_{xy}(k) = \mathscr{F}(x(n))\cdot\mathscr{F}^*(y(n)), \quad k = 0, 1, 2, \ldots, (N-1) \tag{4.18}$$

where $\mathscr{F}^*(y(n))$ is the complex conjugate of $\mathscr{F}(y(n))$; and $\Delta\omega = 2\pi/NT$.

Recall (Eqns (3.16) and (3.17)) that

$$\bar{r}_{xy}(k) = \mathscr{F}^{-1}(S_{xy}(k)) \tag{4.19}$$

This suggests that to evaluate $\bar{r}_{nxy}(k)$, it is necessary only to evaluate $\mathscr{F}(x(n))$, $\mathscr{F}^*(y(n))$, multiply them (Eqn (4.18)), and calculate the inverse Fourier transform of the result (Eqn (4.19). There are, however, two complications.

1. $S_{xy}(k)$ evaluated in this way assumes periodicity of both $x(n)$, and $y(n)$ over the N samples considered, and since the minimum period of $(x(n),\ y(n+k))$, $0 \leqslant k \leqslant (N-1)$ is actually $2N$, the calculation as described is in fact incorrect.

2. $S_{xy}(k)$ is an N-point sequence, $k = 0, 1, \ldots, (N-1)$, rather than an N-point sequence $k = (-N/2), ((-N/2)+1), \ldots, ((N/2)-1)$ which would describe the autocorreiation more exactly.

These problems are largely overcome by 'padding' the sequences $x(n)$, $y(n)$ with 0s to sequences of $2N$ elements, and using the scheme for all $2N$ elements in each case. The first N points of $\bar{r}_{xy}(k)$ (Eqn (4.17)) are then interpreted in a 'folded-over' manner to yield the true cross-correlation:

$$\bar{r}_{xy}(k) = \begin{cases} \bar{r}_{xyc}(N-k), & k = 0, 1, 2, \ldots, \left(\dfrac{N}{2} - 1\right) \\ \bar{r}_{xyc}(k), & k = \dfrac{N}{2}, \dfrac{N}{2} + 1, \ldots, (N-1) \end{cases} \tag{4.20}$$

where $\bar{r}_{xyc}(k)$ represents the calculated values, and $r_{xy}(k)$ the corrected values. $\bar{r}_{xyc}(k)$, $k = N, N+1, \ldots, 2N$, are discarded. This procedure yields reasonably accurate cross-spectral densities and cross-correlation functions, and, by setting $y(n) = x(n)$, PSDs and autocorrelation functions.

The main inaccuracies arise from 'end effects'. The sharp commencement and termination of data, $n = 0$ and $n = (N-1)$, generate frequencies in the Fourier transform which are not present in the original processes. This is tackled by 'smoothing' or 'windowing' the data [ref. 4] before subjecting it to the transform procedure, a topic that is beyond the scope of this book.

4.4 Computer modelling of stochastic systems

Computer modelling of dynamical systems depends on being able to solve differential equations numerically. Runge–Kutta algorithms are the most popular, with predictor corrector and Bulirsh–Stoer algorithms (ref. 3] being used where high accuracy is required, or where the equations are 'stiff' and not susceptible to the relatively simple Runge–Kutta solution.

The Runge–Kutta algorithm is described in this section. Comments on its use in the simulation of stochastic systems, and on variations on the basic algorithm, are also made.

4.4.1 Runge–Kutta algorithms

Consider the first order vector differential equation:

$$\frac{d\mathbf{x}}{dt} = \dot{\mathbf{x}} = \mathbf{f}(\mathbf{x}, t) \tag{4.21}$$

The numerical solution of this, which is the task to be undertaken if the solution (i.e. the behaviour) of the dynamical system model of Eqn (2.67) is to be generated, amounts to finding $\mathbf{x}((n+1)h)$, given $\mathbf{x}(nh)$, where h is a 'calculation interval' which is 'short' compared with the natural dynamical time constants of the system. Then $\mathbf{x}((n+2)h)$, $\mathbf{x}(\mathbf{n}+3)h)$, ..., can be generated by the same process: $\mathbf{x}((\mathbf{n}+1)h)$ is expressed as a Taylor series

$$\mathbf{x}((n+1)h) = \mathbf{x}(nh) + h\mathbf{f}_n + \frac{h^2}{2!}\mathbf{f}'_n + \frac{h^3}{3!}\mathbf{f}''_n + \cdots \tag{4.22}$$

where

$$\mathbf{f}_n = \mathbf{f}(\mathbf{x}(nh), nh)$$

and

$$\mathbf{f}'_n = \left.\frac{d\mathbf{f}}{dt}\right|_{\substack{\mathbf{x}=\mathbf{x}(nh)\\ t=nh}} = \left.\left(\frac{\partial \mathbf{f}}{\partial \mathbf{x}}\cdot \mathbf{f}_n + \frac{\partial \mathbf{f}}{\partial t}\right)\right|_{\substack{\mathbf{x}=\mathbf{x}(nh)\\ t=nh}} \tag{4.23}$$

and so on for the higher derivatives.

An rth order algorithm is devised such that the value of $\mathbf{x}((n+1)h)$, which it generates, matches Eqn (4.22) up to the term in h^r, and such that it avoids the need for evaluating derivatives such as that in Eqn (4.23).

1. The (obvious) first order algorithm (Euler's formula), which is generally rather inaccurate, is

$$\mathbf{x}((n+1)h) = \mathbf{x}(nh) + h\cdot\mathbf{f}(x(nh), nh) \tag{4.24}$$

2. The considerably more accurate second order algorithm is

$$\left.\begin{aligned} \mathbf{k}_1 &= h\cdot\mathbf{f}(\mathbf{x}(nh), nh) \\ \mathbf{k}_2 &= h\cdot\mathbf{f}(\mathbf{x}(nh) + 0.5\mathbf{k}_1, (n+0.5)h) \\ \mathbf{x}((n+1)h) &= \mathbf{x}(nh) + \mathbf{k}_2 \end{aligned}\right\} \tag{4.25}$$

These equations are evaluated in order, yielding k_1, then k_2 and finally $\mathbf{x}((n+1)h)$.

It is worth noting that during this sequence, $\mathbf{f}(\mathbf{x}, t)$ is evaluated at the intermediate point $(\mathbf{x}(nh) + 0.5\mathbf{k}_1, (n+0.5)h)$.

3. The most commonly used algorithm is perhaps the fourth order one:

$$\left.\begin{aligned} \mathbf{k}_1 &= h\cdot\mathbf{f}(\mathbf{x}(nh), nh) \\ \mathbf{k}_2 &= h\cdot\mathbf{f}(\mathbf{x}(nh) + 0.5\mathbf{k}_1, (n+0.5)h) \\ \mathbf{k}_3 &= h\cdot\mathbf{f}(\mathbf{x}(nh) + 0.5\mathbf{k}_2, (n+0.5)h) \\ \mathbf{k}_4 &= h\cdot\mathbf{f}(\mathbf{x}(nh) + \mathbf{k}_3, (n+1)h) \\ \mathbf{x}((n+1)h) &= \mathbf{x}(nh) + \frac{1}{6}(\mathbf{k}_1 + \mathbf{k}_2 + \mathbf{k}_3 + \mathbf{k}_4) \end{aligned}\right\} \tag{4.26}$$

This involves two intermediate-point calculations, and is often felt to give the best compromise between accuracy and complication.

Example

The behaviour of a pendulum (Figure 4.5(a)) is described by the second order equation:

$$\ddot{\theta} + 0.05\dot{\theta} + \sin\theta = 0, \quad \theta(0) = 0, \quad \dot{\theta}(0) = 1 \tag{4.27}$$

where θ is the angular position (rad).

Casting Eqn (4.27) in state equation form (Eqns (2.67) and (2.68)):

$$\dot{\mathbf{x}} = \begin{pmatrix} \dot{x}_1 \\ \dot{x}_2 \end{pmatrix} = \begin{pmatrix} x_2 \\ -\sin x_1 - 0.05x_2 \end{pmatrix}, \quad \mathbf{x}(0) = \begin{pmatrix} 0 \\ 1 \end{pmatrix}$$

where

$$x_1 = \theta$$

$$x_2 = \dot{\theta}$$

A fourth order Runge–Kutta solution, using the rather long calculation interval, $h = 1.0$ (s), yields the figures and graphical result shown in Figure 4.5(b). A second order algorithm applied to the same problem gives results virtually indistinguishable from these.

In such calculations, it is important to choose h well, since too large a value generates faulty results, whilst too small a value leads to a heavy computational load.

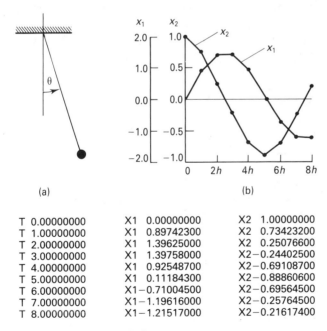

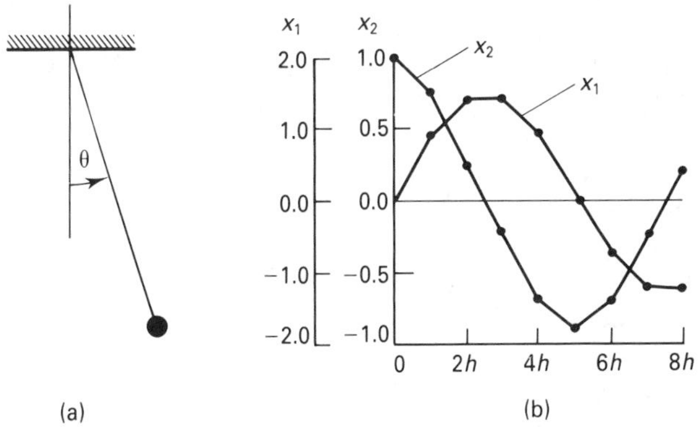

T 0.00000000	X1 0.00000000	X2 1.00000000
T 1.00000000	X1 0.89742300	X2 0.73423200
T 2.00000000	X1 1.39625000	X2 0.25076600
T 3.00000000	X1 1.39758000	X2−0.24402500
T 4.00000000	X1 0.92548700	X2−0.69108700
T 5.00000000	X1 0.11184300	X2−0.88860600
T 6.00000000	X1−0.71004500	X2−0.69564500
T 7.00000000	X1−1.19616000	X2−0.25764500
T 8.00000000	X1−1.21517000	X2−0.21617400

Figure 4.5 Pendulum behaviour.

Some experimentation is usually needed; solutions are generated for progressively larger values of h, and a value of h compatible with no significant change in these solutions is chosen for the final model.

In the example above, $h = 0.5$ and $h = 1.0$ generate virtually identical solutions, but $h = 1.5$ generates very different, and incorrect, results. A calculation interval of 1.0, or shorter, is therefore chosen for the model.

The best choice of h can vary over time for the same problem, a short interval being necessary where violent changes are occurring, and a long interval being more efficient where rates of change are small. Schemes for adapting h automatically according to the perceived conditions are described in the literature [ref. 3] and proprietary simulation packages based on the Runge–Kutta algorithm usually incorporate such features.

4.4.2 Runge–Kutta methods for stochastic models

As explained in Section 4.4.1, the Runge–Kutta calculation interval, h, must be 'short' compared with the natural time constants of the model equations being solved. Clearly, since white noise is involved in the heuristic models of stochastic systems (Eqns (2.73) and (2.75)), and this involves infinite frequencies, the choice of a suitable calculation integral become impossible. If, however, $w(t)$ in these equations is interpreted as band-limited white noise (Section 4.2) with a cut-off frequency that is high compared with the system's natural dynamics but low compared with the calculation interval frequency, then the Runge–Kutta algorithm can yield useful sample functions results.

The examples in Sections 2.10–2.12 are evaluated using this technique. In Figure 2.31, for example, a fourth order Runge–Kutta algorithm is used, $h = 0.02$, and band-limited white noise, cut-off frequency 10 Hz, simulates the 'white noise' vertical position of the vibrating table.

It is worth recalling here that Eqns (2.73) and (2.75) are mathematically invalid in any case if $w(t)$ approaches true white noise, and, furthermore, that there is no guarantee that solutions of non-linear equations of this type are Markov. The results generated by such methods must therefore be treated with a certain caution, though they are generally useful.

It is worth noting, too, that linear models can be cast in discrete time form (Eqns (3.45)–(3.51)) and sample function solutions can be generated by solving these difference equations without resource to Runge–Kutta approximation. This can be an attractive alternative yielding solutions which are theoretically exact at sampling instants.

4.4.3 More elaborate techniques

The shortcomings of the straightforward Runge–Kutta approach described in Section 4.4.2 can be tackled for heuristic models of the structure of Eqn (2.90), which are mathematically interpreted as Eqn (2.92), using Itô or Stratonovich integration as the basis for dealing with the noise corruption.

One idea [ref. 5] is to extend the Runge–Kutta algorithm by converting it to yield results which are statistically correct to a degree that is dependent on the order of the algorithm. A second order algorithm is proposed, but higher order ones are found to be extremely complex, or perhaps even impossible, to develop.

Other approaches [refs. 2, 6] are based on the evaluation of Itô or Stratonovich integrals in the mathematically valid model of Eqn (2.92).

All these methods could fairly be described as elaborate.

EXERCISES

1 Write a program (BASIC or FORTRAN or 'C') to realize the algorithm of Eqn (4.1), allowing α, η, γ, p to be specified by the operator, and use it to generate and store 1000 random numbers for:

α	η	γ	p
23	0	10	8
7	3	10	5
11	3	10	5

2 Write a program (BASIC or FORTRAN or 'C') to generate the PDF, mean and variance of a file of data (Eqns (4.10), (4.11) and (4.12)), and use it to process the data produced in (1).

3 Use the shuffling algorithm described in Section 4.2.1, with a data buffer of 100 words, to generate more files of data, and generate the PDFs of these using the program developed in (2).

4 Write a program using the algorithm of Eqn (4.3) with a buffer of 100 words to generate files of 1000 random numbers in the ranges ± 50, $\pm 10\,000$, $\pm 5\,000\,000$. Use the program developed in (2) to generate the PDFs of these.

5 Write a program to generate 1000 words of Gaussian white noise using the algorithm of Eqn (4.5), and the results generated in (1), for $N = 5, 10, 20$. Use the program developed in (2) to find the PDFs of these.

6 Design a discrete time Gaussian white noise generator using the scheme outlined in Figure 4.3, including the design of a third order discrete Butterworth filter with cut-off frequency 3 rad per sample (Appendix B, Section B2).

7 Write a program realizing the generator designed in (6) and generate 1000 words of noise using it. Find the PDF of this data using the program developed in (2).

8 A well-known FFT algorithm is the Cooley, Lewis, Welch algorithm*:

* Cooley, J. W., P. Lewis and P. D. Welch, 'The finite Fourier transform', *IEEE Trans. on Audio and Electro-acoustics*, **17**, (2), pp. 77–86, 1969.

```
      SUBROUTINE FFT(A,M,N)
      COMPLEX A(N),U,W,T
      N=2**M
      NV2=N/2
      NM1=N-1
      J=1
      DO 7 I=1,NM1
      If(I .GE. J) GO TO 5
      T=A(J)
      A(J)=A(I)
      A(I)=T
5     K=NV2
6     IF(K .GE. J) GO TO 7
      J=J-K
      K=K/2
      GO TO 6
7     J=J+K
      PO=3.141592653589793
      DO 20 L=1,M
      LE=2**L
      LE1=LE/2
      U=(1.0,0.)
      W=CMPLX(COS(PI/LE1),SIN(PI/LE1))
      DO 20 J=1,LE1
      DO 10 I=J,N,LE
      IP=I+LE1
      T=A(IP) * U
      A(IP)=A(I)-T
10    A(I)=A(I)+T
20    U=U * W
      RETURN
      END
```

The time domain data is written to a complex array, A, of $N - 2^m$ elements, and the subroutine, which contains three nested loops, leaves the frequency domain data in A.

Write a program (BASIC or FORTRAN or 'C') utilizing this algorithm and use it to find the Fourier transforms of:

$x = \sin 3t$

$x = \sin t + \cos 4t$

9 Use the algorithm of (8) to evaluate the autocorrelation and PSD of the noise data generated in (5) and (7), according to the algorithms of Eqns (4.20) and (4.18), $x(n) = y(n)$.

10 Write a program (BASIC or FORTRAN or 'C') to realize the second order

Runge–Kutta algorithm of Eqn (4.25) for a second order system:

$$\dot{\mathbf{x}} = \begin{pmatrix} a_{11} & a_{12} & a_{13} \\ a_{21} & a_{22} & a_{23} \\ a_{31} & a_{32} & a_{33} \end{pmatrix} \mathbf{x} + \begin{pmatrix} b_1 \\ b_2 \\ b_3 \end{pmatrix} u$$

so that the operator can specify the calculation interval, h, and a_{11}, $a_{12}, \ldots, a_{33}, b_1, \ldots, b_3$, and u.
Use this to model the systems

(a) $$\dot{\mathbf{x}} = \begin{pmatrix} 0 & 1 \\ -101 & -2 \end{pmatrix} \mathbf{x} + \begin{pmatrix} 0 \\ 101 \end{pmatrix} u, \qquad h = 0.05, \mathbf{x}(0) = 0$$

$$u = \begin{cases} 0, & t < 0 \\ 1, & t \geqslant 0 \end{cases}$$

(b) $$\mathbf{x} = \begin{pmatrix} 0 & 1 & 0 \\ 0 & 0 & 1 \\ -0.404 & -4.080 & -0.500 \end{pmatrix} \mathbf{x} + \begin{pmatrix} 0 \\ 0 \\ 0.404 \end{pmatrix} u,$$

$$h = 0.2, \mathbf{x}(0) = \begin{pmatrix} 0 \\ 0 \\ 0 \end{pmatrix},$$

$$u = \begin{cases} 0, & t < 0 \\ 1, & t \geqslant 0 \end{cases}$$

Find by experiment the maximum values of h which yield tolerable results.

11 Write a program (BASIC or FORTRAN or 'C') to realize the fourth order Runge–Kutta algorithm of Eqn (4.26), and repeat the modelling exercise of (10) using this. Compare the results with those generated in (10).

REFERENCES

[1] Ralston, A. and Wilfe, H. S., *Mathematical Methods for Digital Computers*, Vol II, Wiley, New York, 1967.

[2] Klauder, J. R. and Petersen, W. P., 'Numerical integration of multiplicative noise stochastic differential equations', *SIAM Journal of Numerical Analysis*, **22**, (6), 1985.

[3] Press, W. H., Flannery, B. I., Teukolsky, S. A. and Vetterling, W. T., *Numerical Recipes: The art of scientific computing*, CUP, Cambridge, 1986.

[4] Rabiner, L. R. and Gold, B., *Theory and Application of Digital Signal Processing*, Prentice Hall, Englewood Cliffs, NJ, 1975.

[5] Greenside, H. S. and Helfland, E., 'Numerical integration of stochastic differential equations—II', *Bell System Technical Journal*, **60**, (8), 1981.

[6] Rumelin, W., 'Numerical treatment of stochastic differential equations', *SIAM Journal of Numerical Analysis*, **19**, (3), 1982.

5 Kalman filters

5.1 Introduction

Estimation of the internal states of a stochastic dynamical system (using the system inputs and outputs as a basis for that estimation) is a topic with important applications, not least in the realization of modern control strategies where the states are not accessible for measurement (Section 6.8). The Kalman filter, which applies to linear and some non-linear systems, and is the most promising estimator presently available, is the subject of this chapter.

The underlying ideas [refs 1, 2] are described in Sections 5.2–5.4, and variants, including extended and adaptive filters, in Sections 5.5–5.8. Algorithms and efficient microprocessor realizations of filters are examined in Section 5.9.

5.2 State estimators for deterministic systems – continuous time

As explained in Section 2.10 and Appendix A, Section A2, the behaviour of a linear time-invariant (LTI) deterministic system, S, can be modelled by the dynamical equations (Eqns (2.71) and (2.72)):

$$\dot{\mathbf{x}}(t) = \mathbf{A}\mathbf{x}(t) + \mathbf{B}\mathbf{u}(t) \tag{5.1}$$

$$\mathbf{y}(t) = \mathbf{C}\mathbf{x}(t) + \mathbf{D}\mathbf{u}(t) \tag{5.2}$$

It is generally vauable – for control strategy realization, for example – to measure the states, $\mathbf{x}$, directly, but in practice this may be difficult, or even impossible. Provided

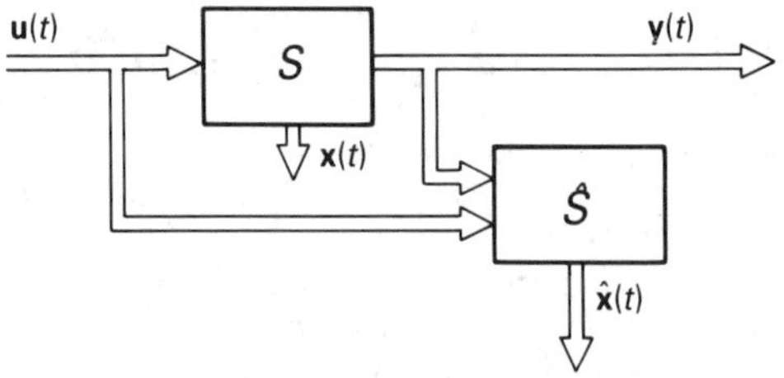

Figure 5.1 System and state estimator.

S is observable (Appendix A, Section A2.3), however, a state estimator, or 'observer', $\hat{S}$, can usually be designed to generate an estimate of $\mathbf{x}$ using $\mathbf{u}$ and $\mathbf{y}$ as the basis for that estimation. This is shown in block form in Figure 5.1.

The two most common types of estimator, the asymptotic state estimator and the Luenberger observer, are examined in this section; they provide the basic structures for Kalman–Bucy filters.

5.1.1 The asymptotic state estimator

An asymptotic states estimator, $\hat{S}$ (Figure 5.1), is an analog simulator of S which seeks to correct any discrepancy between the dynamic behaviours of S and $\hat{S}$. Its model equation is

$$\begin{aligned}\dot{\hat{\mathbf{x}}} &= \mathbf{A}\hat{\mathbf{x}} + \mathbf{B}\mathbf{u} + \mathbf{M}(\mathbf{y} - \hat{\mathbf{y}}) \\ &= (\mathbf{A} - \mathbf{M}\mathbf{C})\hat{\mathbf{x}} + (\mathbf{B} - \mathbf{M}\mathbf{D})\mathbf{u} + \mathbf{M}\mathbf{y} \qquad (5.3)\end{aligned}$$

where

$\hat{\mathbf{x}}$ represents the (m) estimated states,
$\hat{\mathbf{y}}$ represents the (r) estimated outputs,
$\mathbf{M}$ is a constant $(m \times r)$ 'gain' matrix.

Eqn (5.3) replicates the system model (Eqn (5.1)), but includes a corrective term, $\mathbf{M}(\mathbf{y} - \hat{\mathbf{y}})$, dependent on any discrepancy between the system and estimator outputs, $(\mathbf{y} - \hat{\mathbf{y}})$.

The behaviour of any discrepancy between the system and estimator states, i.e. the estimator error, $(\mathbf{x} - \hat{\mathbf{x}})$, is found by subtracting Eqn (5.3) from (5.1):

$$(\dot{\mathbf{x}} - \dot{\hat{\mathbf{x}}}) = (\mathbf{A} - \mathbf{M}\mathbf{C})(\mathbf{x} - \hat{\mathbf{x}})$$

Being of standard state form (Appendix A, Eqn (A2.1)), this signifies that the estimator error behaviour is determined by the eigenvalues of $(\mathbf{A} - \mathbf{M}\mathbf{C})$. Designing the estimator thus amounts to choosing an estimator gain matrix, $\mathbf{M}$, such that the eigenvalues of $(\mathbf{A} - \mathbf{M}\mathbf{C})$ are stable and 'fast' compared with those of $\mathbf{A}$. Any error that develops in $\hat{\mathbf{x}}$ is therefore removed relatively quickly compared with the dynamics of S.

Example

Consider an experimental equipment consisting of a local angular position servo (amplifier, motor, gearbox, potentiometer), a flexible torsion bar, and a flywheel with angular displacement potentiometer (Figure 5.2(a)). The system input, u, is the voltage applied to the servo amplifier, its output, y, the voltage at the flywheel potentiometer.

The system behaviour is modelled by (Eqns (5.1) and (5.2)):

$$\dot{\mathbf{x}} = \begin{pmatrix} 0 & 10.00 & 0 \\ 0 & 0 & 10.00 \\ -10.10 & -12.10 & -13.00 \end{pmatrix} \mathbf{x} + \begin{pmatrix} 0 \\ 0 \\ 10.10 \end{pmatrix} u \tag{5.4}$$

$$y = (1.00 \quad 0 \quad 0)\mathbf{x} \tag{5.5}$$

where

u represents the input to the servo amplifier (volts),
y represents the potentiometer output (volts),
and $x_1 = y$, $x_2 = 0.1\dot{y}$, $x_3 = 0.01\ddot{y}$.

x_1, x_2, x_3 thus represent the (scaled) position, velocity and acceleration of the flywheel respectively; x_2, x_3 are not measured directly, however, and it is required to construct an estimator which generates estimates of x_1, x_2, x_3.

The eigenvalues of the state matrix, $\mathbf{A}$, which describe the natural modes of

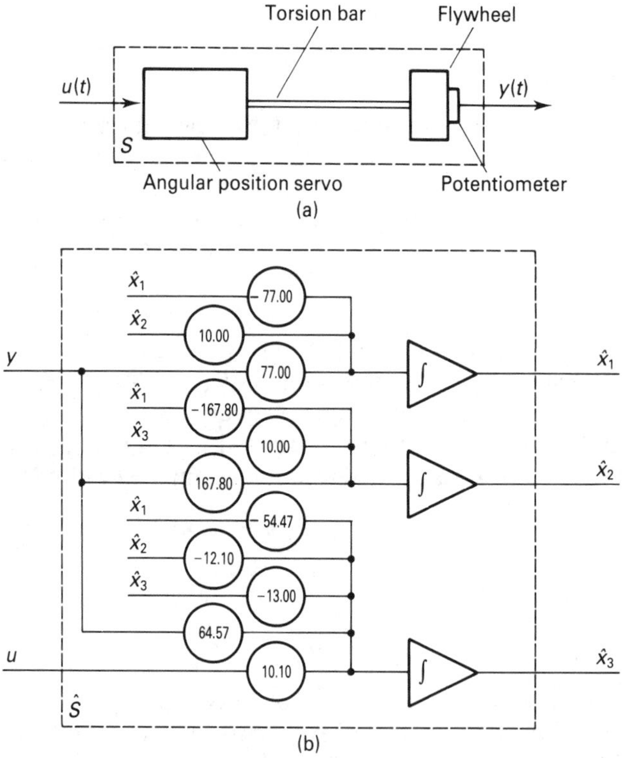

Figure 5.2 LTI system and asymptotic state estimator.

S are easily found (Appendix A, Eqn (A2.13)):

$$\lambda = -10.58, \quad \lambda = -1.21 + j9.69, \quad \lambda = -1.21 - j9.69$$

The state estimator model (Eqn (5.3)), is

$$\dot{\hat{\mathbf{x}}} = \begin{pmatrix} -m_1 & 10.00 & 0 \\ -m_2 & 0 & 10.00 \\ -10.10 - m_3 & -12.10 & -13.00 \end{pmatrix} \hat{\mathbf{x}} + \begin{pmatrix} 0 \\ 0 \\ 10.10 \end{pmatrix} u + \begin{pmatrix} m_1 \\ m_2 \\ m_3 \end{pmatrix} y$$

where

$$\mathbf{M} = (m_1 \quad m_2 \quad m_3)^T.$$

$\mathbf{M}$ is chosen so that eigenvalues of $(\mathbf{A} - \mathbf{MC})$ are fast compared with those of $\mathbf{A}$, say:

$$\lambda = -50, \quad \lambda = -20 + j20, \quad \lambda = -20 - j20$$

Thus:

$$\det(\lambda\mathbf{I} - (\mathbf{A} - \mathbf{MC})) = \det \begin{pmatrix} \lambda + m_1 & -10.00 & 0 \\ m_2 & \lambda & -10.00 \\ 10.10 + m_3 & 12.10 & \lambda + 13.00 \end{pmatrix}$$
$$= (\lambda + 50)(\lambda + 20 + j20)(\lambda + 20 - j20) = 0$$

Equivalently,

$$\lambda^3 + \lambda^2(m_1 + 13) + \lambda(13m_1 + 10m_2 + 121) + (121m_1 + 100m_3 + 1140)$$
$$= \lambda^3 + \lambda^2(90) + \lambda(2800) + 4000 = 0$$

Equating coefficients of λ^0, λ^1, λ^2 and solving for m_1, m_2, m_3 yields

$$\mathbf{M} = (m_1 \quad m_2 \quad m_3)^T = (77.00 \quad 167.80 \quad -64.57)^T$$

The estimator model is thus:

$$\dot{\hat{\mathbf{x}}} = \begin{pmatrix} -77.00 & 10.00 & 0 \\ -167.80 & 0 & 10.00 \\ +54.47 & -12.10 & -13.00 \end{pmatrix} \hat{\mathbf{x}} + \begin{pmatrix} 0 \\ 0 \\ 10.10 \end{pmatrix} u + \begin{pmatrix} 77.00 \\ 167.80 \\ -64.57 \end{pmatrix} y$$

A schematic realization of this is shown in Figure 5.2(b). Step function responses for $\mathbf{x}$ and $\hat{\mathbf{x}}$ are given in Figure 5.3, where it can be seen that the estimator error is very small.

5.2.2 The Luenberger observer

The Luenberger observer (Figure 5.4) uses the r system outputs to calculate, algebraically, r 'pseudo-states', and asymptotic state estimation to estimate the remaining $(m - r)$ pseudo-states. The resulting m pseudo-state estimates are combined algebraically to yield m state estimates.

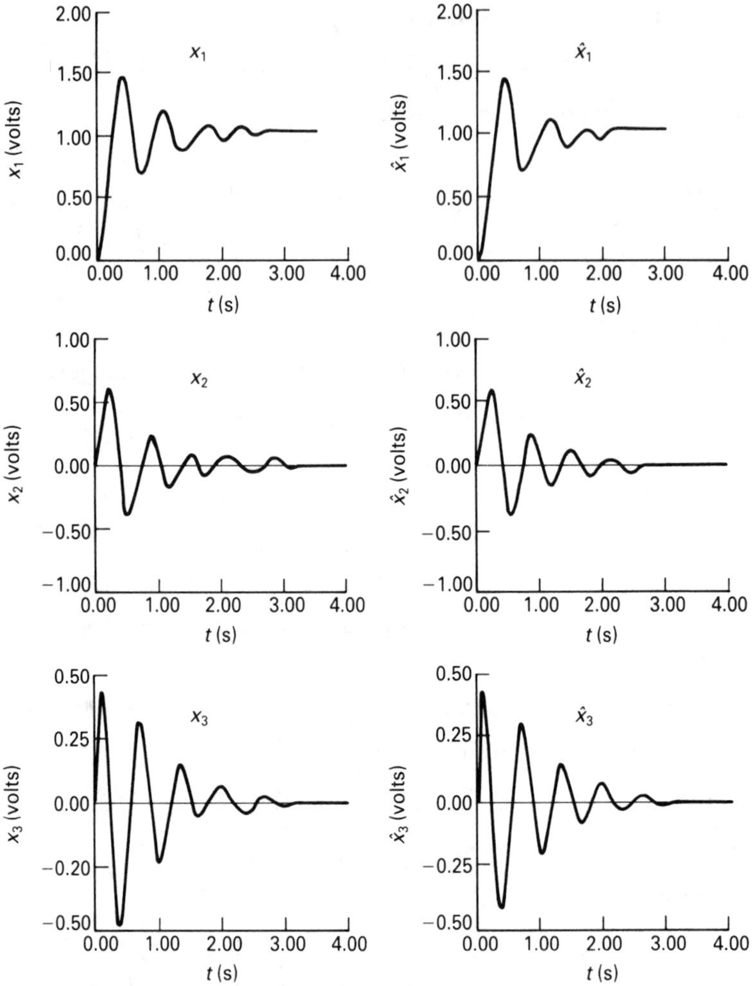

Figure 5.3 System and estimator behaviour.

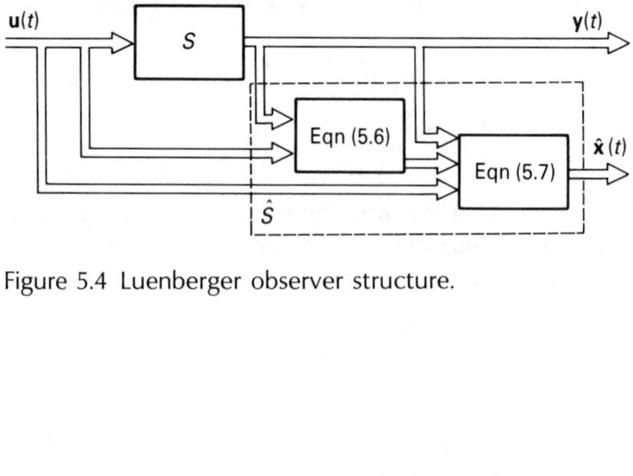

Figure 5.4 Luenberger observer structure.

The Luenberger observer model equations are

$$\dot{\boldsymbol{\eta}} = \mathbf{L}\boldsymbol{\eta} + (\mathbf{G} - \mathbf{HD})\mathbf{u} + \mathbf{Hy} \qquad (5.6)$$

$$\hat{\mathbf{x}} = \mathbf{N}\boldsymbol{\eta} - \mathbf{MDu} + \mathbf{My} \qquad (5.7)$$

where $\boldsymbol{\eta}$ represents the $(m - r)$ pseudo-state estimates, and **L**, **G**, **H**, **N**, **M** are constant matrices found by the following design procedure.

1. Define an $(m - r) \times (m - r)$ matrix **L** with eigenvaues which are 'faster' than, and in no case equal to, those of **A**; **L** is typically chosen to be diagonal.
2. Define arbitrarily an $(m - r) \times r$ matrix **H**, and find the $(m - r) \times m$ matrix **T** which satisfies:

$$\mathbf{TA} - \mathbf{LT} = \mathbf{HC} \qquad (5.8)$$

   T must be such that the $(m \times m)$ matrix

$$\mathbf{R} = \begin{pmatrix} \mathbf{C} \\ \text{---} \\ \mathbf{T} \end{pmatrix} \qquad (5.9)$$

   is not singular. If it is, **H** must be redefined.
3. Calculate

$$\mathbf{G} = \mathbf{TB}, \ (\mathbf{M} \,\vdots\, \mathbf{N}) = \mathbf{R}^{-1},$$

   where **M** is $(m \times r)$ and **N** is $(m \times (m - r))$.
4. Then the estimator is modelled by Eqns (5.6) and (5.7).

5.3 The Kalman–Bucy filter

The ideas of Section 5.2 extend fairly readily to linear time-variant systems with process and measurement noise corruption. Such a system, S, and its estimator, $\hat{S}$, are shown in Figure 5.5.

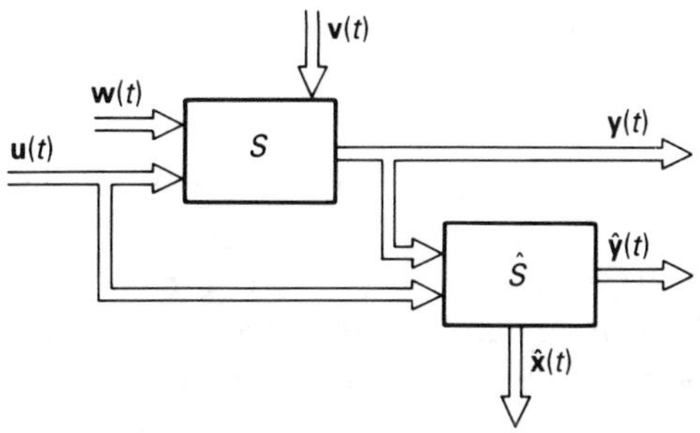

Figure 5.5 Linear stochastic system and state estimator.

The estimator, $\hat{S}$, has no direct knowledge of the noise signals, $\mathbf{w}$, $\mathbf{v}$, but its estimates, $\hat{\mathbf{x}}$, are required to take account of the presence of $\mathbf{w}$, and to attenuate the effects of $\mathbf{v}$. The quality of the results naturally depends on the choice of the estimator gain matrix $\mathbf{M}$ (Eqn (5.3)). If the choice is optimal in a particular sense, the estimator is a Kalman–Bucy filter [ref. 3].

The system is modelled (heuristically) by Eqns (2.75) and (2.76):

$$\dot{\mathbf{x}}(t) = \mathbf{A}(t)\mathbf{x}(t) + \mathbf{B}(t)\mathbf{u}(t) + \mathbf{H}(t)\bar{\mathbf{w}}(t) \tag{5.10}$$

$$\mathbf{y}(t) = \mathbf{C}(t)\mathbf{x}(t) + \mathbf{D}(t)\mathbf{u}(t) + \bar{\mathbf{v}}(t) \tag{5.11}$$

where

$\mathbf{x}(t)$ represents the (m) system states,
$\mathbf{u}(t)$ represents the (q) system inputs,
$\mathbf{y}(t)$ represents the (r) system outputs,
$\mathbf{w}(t)$ represents an s-vector of Gaussian extended white noise, strength $\mathbf{Q}(t)$,
$\bar{\mathbf{v}}(t)$ represents an r-vector of Gaussian extended white noise, strength $\mathbf{R}(t)$, uncorrelated with $\bar{\mathbf{w}}(t)$,
$\mathbf{A}(t)$, $\mathbf{B}(t)$, $\mathbf{C}(t)$, $\mathbf{D}(t)$, $\mathbf{H}(t)$ are time-dependent matrices.

As explained in Section 2.10, for deductive purposes, including the derivation of Kalman–Bucy filter equations, Eqn (5.10) must be interpreted as the stochastic differential equation (Eqn (2.77)):

$$\mathrm{d}\mathbf{x}(t) = \mathbf{A}(t)\mathbf{x}(t)\,\mathrm{d}t + \mathbf{B}(t)\mathbf{u}(t)\,\mathrm{d}t + \mathbf{H}(t)\,\mathrm{d}\bar{\boldsymbol{\beta}}(t)$$

where $\bar{\boldsymbol{\beta}}(t)$ is an m-vector of extended Wiener processes, diffusion $\mathbf{Q}(t)$.

The Kalman–Bucy filter is an asymptotic state estimator with its time-variant gain $\mathbf{M}(t)$ selected in such a way that it minimizes the scalar measure of the filter accuracy:

$$\operatorname{trace}(\mathbf{G}(t)) = \varepsilon[(x_1(t) - \hat{x}_1(t))^2 + (x_2(t) - \hat{x}_2(t))^2 + \cdots + (x_m(t) - \hat{x}_m(t))^2 \tag{5.12}$$

where

$$\begin{aligned}\mathbf{G}(t) &= \varepsilon[(\mathbf{x}(t) - \hat{\mathbf{x}}(t))(\mathbf{x}(t) - \hat{\mathbf{x}}(t))^T] \\ &= \operatorname{cov}(\mathbf{x}(t) - \hat{\mathbf{x}}(t))|_{\tau=0}\end{aligned} \tag{5.13}$$

$\mathbf{G}(t)$, slightly loosely called the 'error covariance matrix', is positive definite and symmetrical.

If $\bar{\mathbf{w}}(t)$ and $\bar{\mathbf{v}}(t)$ are uncorrelated, $\mathbf{G}(t)$ can be shown to be positive definite and symmetrical. Furthermore, if S is observable (Appendix A, Section A2.3), if $\mathbf{Q}(t)$ is positive semidefinite (implying that at least some states are corrupted with process noise), and if $\mathbf{R}(t)$ is positive definite (implying that all outputs are corrupted with measurement noise), then (trace $\mathbf{G}(t)$) is minimized if $\mathbf{G}(t)$ satisfies the matrix Riccati equation:

$$\dot{\mathbf{G}}(t) = \mathbf{A}(t)\mathbf{G}(t) + \mathbf{G}(t)\mathbf{A}^T(t) + \mathbf{H}(t)\mathbf{Q}(t)\mathbf{H}^T(t) - \mathbf{G}(t)\mathbf{C}^T(t)\mathbf{R}^{-1}(t)\mathbf{C}(t)\mathbf{G}(t) \tag{5.14}$$

The fact that $\mathbf{R}(t)$ is positive definite ensures that $\mathbf{R}^{-1}(t)$ exists, and guarantees that $\mathbf{G}(t)$ converges to a unique solution (after an initial transient) from any initial condition:

$$\mathbf{G}(0) = \varepsilon[(\mathbf{x}(0) - \hat{\mathbf{x}}(0))(\mathbf{x}(0) - \hat{\mathbf{x}}(0))^T]$$

The estimator gain matrix is

$$\mathbf{M}(t) = \mathbf{G}(t)\mathbf{C}^T(t)\mathbf{R}^{-1}(t) \tag{5.15}$$

The Kalman–Bucy filter thus consists of a network (or computer algorithm) which solves Eqn (5.14), calculates $\mathbf{M}(t)$ from Eqn (5.15) and uses this result in the estimator:

$$\dot{\hat{\mathbf{x}}}(t) = (\mathbf{A}(t) - \mathbf{M}(t)\mathbf{C}(t))\hat{\mathbf{x}}(t) + (\mathbf{B}(t) - \mathbf{M}(t)\mathbf{D}(t))\mathbf{u}(t) + \mathbf{M}(t)\mathbf{y}(t) \tag{5.16}$$

This is shown in Figure 5.6.

All this is difficult to achieve by analog methods – not least because Eqn (5.14) is non-linear. Realization using numerical methods is always possible (Section 4.4), but this involves the use of an on-line digital computer, and in these circumstances it is far more straightforward to cast the original system model in discrete time form (Sections 3.8–3.10), and design a filter directly in the discrete time domain. This yields the Kalman (as opposed to the Kalman–Bucy) filter [ref. 4], which is described in the bulk of this chapter.

An easy realization of the Kalman–Bucy filter is, however, for the linear time-invariant system, where the matrices of the model, including the process and measurement noise strengths $\mathbf{Q}$, $\mathbf{R}$, are constant (i.e. $\mathbf{w}$, $\mathbf{v}$ are pure (not extended) white noise).

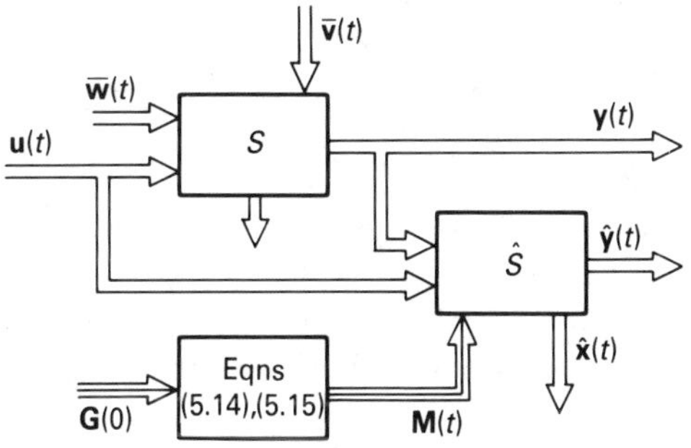

Figure 5.6 Linear stochastic system and Kalman–Bucy filter.

In this case the filter devolves into an asymptotic state estimator (Eqn (5.3)) whose gain matrix is

$$\mathbf{M} = \mathbf{G}(\infty)\mathbf{C}^T\mathbf{R}^{-1} \tag{5.17}$$

where $\mathbf{G}(\infty)$ is the constant value to which $\mathbf{G}(t)$ converges (Eqn (5.14)), $t \to \infty$:

$$\dot{\mathbf{G}}(\infty) = 0 = \mathbf{A}\mathbf{G}(\infty) + \mathbf{G}(\infty)\mathbf{A}^T + \mathbf{H}\mathbf{Q}\mathbf{H}^T - \mathbf{G}(\infty)\mathbf{C}^T\mathbf{R}^{-1}\mathbf{C}\mathbf{G}(\infty) \tag{5.18}$$

A state estimator designed in this way yields $\mathbf{x}(t)$ such that trace $\mathbf{G}(\infty)$ is minimized. The estimator, or filter, is thus optimal except during an initial transient at switch-on: strictly, the filter is 'suboptimal'.

Example

An example is provided by the experimental LTI system considered in Section 5.2, extended to include process and measurement noise, as illustrated in Figure 5.7(a). In this case the flywheel is subjected to coloured process noise via a torque servo and belt drive, and the flywheel position output is corrupted by measurement noise.

The task of the Kalman–Bucy filter is to estimate the system states, in this case the position, velocity, and acceleration of the flywheel, given the deterministic input, $u(t)$, the noise corrupted output, $y(t)$, and the strengths of the process and measurement noise signals. Ideally, the estimates would fully reflect the effects of $w(t)$, and altogether ignore the effects of $v(t)$, but this is clearly impossible and the filter is merely optimal in the sense that trace $\mathbf{G}(t)$ (Eqn (5.12)) is minimized.

The heuristic system model equations are:

$$\dot{\mathbf{x}} = \begin{pmatrix} 0 & 10.00 & 0 & 0 \\ 0 & 0 & 10.00 & 0 \\ -10.10 & -12.10 & -13.00 & 5.00 \\ 0 & 0 & 0 & -5.00 \end{pmatrix}\mathbf{x} + \begin{pmatrix} 0 \\ 0 \\ 10.10 \\ 0 \end{pmatrix} u + \begin{pmatrix} 0 \\ 0 \\ 5 \\ 5 \end{pmatrix} w \tag{5.19}$$

$$y = (1 \quad 0 \quad 0 \quad 0)\mathbf{x} + v \tag{5.20}$$

where

u, $\mathbf{x}$, are as in Eqns (5.4) and (5.5),
w is the input voltage to the torque servo (volts),
v is the measurement noise (volts), and
y is the noise corrupted potentiometer output (volts),
w, v, have strengths: $Q = 0.01$ (volts)^2, $R = 0.01$ (volts)^2.

The suboptimal filters is designed by solving Eqn (5.14) for $\mathbf{G}(\infty)$, checking that the solution satisfies Eqn (5.18), finding $\mathbf{M}$ from Eqn (5.17), and using this in Eqn (5.16)

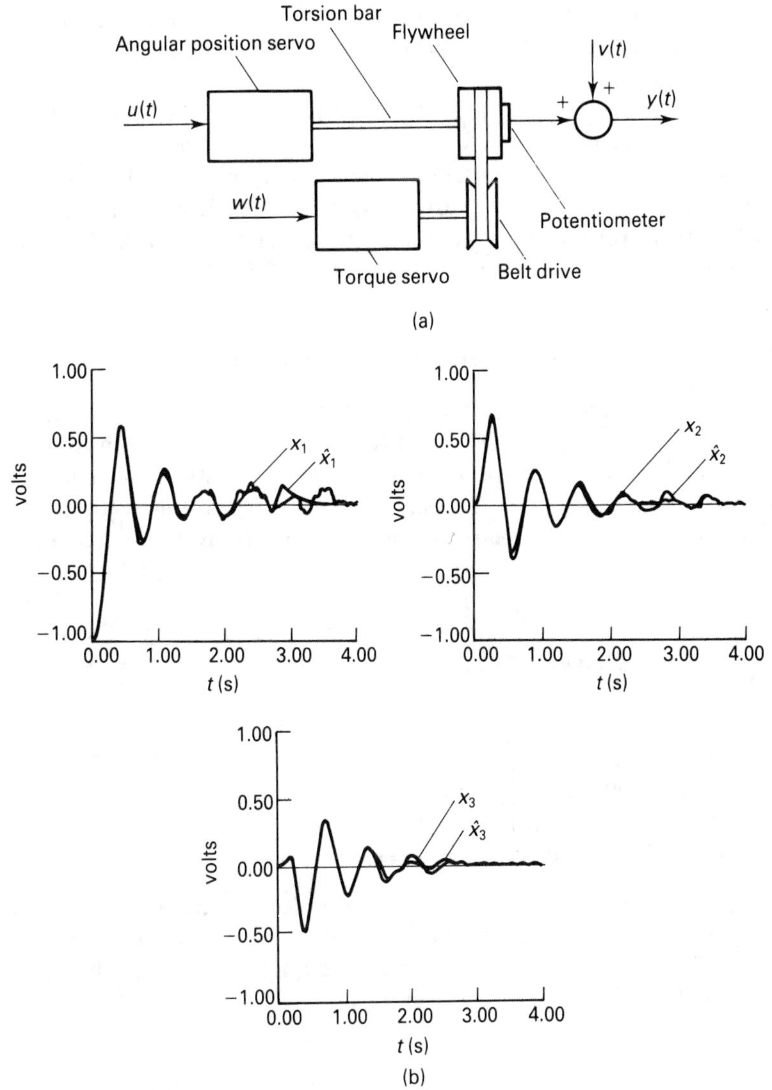

Figure 5.7 Experimental stochastic system and suboptimal Kalman filter.

(i.e. in this case, Eqn (5.3)):

$$\dot{\hat{\mathbf{x}}} = \begin{pmatrix} -16.047 & 10.000 & 0 & 0 \\ -12.876 & 0 & 10.000 & 0 \\ -12.046 & -12.100 & -13.000 & 5.000 \\ -58.301 & 0 & 0 & -5.000 \end{pmatrix} \hat{\mathbf{x}} + \begin{pmatrix} 0 \\ 0 \\ 10.100 \\ 0 \end{pmatrix} u + \begin{pmatrix} 16.047 \\ 12.876 \\ 1.946 \\ 58.301 \end{pmatrix} y$$

A realization of this has the structure shown, with suitably modified gains, in Figure 5.2(b), and some results comparing $\mathbf{x}(t)$ and $\hat{\mathbf{x}}(t)$ for a step function input are shown in Figure 5.7(b).

This technique deals with the specific case where $\mathbf{w}(t)$, $\mathbf{v}(t)$ are white uncorrelated Gaussian noise vectors, with at least some elements of $\mathbf{w}(t)$, and all elements of $\mathbf{v}(t)$ non-zero. Variations of the method are available for cases where $\mathbf{w}(t)$, $\mathbf{v}(t)$ are not white, where they are correlated, or where elements of $\mathbf{v}(t)$ are zero, but since practical realizations of such strategies are almost invariably in the discrete time domain, these topics are explored in Sections 5.5–5.9.

5.4 State estimators for deterministic systems – discrete time

As explained in Section 5.3, Kalman filters are generally realized in on-line digital computers, usually microprocessors, and their design is therefore most conveniently based on discrete time stochastic models. The deterministic state estimators which provide the structures for Kalman filters are considered in this section.

The behaviour of a linear time-invariant deterministic discrete system, Σ, is modelled by Eqns (3.34) and (3.35)):

$$\mathbf{x}(n+1) = \varphi\mathbf{x}(n) + \psi\mathbf{u}(n)$$

$$\mathbf{y}(n) = \mathbf{C}_d\mathbf{x}(n) + \mathbf{D}_d\mathbf{u}(n)$$

where

$\mathbf{x}(n)$ represents the (m) states,
$\mathbf{u}(n)$ represents the (q) inputs,
$\mathbf{y}(n)$ represents the (r) outputs,
φ, ψ, $\mathbf{C}_d$, $\mathbf{D}_d$ are constant matrices.

As in the continuous time case, a state estimator, $\hat{\Sigma}$, can readily be constructed, this time in algorithm form, using $\mathbf{u}(n)$, $\mathbf{y}(n)$ as data, as shown in Figure 5.8.

The design procedures for asymptotic estimators and Luenberger observers closely follow those for continuous time systems.

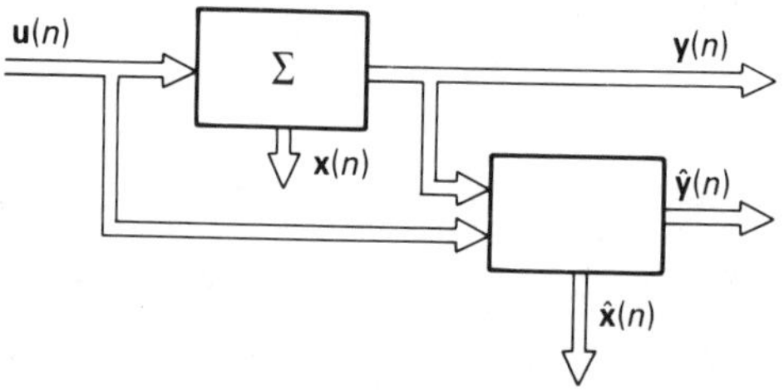

Figure 5.8 Discrete time state estimator.

5.4.1 The asymptotic state estimator

The model equations of the asymptotic state estimator are (cf. Eqn (5.3)):

$$\hat{\mathbf{x}}(n+1) = (\boldsymbol{\varphi} - \mathbf{M}\mathbf{C}_d)\hat{\mathbf{x}}(n) + (\boldsymbol{\psi} - \mathbf{M}\mathbf{D}_d)\mathbf{u}(n) + \mathbf{M}\mathbf{y}(n) \tag{5.21}$$

where

$\hat{\mathbf{x}}(n)$ represents the (m) estimated states, and
$\mathbf{M}$ is the $(m \times r)$ estimator gain matrix.

The behaviour of the estimator error $(\mathbf{x}(n) - \hat{\mathbf{x}}(n))$ is governed by

$$(\mathbf{x}(n+1) - \hat{\mathbf{x}}(n+1)) = (\boldsymbol{\varphi} - \mathbf{M}\mathbf{C}_d)(\mathbf{x}(n) - \hat{\mathbf{x}}(n))$$

Consequently (Appendix A, Section A3), the estimator design consists of selecting $\mathbf{M}$ so that $(\boldsymbol{\varphi} - \mathbf{M}\mathbf{C}_d)$ has eigenvalues which are stable and 'fast' compared with those of $\boldsymbol{\varphi}$.

Example

Consider the discrete time (sampled data) system consisting of a DAC and ADC and the continuous time system, *S*, of Figure 5.2(a), as shown in Figure 5.9(a). The sample time of the converters is 0.050 s.

The continuous time model (Eqns (5.4) and (5.5)) and the continuous–discrete time conversion equations (Appendix A, Eqns (A3.20)–(A3.23)) yield

$$\mathbf{x}(n+1) = \begin{pmatrix} 0.982 & 0.476 & 0.099 \\ -0.100 & 0.862 & 0.347 \\ -0.351 & -0.521 & 0.410 \end{pmatrix}\mathbf{x}(n) + \begin{pmatrix} 0.018 \\ 0.100 \\ 0.351 \end{pmatrix}\mathbf{u}(n) \tag{5.22}$$

$$y(n) = (1 \quad 0 \quad 0)\mathbf{x}(n) \tag{5.23}$$

The eigenvalues of $\boldsymbol{\varphi}$ are

$$\lambda = 0.833 + \mathrm{j}0.438, \quad \lambda = 0.833 - \mathrm{j}0.438, \quad \lambda = 0.589$$

The matrix $\mathbf{M}$ is chosen so that the eigenvalues of $(\boldsymbol{\varphi} - \mathbf{M}\mathbf{C}_d)$ are 'fast', compared with these. Repeating the design procedure of Section 5.2.1 (best performed using a CAD facility) yields the following.

For eigenvalues of $(\boldsymbol{\varphi} - \mathbf{M}\mathbf{C}_d) = \lambda = 0.1 + \mathrm{j}0.1, \lambda = 0.1 - \mathrm{j}0.1, \lambda = 0.1$,

$$\mathbf{M} = \begin{pmatrix} 1.954 \\ 1.661 \\ -1.424 \end{pmatrix} (\boldsymbol{\varphi} - \mathbf{M}\mathbf{C}_d) = \begin{pmatrix} -0.972 & 0.476 & 0.099 \\ -1.761 & 0.862 & 0.347 \\ 1.073 & -0.521 & 0.410 \end{pmatrix}$$

The estimator algorithm (Eqn (5.21)) is thus

$$\hat{\mathbf{x}}(n+1) = \begin{pmatrix} -0.972 & 0.476 & 0.099 \\ -1.761 & 0.862 & 0.347 \\ 1.073 & -0.521 & 0.410 \end{pmatrix} \hat{\mathbf{x}}(n) + \begin{pmatrix} 0.018 \\ 0.100 \\ 0.351 \end{pmatrix} u(n) + \begin{pmatrix} 1.954 \\ 1.661 \\ -1.421 \end{pmatrix} y(n)$$

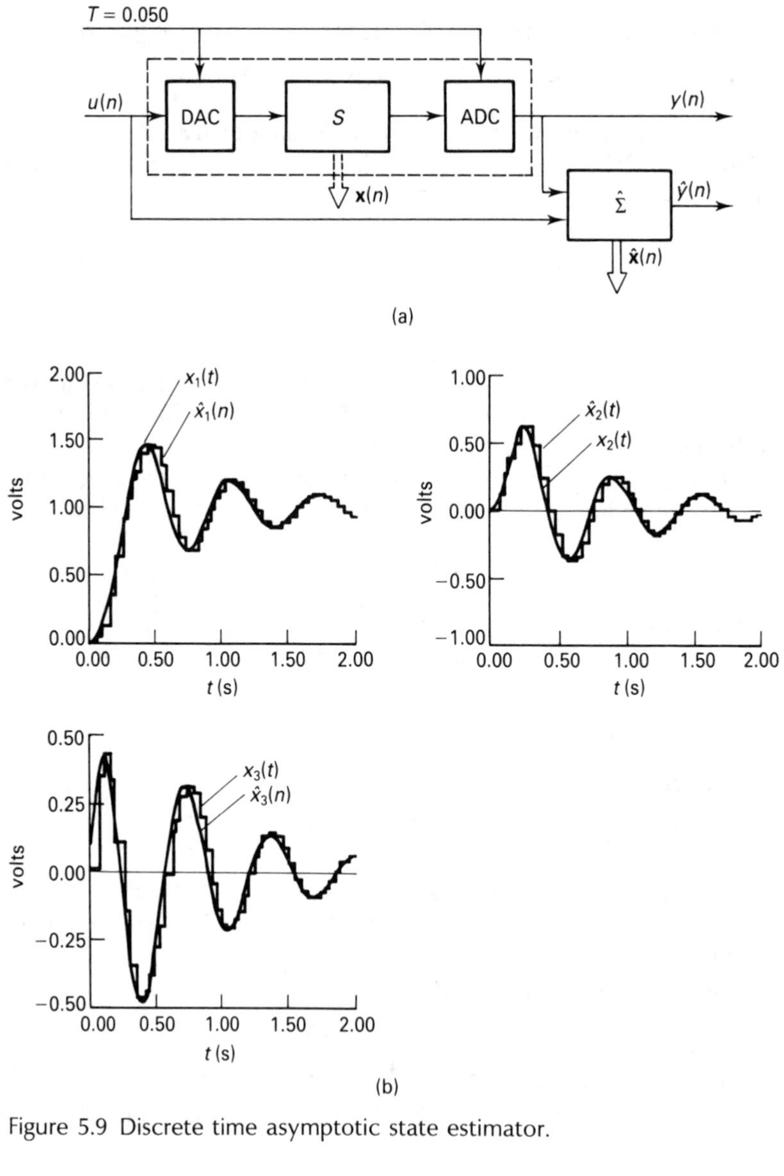

Figure 5.9 Discrete time asymptotic state estimator.

Some waveforms demonstrating the behaviour of this estimator are shown in Figure 5.9(b). The estimated states, at the sampling instants, are accurate, and compare well with those generated by the continuous time estimator in Figure 5.3.

5.4.2 The discrete time Luenberger observer

Since the algebraic structures of the continuous and discrete time models are identical, a discrete time Luenberger observer can be designed by following the procedure outlined in Section 5.2.2, with, of course, eigenvalues appropriate to the discrete time domain. The observer structure provides the basis for certain types of Kalman filter and this, rather than the small economy of software it offers compared with the asymptotic state estimator, makes it valuable.

The Luenberger observer equations, or algorithm, have the structure of Eqns (5.6) and (5.7):

$$\boldsymbol{\eta}(n+1) = \mathbf{L}\boldsymbol{\eta}(n) + (\mathbf{G} - \mathbf{H}\mathbf{D}_d)\mathbf{u}(n) + \mathbf{H}\mathbf{y}(n) \tag{5.24}$$

$$\hat{\mathbf{x}}(n) = \mathbf{N}\eta(n) - \mathbf{M}\mathbf{D}_d\mathbf{u}(n) + \mathbf{M}\mathbf{y}(n) \tag{5.25}$$

where **L**, **G**, **H**, **N**, **M** are designed as in Section 5.2.2, **A**, **B**, **C**, **D** being replaced with $\boldsymbol{\varphi}$, $\boldsymbol{\psi}$, $\mathbf{C}_d$, $\mathbf{D}_d$.

5.5 The Kalman filter

The Kalman filter is a discrete time state estimator designed for linear time-variant stochastic systems, with the structure of the asymptotic estimator, or, in certain cases, the Luenberger observer.

A discrete time-variant stochastic system is modelled by Eqns (3.36) and (3.37):

$$\mathbf{x}(n+1) = \boldsymbol{\varphi}(n)\mathbf{x}(n) + \boldsymbol{\psi}(n)\mathbf{u}(n) + \boldsymbol{\Gamma}(n)\bar{\mathbf{w}}(n) \tag{5.26}$$

$$\mathbf{y}(n) = \mathbf{C}_d(n)\mathbf{x}(n) + \mathbf{D}_d(n)\mathbf{u}(n) + \bar{\mathbf{v}}(n) \tag{5.27}$$

The strengths of $\bar{\mathbf{w}}(n)$, $\bar{\mathbf{v}}(n)$, which are specified as uncorrelated, are $\mathbf{Q}_d(n)$, $\mathbf{R}_d(n)$ respectively.

A positive-definite, symmetrical 'error covariance matrix' is defined (rather loosely, cf. Eqn (5.13)):

$$\mathbf{G}(n) = \varepsilon[(\mathbf{x}(n) - \hat{\mathbf{x}}(n))(\mathbf{x}(n) - \hat{\mathbf{x}}(n))^T) \tag{5.28}$$

A scalar measure of the estimator accuracy is given by

$$\begin{aligned}\text{trace}(\mathbf{G}(n)) &= \varepsilon[(x_1(n) - \hat{x}_1(n))^2 + (x_2(n) - \hat{x}_2(n))^2 + \cdots (x_m(n) - \hat{x}_m(n))^2] \\ &= \varepsilon[(\mathbf{x}(n) - \hat{\mathbf{x}}(n))^T(\mathbf{x}(n) - \hat{\mathbf{x}}(n))]\end{aligned} \tag{5.29}$$

The Kalman filter equation is (cf. Eqn (5.21))

$$\hat{\mathbf{x}}(n+1) = (\varphi(n) - \mathbf{M}(n)\mathbf{C}_d(n))\hat{\mathbf{x}}(n) + (\psi(n) - \mathbf{M}(n)\mathbf{D}_d(n))\mathbf{u}(n) + \mathbf{M}(n)\mathbf{y}(n) \tag{5.30}$$

where the $(m \times r)$ gain matrix $\mathbf{M}(n)$, which minimizes trace $(\mathbf{G}(n))$ is calculated by the following routine. Let:

$$\mathbf{P}(n) = \mathbf{R}_d(n) + \mathbf{C}_d(n)\mathbf{G}(n)\mathbf{C}_d^T(n) \tag{5.31}$$

Then

$$\mathbf{M}(n) = \varphi(n)\mathbf{G}(n)\mathbf{C}_d^T(n)\mathbf{P}^{-1}(n) \tag{5.32}$$

$$\mathbf{G}(n+1) = (\varphi(n) - \mathbf{M}(n)C_d(n)\mathbf{G}(n)\varphi^T(n) + \boldsymbol{\Gamma}(n)\mathbf{Q}_d(n)\boldsymbol{\Gamma}^T(n) \tag{5.33}$$

The filter algorithm thus consists of Eqns (5.31), (5.32), (5.30) and (5.33), executed in that order, repeatedly, once per sampling period.

Formally, an initial value $\mathbf{G}(0)$ is required for the solution of these equations, but convergence to the correct $\mathbf{G}(n)$, after an initial transient, is guaranteed for any $\mathbf{G}(0)$. Thus, for convenience, $\mathbf{G}(0)$ is often set to (0), which is usually incorrect, since, almost inevitably, $\hat{\mathbf{x}}(0) \neq \mathbf{x}(0)$, and the filter generates an initial transient after switch-on, before settling down to steady state, optimal operation. Naturally, if it can be arranged that $\hat{\mathbf{x}}(0) = \mathbf{x}(0)$, i.e. if $\mathbf{G}(0) = (0)$, this transient is avoided, but since $\mathbf{x}(0)$ is usually unknown, this is seldom practical.

A discrete time system and Kalman filter are shown in Figure 5.10. The algorithm simplifies considerably for the linear time-invariant case (Eqns (3.38) and (3.39)), $\mathbf{Q}_d$, $\mathbf{R}_d$ constant:

$$\hat{\mathbf{x}}(n+1) = (\varphi - \mathbf{M}(n)\mathbf{C}_d)\hat{\mathbf{x}}(n) + (\psi - \mathbf{M}(n)\mathbf{D}_d)\mathbf{u}(n) + \mathbf{M}(n)\mathbf{y}(n) \tag{5.34}$$

$$\mathbf{P}(n) = \mathbf{R}_d + \mathbf{C}_d\mathbf{G}(n)\mathbf{C}_d^T \tag{5.35}$$

$$\mathbf{M}(n) = \varphi\mathbf{G}(n)\mathbf{C}_d^T\mathbf{P}^{-1}(n) \tag{5.36}$$

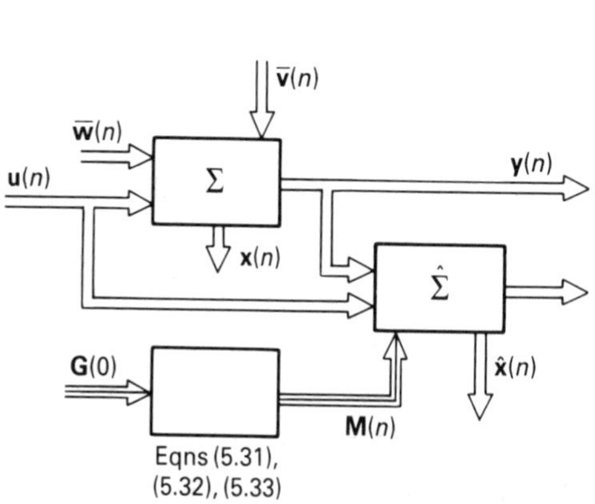

Figure 5.10 Discrete time system and Kalman filter.

$$\mathbf{G}(n+1) = (\boldsymbol{\varphi} - \mathbf{M}(n)\mathbf{C}_d)\mathbf{G}(n)\boldsymbol{\varphi}^T + \boldsymbol{\Gamma}\mathbf{Q}_d\boldsymbol{\Gamma}^T \tag{5.37}$$

The filter algorithm here consists of Eqns (5.35), (5.36), (5.34) and (5.37) in that order. After an initial transient, $\mathbf{G}(n)$ and $\mathbf{M}(n)$ settle down to constant matrices $\mathbf{G}(\infty)$, $\mathbf{M}(\infty)$ and the algorithm becomes a state estimator consisting of Eqn (5.34), $\mathbf{M}(n) = \mathbf{M}(\infty)$.

Since $\mathbf{G}(\infty)$, $\mathbf{M}(\infty)$ are independent of $\mathbf{G}(0)$, it is usually acceptable to precalculate them off-line and construct a suboptimal filter using only Eqn (5.34), $\mathbf{M}(n) = \mathbf{M}(\infty)$. This is, of course, simply a state estimator with a gain selected by the Kalman filter algorithm. Its performance is suboptimal during the initial transient, but is subsequently identical to that of the optimal filter.

Example

Consider the linear time-invariant system of Figure 5.7, buffered by DACs and ADCs, as indicated in Figure 5.11(a). The 'virtual' DACs and ADCs, indicated by dotted outlines, serve to define $\mathbf{w}(n)$, $\mathbf{v}(n)$ in such a way that their effects on Σ are identical (at sampling instants) to those of the continuous time noise processes $\mathbf{w}(t)$, $\mathbf{v}(t)$. S is modelled by Eqns (5.19) and (5.20), $w(t)$, $v(t)$ uncorrelated, with strengths

$$Q = 0.01, \quad R = 0.01$$

For a sampling period $T = 0.050$ s, the system is modelled by (Eqns (3.38) and (3.39)):

$$\mathbf{x}(n+1) = \begin{pmatrix} 0.982 & 0.476 & 0.099 & 0.008 \\ -0.100 & 0.862 & 0.347 & 0.046 \\ -0.351 & -0.521 & 0.410 & 0.151 \\ 0 & 0 & 0 & 0.779 \end{pmatrix} \mathbf{x}(n) + \begin{pmatrix} 0.018 \\ 0.100 \\ 0.351 \\ 0 \end{pmatrix} u(n) + \begin{pmatrix} 0.009 \\ 0.054 \\ 0.196 \\ 0.221 \end{pmatrix} w(n) \tag{5.38}$$

$$y(n) = (1 \quad 0 \quad 0 \quad 0)\mathbf{x}(n) + v(n) \tag{5.39}$$

where the strengths of $w(n)$, $v(n)$ are

$$Q_d = 5 \times 10^{-4}, \quad R_d = 10^{-2}$$

The sequence of Eqns (5.35), (5.36) and (5.37) repeated about twenty times yields

$$\mathbf{M} = \mathbf{M}(\infty) = \begin{pmatrix} 0.0131 \\ 0.0002 \\ -0.0084 \\ 0.0044 \end{pmatrix}$$

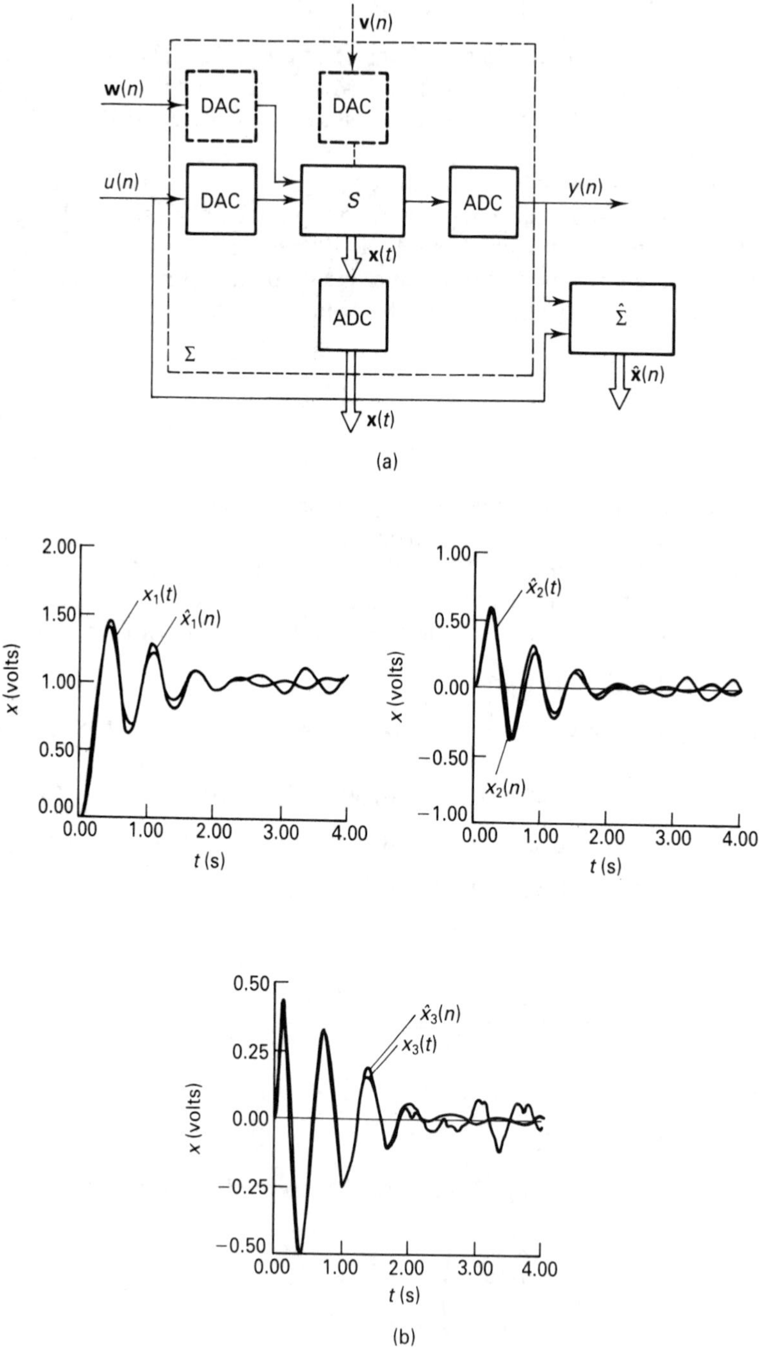

Figure 5.11 Discrete LTI stochastic system and Kalman filter.

This yields the suboptimal filter algorithm (Eqn (5.34), $\mathbf{M}(n) = \mathbf{M}(\infty)$):

$$\hat{\mathbf{x}}(n+1) = \begin{pmatrix} 0.969 & 0.476 & 0.099 & 0.008 \\ -0.100 & 0.862 & 0.347 & 0.046 \\ -0.343 & -0.521 & 0.410 & 0.151 \\ -0.004 & 0 & 0 & 0.779 \end{pmatrix} \mathbf{x}(n) + \begin{pmatrix} 0.018 \\ 0.100 \\ 0.351 \\ 0 \end{pmatrix} u(n) + \begin{pmatrix} 0.0131 \\ 0.0002 \\ -0.0084 \\ 0.0044 \end{pmatrix} y(n)$$

Some step function responses for the system states and estimates using this algorithm are shown in Figure 5.11(b), where it can be seen that the filter detects the effects of the process noise and attenuates those of the measurement noise.

The values of $\mathbf{Q}_d, \mathbf{R}_d$ determine $\mathbf{M}(\infty)$ and, via this, the eigenvalues of $(\varphi - \mathbf{MC}_d)$. These in turn determine the behaviour of the estimator error $(\mathbf{x}(n) - \hat{\mathbf{x}}(n))$.

Large $\mathbf{Q}_d$ values encourage the filter to follow the states closely but allow more measurement noise through; large $\mathbf{R}_d$ values encourage the filter to suppress measurement noise, but impair its ability to follow state changes closely. The optimal values of $\mathbf{Q}_d, \mathbf{R}_d$ usually yield reasonable results, but there can be a distinct advantage in tuning the Kalman filter by altering $\mathbf{Q}_d$, $\mathbf{R}_d$ empirically – after all, the optimal $\mathbf{Q}_d$, $\mathbf{R}_d$ simply specify a filter which minimizes a rather arbitrary (but mathematically tractable) measure of quality (Eqn 5.29).

It is, of course, easiest to alter the diagonal elements of $\mathbf{Q}_d$, $\mathbf{R}_d$ empirically and to test the effects, usually in simulation, but it can prove very advantageous to tune the off-diagonal elements – which imply cross-correlation between the elements within the two noise vectors. However, this can be time consuming and difficult.

In Figure 5.11(b), the estimates are reasonable where $u(n)$ is the dominant input, but poor where $w(n)$ become dominant ($t > 2$). Arbitrarily redesigning the filter by setting $\mathbf{Q}_d = 1$ forces it to take more account of the effects of $w(n)$ generating results which are in many respects better than those of the optimal filter.

5.6 Variations on the Kalman filter [refs 5, 6, 10]

Variations on the algorithms of Section 5.5 deal with practical deviations from ideal conditions, such as non-white noise, correlation between process and measurement noise, and absence of measurement noise on some system outputs. These are considered in this section.

5.6.1 Systems with non-white process noise

In many problems the process noise cannot reasonably be regarded as white, and it may be necessary to incorporate a shaping filter in the state equations to model this (Section 3.10). This can apply to systems where noise inputs are quite unspecified and are simply regarded as non-white noise for want of any attractive alternative. Many target tracking problems fall into this category.

Example

A laboratory equipment, which simulates an airborne target and a tracker, consists of an array of randomly illuminated target lights mounted on a vertical rotatable arm situated about 2 metres from a target tracker, which carries a lens, photocell array, appropriate electronics and elevation and azimuth drive servos (Figure 5.12(a)). The tracker directionally locks onto the dynamically varying 'centre of gravity' of the illuminated target lights. The target is driven manually and its position and velocity are estimated by a Kalman filter realized in a microprocessor attached to the tracker. The target, position and velocity are also directly measured by a potentiometer and

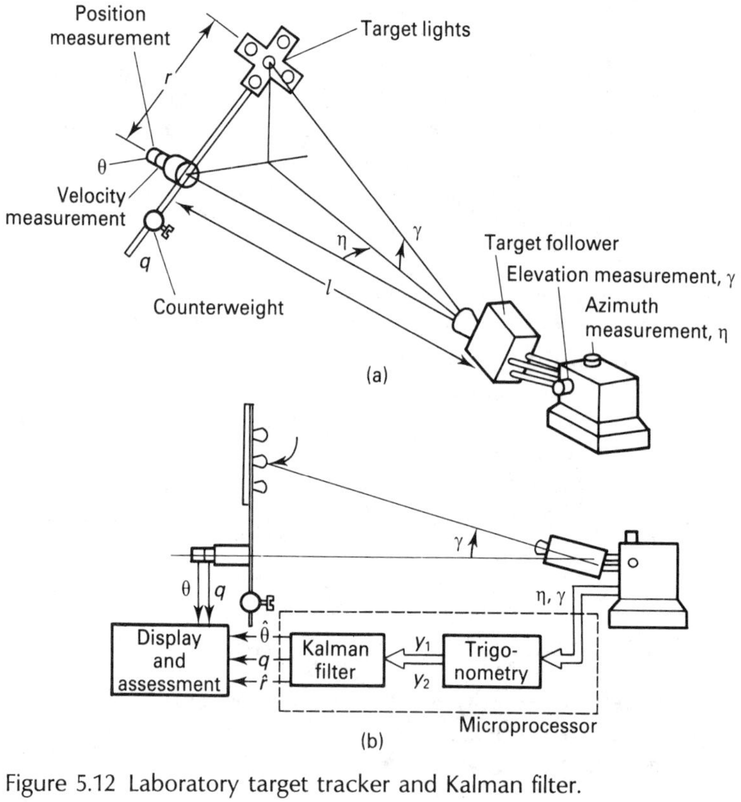

Figure 5.12 Laboratory target tracker and Kalman filter.

tachogenerator so that the Kalman filter performance can be assessed. The manual drive is quite unspecified, and is regarded as coloured (non-white) noise in the Kalman filter design for want of any easy alternative.

The tracker elevation and azimuth are converted to target angular displacement and radius:

$$y_1 = \theta + v_\theta = \tan^{-1}\left(\frac{\sin\eta}{\tan\gamma}\right)$$

$$y_2 = r + v_r = l\left(\tan^2\eta + \frac{\tan^2\gamma}{\cos^2\eta}\right)$$

where

θ is the target angular position (rad),
r is the target radial position (m),
v_θ is the target angular position measurement noise (rad),
v_r is the target position measurement noise (m),
y_1, y_2 are the measured target angular and radial positions measured by the tracker (rad, m),
η, γ are the target azimuth and elevation (rad),
l is the target-tracker axial displacement (m).

The target manual drive is modelled – rather arbitrarily – as Gaussian white noise seen through a first order filter (Section 2.11) of transfer function·

$$F(s) = \frac{1}{1+s}$$

Together with the target equations of motion, this gives an augmented heuristic state space model:

$$\begin{pmatrix} \dot{\theta} \\ \dot{q} \\ \dot{r} \\ \dot{w}_f \end{pmatrix} = \begin{pmatrix} 0 & 1.00 & 0 & 0 \\ 0 & -0.07 & 0 & 1.00 \\ 0 & 0 & 0 & 0 \\ 0 & 0 & 0 & -1.00 \end{pmatrix} \begin{pmatrix} \theta \\ q \\ r \\ w_f \end{pmatrix} + \begin{pmatrix} 0 \\ 0 \\ 0 \\ 1.00 \end{pmatrix} w \tag{5.40}$$

$$\mathbf{y} = \begin{pmatrix} 1 & 0 & 0 & 0 \\ 0 & 0 & 1 & 0 \end{pmatrix} \begin{pmatrix} \theta \\ q \\ r \\ w_f \end{pmatrix} + \begin{pmatrix} v_\theta \\ v_r \end{pmatrix} \tag{5.41}$$

where

q is the target angular velocity (rad s^{-1}),
w_f is the filtered noise (rad s^{-2}),
w is Gaussian white noise (rad s^{-2}) whose strength is assessed arbitrarily from the kind of manual input to be expected, say $Q = 2.5$.

The measurement noise strength, calculated from the dimensions of the target array of lights, is

$$\mathbf{R} = \begin{pmatrix} 0.01 & 0 \\ 0 & 0.01 \end{pmatrix}$$

The first step is to synthesize the equivalent discrete time model (Eqns (3.52)–(3.56)). A sampling time $T = 0.01$ is chosen:

$$\mathbf{x}(n+1) = \begin{pmatrix} 1.00 & 0.10 & 0 & 0 \\ 0 & 1.00 & 0 & 0.10 \\ 0 & 0 & 1.00 & 0 \\ 0 & 0 & 0 & 0.90 \end{pmatrix} \mathbf{x}(n) + \begin{pmatrix} 0 \\ 0 \\ 0 \\ 0.01 \end{pmatrix} w(n)$$

$$\mathbf{y}(n) = \begin{pmatrix} 1 & 0 & 0 & 0 \\ 0 & 0 & 1 & 0 \end{pmatrix} \mathbf{x}(n) + \begin{pmatrix} v_\theta(n) \\ v_r(n) \end{pmatrix}$$

where

$$\mathbf{x}(n) = (\theta(n) \quad q(n) \quad r(n) \quad w_f(n))^T$$

$$Q_d \doteqdot QT = 0.025$$

$$\mathbf{R}_d = \mathbf{R} = \begin{pmatrix} 0.01 & 0 \\ 0 & 0.01 \end{pmatrix}$$

Eqns (5.35)–(5.37) yield (any $\mathbf{G}(0)$)

$$\mathbf{M}(\infty) = \begin{pmatrix} 0.41 & 0 \\ 0.73 & 0 \\ 0 & 0.77 \\ 0.53 & 0 \end{pmatrix}$$

The suboptimal Kalman filter is (Eqn (5.34))

$$\hat{\mathbf{x}}(n+1) = \begin{pmatrix} 0.59 & 0.10 & 0 & 0 \\ -0.73 & 1.00 & 0 & 0.10 \\ 0 & 0 & 0.23 & 0 \\ -0.53 & 0 & 0 & 0.90 \end{pmatrix} \mathbf{x}(n) + \begin{pmatrix} 0.41 & 0 \\ 0.73 & 0 \\ 0 & 0.77 \\ 0.53 & 0 \end{pmatrix} \mathbf{y}(n)$$

Responses of the system to a manually applied target input are shown in Figure 5.13, where it can be seen that good target state estimates are generated.

Figure 5.13 also shows the noise corruption of $\theta(t)$, $r(t)$ to give $y_1(t)$, $y_2(t)$. The effect of this on $\hat{x}_3(n)$ is considerable, on $\hat{x}_2(n)$ somewhat less, and on $\hat{x}_1(n)$ almost negligible.

Increasing the elements of $\mathbf{R}_d$ arbitrarily and redesigning the filter accordingly reduces the effect of v_θ, v_r, but impairs the capacity of the filter to follow state changes rapidly. Increasing elements in $\mathbf{Q}_d$ has the opposite effects. Experimentation along these lines can prove very worthwhile.

5.6.2 Systems with zero measurement noise vector elements

A mathematical requirement for the convergence of the Kalman filter algorithm (Eqns (5.31)–(5.33)) is that $\mathbf{R}_d$ should be positive definite, which implies that no element

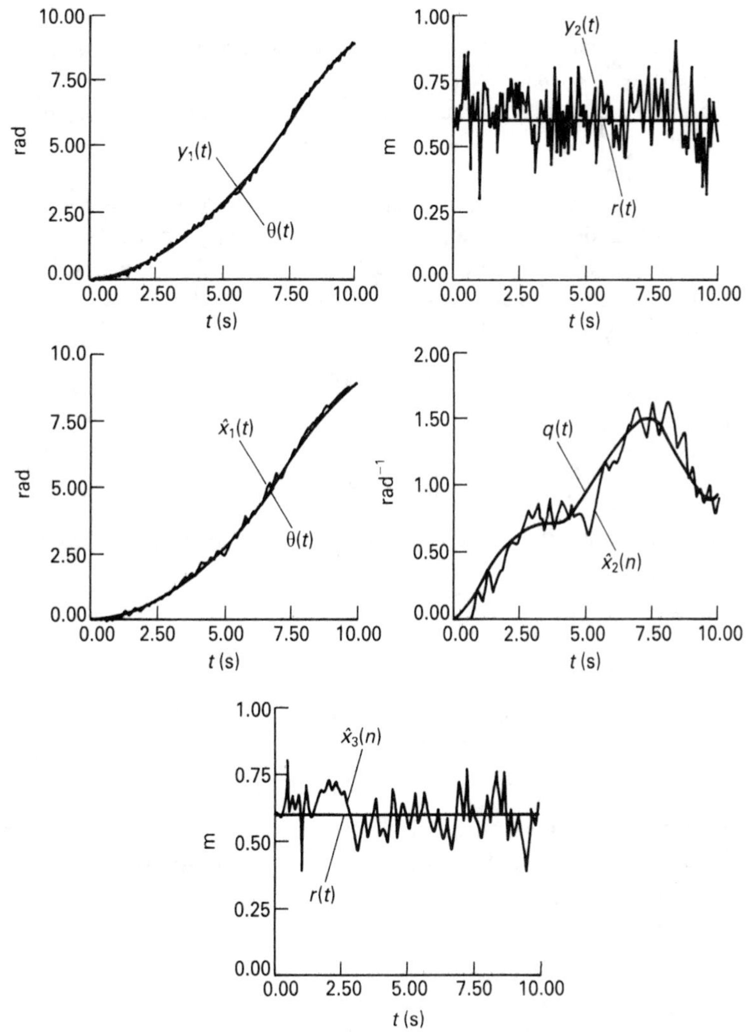

Figure 5.13 Kalman filter results.

of the measurement noise vector, $\mathbf{v}(n)$ (Eqn (5.27)), is zero. This condition is not infrequently breached. Kalman filters can be designed for some such cases using the Luenberger observer structure, though not for those where every measurement noise element is zero.

Consider a linear time-variant system (Eqns (5.26) and (5.27)) where $\mathbf{v}(n)$ has zero elements. Eqn (5.27) is partitioned:

$$\mathbf{y}(n) = \begin{pmatrix} \mathbf{y}_1(n) \\ \mathbf{y}_2(n) \end{pmatrix} = \begin{pmatrix} \mathbf{C}_{1d}(n) \\ \mathbf{C}_{2d}(n) \end{pmatrix} \mathbf{x}(n) + \begin{pmatrix} \mathbf{D}_{1d}(n) \\ \mathbf{D}_{2d}(n) \end{pmatrix} \mathbf{u}(n) + \begin{pmatrix} \mathbf{v}(n) \\ \mathbf{0} \end{pmatrix} \tag{5.42}$$

where

$\mathbf{y}_1(n)$, an r_1-vector, represents the r_1 outputs corrupted by measurement noise $\mathbf{v}(n)$, strength $\mathbf{R}_d(n)$,
$\mathbf{y}_2(n)$, an r_2-vector, represents the r_2 outputs not corrupted by measurement noise.

The design procedure follows that of the Luenberger observer (Sections 5.2.2 and 5.4.2). The r_2 pseudo-states are calculated algebraically from $\mathbf{y}_2(n)$; the remaining $(m - r_2)$ pseudo-states are estimated by a Kalman filter; and the results of the two algorithms are combined algebraically to yield m estimated states (cf. Figure 5.4). The detailed procedure is as follows:

1. An $(m - r_2) \times m$ constant matrix, $\mathbf{T}_2$, is found such that

$$\begin{pmatrix} \mathbf{C}_{2d}(n) \\ \mathbf{T}_2 \end{pmatrix} \text{ is non-singular}$$

If it fulfils this non-singularity condition, $\mathbf{T}_2$ is most conveniently chosen:

$$\mathbf{T}_2 = \begin{pmatrix} 1 & 0 & 0 & 0 & \dots & & & 0 \\ 0 & 1 & 0 & 0 & & & & 0 \\ 0 & 0 & 1 & 0 & & & & 0 \\ \vdots & & & & & & & \vdots \\ 0 & 0 & 0 & 0 & \dots & 1 & \dots & 0 \end{pmatrix}$$

This simplifies the definition of the pseudo-states considerably. Alternatively, with some added complication, matrices $\mathbf{L}$, $\mathbf{H}$, can be specified, and $\mathbf{T}_2$ derived using Eqns (5.8) and (5.9).

2. Calculate

$$(\mathbf{M}_2(n) \,\vdots\, \mathbf{N}_2(n)) = \begin{pmatrix} \mathbf{C}_{2d}(n) \\ \text{---------} \\ \mathbf{T}_2 \end{pmatrix}^{-1} \tag{5.43}$$

where

$\mathbf{M}_2(n)$ is $(m \times r_2)$ and $\mathbf{N}_2(n)$ is $(m \times r_1)$

3. Then the Kalman filter which estimates $(m - r_2)$ pseudo-states is

$$\boldsymbol{\eta}(n+1) = [\mathbf{T}_2\boldsymbol{\varphi}(n)\mathbf{N}_2(n) - \mathbf{M}_1(n)\mathbf{C}_{1d}(n)]\boldsymbol{\eta}(n)$$
$$+ [\mathbf{T}_2\boldsymbol{\varphi}(n)(\mathbf{I} - \mathbf{M}_2(n)\mathbf{D}_2(n)) + \mathbf{M}_1(n)(\mathbf{D}_1(n) - \mathbf{C}_1(n)\mathbf{M}_2(n)\mathbf{D}_2(n)]\mathbf{u}(\mathrm{n})$$
$$+ [\mathbf{T}_2\boldsymbol{\varphi}(n)\mathbf{M}_2(n) + \mathbf{M}_1(n)\mathbf{C}_1(n)\mathbf{M}_2(n)]\mathbf{y}_2(n) + \mathbf{M}_1(n)y_1(n) \tag{5.44}$$

where $\mathbf{M}_1(n)$ is the $((m - r_2) \times r_1)$ gain matrix derived from:

$$\mathbf{P}(n) = \mathbf{R}_d(n) + \mathbf{C}_{1d}(n)\mathbf{N}_2(n)\mathbf{G}(n)\mathbf{N}_2^T(n)\mathbf{C}_{1d}^T(n) \tag{5.45}$$

$$\mathbf{M}_1(n) = \mathbf{T}_2\boldsymbol{\varphi}(n)\mathbf{N}_2(n)\mathbf{G}(n)\mathbf{C}_{1d}^T(n)\mathbf{P}^{-1}(n) \tag{5.46}$$

$$\mathbf{G}(n+1) = (\mathbf{T}_2\mathbf{N}_2(n)\boldsymbol{\varphi}(n) - \mathbf{M}_1(n)\mathbf{C}_{1d}(n)\mathbf{N}_2(n))\mathbf{G}(n)\mathbf{N}_2^T(n)\boldsymbol{\varphi}^T(n)\mathbf{T}_2^T$$
$$+ \mathbf{T}_2\boldsymbol{\Gamma}(n)\mathbf{Q}d(n)\boldsymbol{\Gamma}^T(n)\mathbf{T}_2 \tag{5.47}$$

4. The filter output is

$$\hat{\mathbf{x}}(n) = \mathbf{N}_2(n)\boldsymbol{\eta}(n) - \mathbf{M}_2(n)\mathbf{D}_{2d}(n)\mathbf{u}(n) + \mathbf{M}_2(n)\mathbf{y}(n) \tag{5.48}$$

The Kalman filter thus consists of Eqns (5.45), (5.46), (5.47), (5.44), (5.43) and (5.48) evaluated in that order, repeatedly.

Naturally, if Σ is time invariant, the filter devolves into a Luenberger observer (Eqns (5.24) and (5.25)), with gain matrices calculated off-line from Eqns (5.45)–(5.47) and Eqn (5.43).

5.6.3 Non-white measurement noise

If elements in the measurement noise vector are Gaussian, but non-white, it is usually easy to accommodate this using filter models (Section 3.10), the system state equation models being augmented accordingly. This inevitably gives rise to non-zero elements in the augmented model measurement noise vector, requiring the Kalman filter design treatment described in Section 5.6.2.

5.6.4 Systems with cross-correlation between process and measurement noise

The optimality of the algorithm of Eqns (5.31), (5.32) and (5.33) depends on there being no cross-correlation between $\bar{\mathbf{w}}(n)$, $\bar{\mathbf{v}}(n)$. Where there is such cross-correlation, an optimal algorithm can nevertheless be found, as follows.

Suppose the cross-correlation of $\boldsymbol{\Gamma}(n)\bar{\mathbf{w}}(n)$, $\bar{\mathbf{v}}(n)$ is (Section 3.7)

$$\mathbf{R}_{dwv}(k) = \varepsilon(\boldsymbol{\Gamma}(n)\bar{\mathbf{w}}(n)\bar{\mathbf{v}}^T(n+k)) = \mathbf{S}_d(n)\cdot\delta(k) \tag{5.49}$$

where

$\mathbf{S}_d(n)$ is an $(m \times r)$ matrix,

$\delta(k)$ is the Kronecker delta function (Appendix D, Section D2).

The optimal filter in this case is given by Eqns (5.31)–(5.33), altered slightly as follows:

1. Eqn 5.31 is replaced by

$$\mathbf{G}(n+1) = \varphi(n)\mathbf{G}(n)\varphi^T(n) - \varphi(n)(\mathbf{S}_d(n) + \mathbf{G}(n)\mathbf{C}_d^T(n))\mathbf{P}^{-1}(n)(\mathbf{S}_d^T(n) + \mathbf{C}_d(n)\mathbf{G}(n))\varphi^T(n) + \mathbf{\Gamma}(n)\mathbf{Q}_d(n)\mathbf{\Gamma}^T(n)$$

2. In Eqn (5.32),

$[\mathbf{G}(n)\mathbf{C}_d^T(n)]$ is replaced by $[\mathbf{S}_d(n) + \mathbf{G}(n)\mathbf{C}_d^T(n)]$.

This also applies to the time-invariant case (Eqns (5.34)–(5.37)), and to its associated suboptimal filter.

In practice, at least in the time-invariant case, it can be better simply to ignore cross-correlation between $\mathbf{w}(n)$, $\mathbf{v}(n)$, design the suboptimal filter, and then tune it empirically by altering $\mathbf{Q}_d$, $\mathbf{R}_d$, as suggested in Section 5.5.

5.7 The split Kalman filter [ref. 2]

The Kalman filter of Eqns (5.30)–(5.33) can be recast in a 'split' form which uses the information at time n, namely $\mathbf{u}(n)$, $\mathbf{y}(n)$, to estimate the current state $\mathbf{x}(n)$, rather than predicting $\mathbf{x}(n+1)$ as the basic filter, Eqn (5.30), does. This can yield improved accuracy, particularly for slow sampling rates, and it incidentally permits easy extension of the algorithm to deal with non-linear systems (Section 5.8).

The recast algorithm is:

$$\mathbf{J}(n) = \mathbf{G}^-(n)\mathbf{C}_d^T(n)[\mathbf{R}_d(n) + \mathbf{C}_d(n)\mathbf{G}(n)\mathbf{C}_d^T(n)]^{-1} \tag{5.50}$$

$$\hat{\mathbf{x}}^+(n) = \hat{\mathbf{x}}^-(n) + \mathbf{J}(n)[\mathbf{y}(n) - \mathbf{C}_d(n)\hat{\mathbf{x}}^-(n) - \mathbf{D}_d(n)\mathbf{u}(u)] \tag{5.51}$$

$$\mathbf{G}^+(n) = [\mathbf{I} - \mathbf{J}(n)\mathbf{C}_d(n)]\mathbf{G}^-(n) \tag{5.52}$$

$$\hat{\mathbf{x}}^-(n+1) = \varphi(n)\hat{\mathbf{x}}^+(n) + \psi(n)\mathbf{u}(n) \tag{5.53}$$

$$\mathbf{G}^-(n+1) = \varphi(n)\mathbf{G}^+(n)\varphi^T(n) + \mathbf{\Gamma}(n)\mathbf{Q}_d(n)\mathbf{\Gamma}^T(n) \tag{5.54}$$

where the symbols are as defined in Section 5.5, except that the suffix '$-$' signifies information available at the sampling instant, nT, and '$+$', information available as the result of calculations some time (hopefully very shortly) after that instant. $\mathbf{J}(n)$ is defined by Eqn (5.50).

The equations are realized in the sequence shown, with Eqn (5.51) being realized immediately after the sampling instant, when $\mathbf{u}(n)$, $\mathbf{y}(n)$ have become available; where the Kalman filter is part of some real-time loop, it is clearly desirable that $\hat{\mathbf{x}}^+(n)$ should be generated as quickly as possible after the sampling instant. The remaining equations need be evaluated only in time for the next sampling instant. The ideal initial conditions are $\hat{\mathbf{x}}^-(0) = \mathbf{x}(0)$ and $\mathbf{G}^-(0) = \mathbf{0}$, but in practice $\mathbf{x}(0)$

and $\mathbf{G}(0)$ are usually unknown and, as with the basic Kalman filter, convergence of the algorithm to the same estimate is guaranteed whatever the initial conditions. If these are set inaccurately, only the initial transient is affected.

Eqns (5.51) and (5.52) are known as the 'measurement update' stage, being dependent on $\mathbf{u}(n)$, $\mathbf{y}(n)$ which become available at the measurement instant, while Eqns (5.53), (5.54) and (5.50) are the 'time update' stage. The sequence is illustrated in Figure 5.14.

Eqns (5.30)–(5.33) are readily derived from Eqns (5.50)–(5.54) by eliminating $\hat{\mathbf{x}}^+(n)$ in the latter and using the relationship

$$\mathbf{M}(n) = \varphi(n)\mathbf{J}(n) \tag{5.55}$$

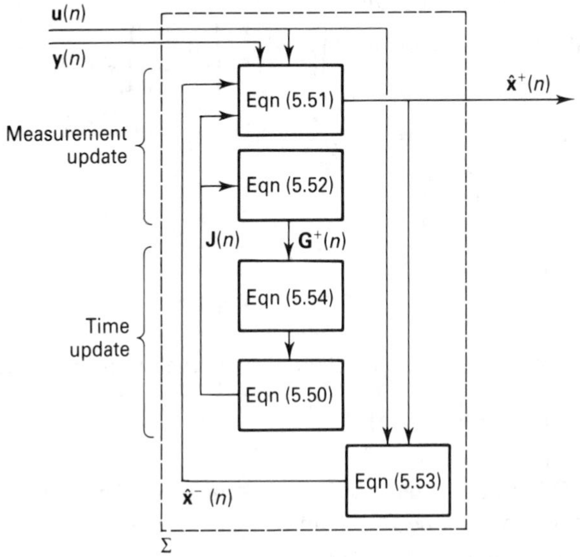

Figure 5.14 Split Kalman filter sequence.

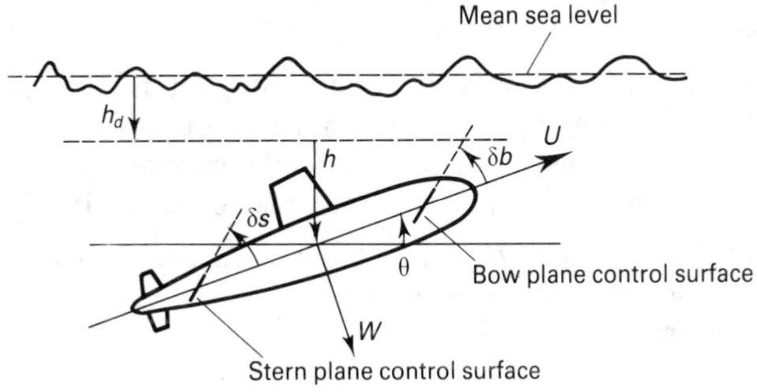

Figure 5.15 Submarine vertical dynamics.

Example

Consider the vertical dynamics of a submarine travelling a few metres below the surface of a choppy sea, as illustrated in Figure 5.15 (see previous page). The vertical plane behaviour of the submarine is modelled by the simplified linear time-variant, heuristic equations:

$$\begin{pmatrix} \dot{W} \\ \dot{q} \\ \dot{\theta} \\ h \end{pmatrix} = \begin{pmatrix} -0.016U & 0.250U & 0 & 0 \\ 0.0002U & -0.077U & -0.008 & -0.0014 \\ 0 & 1 & 0 & 0 \\ 1 & 0 & -U & 0 \end{pmatrix} \begin{pmatrix} W \\ q \\ \theta \\ h \end{pmatrix} + \begin{pmatrix} -0.0033U^2 & -0.0065U^2 \\ 0.0035U^2 & -0.0010U^2 \\ 0 & 0 \\ 0 & 0 \end{pmatrix} \begin{pmatrix} \delta b \\ \delta s \end{pmatrix} + \begin{pmatrix} 1 & 0 \\ 0 & 1 \\ 0 & 0 \\ 0 & 0 \end{pmatrix} \mathbf{w} \tag{5.56}$$

$$\mathbf{y} = \begin{pmatrix} 0 & 0 & 1 & 0 \\ 0 & 0 & 0 & 1 \end{pmatrix} \begin{pmatrix} W \\ q \\ \theta \\ h \end{pmatrix} + \mathbf{v} \tag{5.57}$$

where

W is the downwards velocity at right angles to the submarine main axis (m s^{-1}),
q is the pitch rate (rad s^{-1}),
θ is the pitch (rad),
h is the depth WRT a datum depth, h_d, below mean sea level (m),
U is the forward velocity along the main body axis (m s^{-1}),
δb is the bow control plane angle (rad),
δs is the stern control plane angle (rad),
$\mathbf{y}$ is the output, $\mathbf{y} = (\theta \quad h)^T$,
$\mathbf{w}$ is the process noise of the sea, strength $\mathbf{Q}$,
$\mathbf{v}$ is the measurement noise of the sea, strength $\mathbf{R}$.

A split Kalman filter, for use in a submarine autopilot, is designed using a sampling time $T = 2.0$ s. This entails calculating $\boldsymbol{\varphi}(n)$, $\boldsymbol{\psi}(n)$, $\mathbf{Q}_d(n)$, $\mathbf{R}_d(n)$ (Eqns (3.46)–(3.51)) at each sampling instant, according to the current value of U, and realizing the algorithm of Eqns (5.50)–(5.54) in the order indicated in Figure 5.14. The current values of $\mathbf{Q}(t)$, $\mathbf{R}(t)$ must, of course, be assessed, and the corresponding $\mathbf{Q}_d(n)$, $\mathbf{R}_d(n)$ calculated and input to the algorithm.

Some results generated for

$$\mathbf{Q} = \begin{pmatrix} 0.01 & 0 \\ 0 & 0.001 \end{pmatrix} \quad \mathbf{R} = \begin{pmatrix} 10 & 0 \\ 0 & 0.01 \end{pmatrix}$$

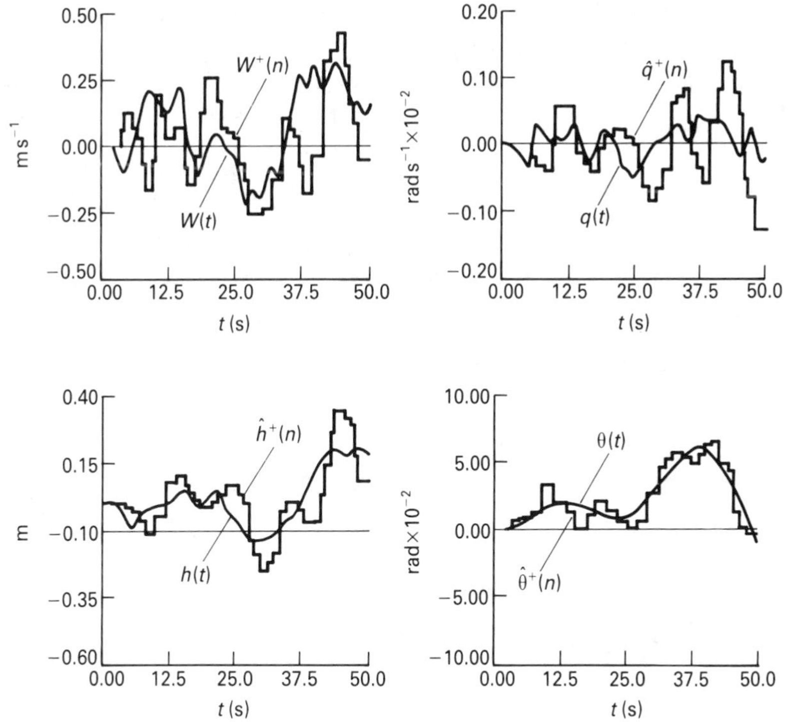

Figure 5.16 Split Kalman filter results.

and a varying forward velocity, $U = (3 + (t/100))$, $\delta b = \delta s = 0$, are shown in Figure 5.16, where it can be seen that the estimates of θ, h are significantly better than those of W, q. Such results are not untypical; their quality can be altered by tuning $\mathbf{Q}_d$, $\mathbf{R}_d$ as described in Section 5.5.

5.8 Kalman filters for non-linear systems [ref. 5]

The split Kalman filter of Section 5.7 can be extended fairly readily to deal with reasonably well-behaved non-linear systems.

As explained in Section 3.8, it is usually impractical to model the behaviour of a non-linear system (whether deterministic or stochastic) by difference equations, and a continuous time model must therefore provide the basis of the appropriate Kalman filter design, which is admittedly rather empirical.

In heuristic form (Eqns (2.90) and (2.91)) this is

$$\dot{\mathbf{x}} = \mathbf{f}(\mathbf{x}, \mathbf{u}, t) + \mathbf{H}(\mathbf{x}, t)\bar{\mathbf{w}}(t) \tag{5.58}$$

$$\mathbf{y} = \mathbf{g}(\mathbf{x}, \mathbf{u}, t) + \bar{\mathbf{v}}(t) \tag{5.59}$$

where $\bar{\omega}(t)$ and $\bar{\mathbf{v}}(t)$ represent additive Gaussian extended white process and measurement noise respectively, strengths $\mathbf{Q}(t)$, $\mathbf{R}(t)$.

Eqn (5.58) must, as usual, be interpreted as a stochastic differential equation (Eqn (2.92)). Eqn (5.59) is written, for the purpose of Kalman filter construction, in sampled form:

$$\mathbf{y}(n) = \mathbf{g}(\mathbf{x}(n), \mathbf{u}(n), n) + \bar{\mathbf{v}}(n) \tag{5.60}$$

where

$$\mathbf{y}(n) = \mathbf{y}(nT),$$
$$\mathbf{x}(n) = \mathbf{x}(nT),$$

and T is the sampling period of the Kalman filter.

Following the format of the split filter (Eqns (5.50)–(5.54)), the filter is as follows:

$$\mathbf{J}(n) = \mathbf{G}^-(n)\mathbf{C}_d^T(n)[\mathbf{R}_d(n) + \mathbf{C}_d(n)\mathbf{G}^-(n)\mathbf{C}_d^T(n)]^{-1} \tag{5.61}$$

where

$$\mathbf{C}_d(n) = \left.\frac{\partial \mathbf{g}}{\partial \mathbf{x}}\right|_{\substack{\mathbf{x} = \hat{\mathbf{x}}^-(n) \\ \mathbf{u} = \mathbf{u}(n) \\ t = nT}} \tag{5.62}$$

$$\hat{\mathbf{x}}^+(n) = \hat{\mathbf{x}}^-(n) + \mathbf{J}(n)[\mathbf{y}(n) - \mathbf{g}(\hat{\mathbf{x}}^-(n), \mathbf{u}(n), n)] \tag{5.63}$$

$$\mathbf{G}^+(n) = [\mathbf{I} - \mathbf{J}(n)\mathbf{C}_d(n)]\mathbf{G}^-(n) \tag{5.64}$$

$\hat{\mathbf{x}}^-(n+1)$ and $\mathbf{G}^-(n+1)$ are found by solving the approximate differential equations (cf. Eqns (5.58) and (5.14)) over the time interval $[nT, (n+1)T]$:

$$\dot{\hat{\mathbf{x}}} = \mathbf{f}(\hat{\mathbf{x}}, \mathbf{u}, t) \tag{5.65}$$

$$\dot{\mathbf{G}}(t) = \mathbf{A}^T\mathbf{G}(t) + \mathbf{G}(t)\mathbf{A}^T + \mathbf{HQH}^T - \mathbf{G}(t)\mathbf{CR}^{-1}\mathbf{CG}(t) \tag{5.66}$$

where

$$\mathbf{A} = \left.\frac{\partial \mathbf{f}}{\partial \mathbf{x}}\right|_{\substack{\mathbf{x} = \hat{\mathbf{x}}^+(n) \\ \mathbf{u} = \mathbf{u}(n) \\ t = nT}}; \quad \mathbf{C} = \mathbf{C}\left.\frac{\partial \mathbf{g}}{\partial \mathbf{x}}\right|_{\substack{\mathbf{x} = \tilde{\mathbf{x}}^+(n) \\ \mathbf{u} = \mathbf{u}(n) \\ t = nT}}; \quad \mathbf{H} = \mathbf{H}(\mathbf{x}, t)|_{\substack{\mathbf{x} = \hat{\mathbf{x}}^+(n) \\ t = nT}} \tag{5.67}$$

with initial conditions:

$$\hat{\mathbf{x}}(nT) = \hat{\mathbf{x}}^+(n), \quad \mathbf{G}(nT) = \mathbf{G}^+(n)$$

and

$$\mathbf{Q} = \mathbf{Q}(nT), \quad \mathbf{R} = R(nT)$$

Eqns (5.65) and (5.66) may be solved using a Runge–Kutta algorithm and a calculation interval which is short compared with T (typically $0.1T$).

Alternatively, if T is short compared with the system time constants, the system can be treated as linear time-invariant over the interval $[nT, (n+1)T]$, so that

$$\hat{\mathbf{x}}^-(n+1) = \boldsymbol{\varphi}\hat{\mathbf{x}}^+(n) + \boldsymbol{\psi}\mathbf{n}(n) \tag{5.68}$$

$$\mathbf{G}^-(n+1) = \boldsymbol{\varphi}\mathbf{G}^+(n)\boldsymbol{\varphi}^T + \boldsymbol{\Gamma}\mathbf{Q}_d\boldsymbol{\Gamma}^T \tag{5.69}$$

where

$$\boldsymbol{\varphi} = e^{\mathbf{A}T} \doteqdot \mathbf{I} + \mathbf{A}T, \quad \boldsymbol{\psi} = \mathbf{B}T$$

$$\boldsymbol{\Gamma} \doteqdot \mathbf{H}T$$

$$\mathbf{Q}_d \doteqdot \mathbf{Q}T$$

Example

The behaviour of an airborne missile and its target are modelled in simplified form in two horizontal dimensions only, as indicated in Figure 5.17, where

a_M is the missile lateral acceleration (m s^{-2}),
a_T is the target lateral acceleration (m s^{-2}),
r is the missile to target distance (m),
θ_M is the missile velocity direction (rad),
θ_T is the target velocity direction (rad),
σ is the 'line-of-sight' angle (rad),
V_M is the missile forward velocity (m s^{-1}),
V_T is the target forward velocity (m s^{-1}),
T is the sample time (s), $T = 1$.

Assuming constant missile and target forward velocities:

$$\dot{r} = V_T\cos(\theta_T - \sigma) - V_M\cos(\theta_M - \sigma)$$

$$r\dot{\sigma} = V_T\sin(\theta_T - \sigma) - V_M\sin(\theta_M - \sigma)$$

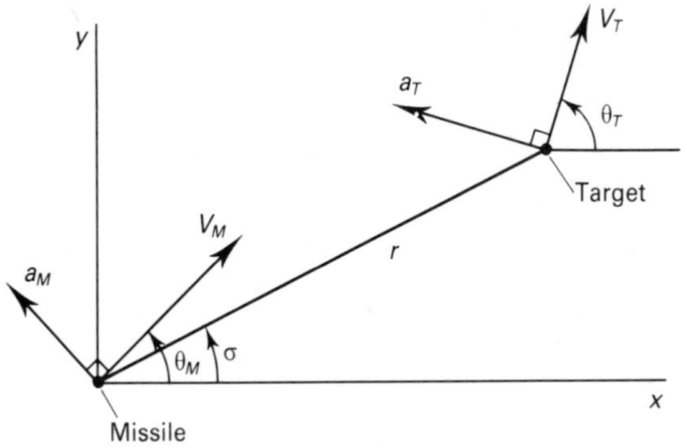

Figure 5.17 Missile-target system.

With the simplifying assumptions that $\theta_M \doteqdot \theta_T \doteqdot \sigma$, and $\ddot{r} = 0$, and modelling the rate of change of target lateral acceleration as white noise, $\dot{a}_T = w_T$ (which, as in the example of Section 5.6.1, is certainly untrue but does allow access to the Kalman filter design procedure), the following heuristic model can be developed (Eqns (5.58) and (5.59)):

$$\dot{\mathbf{x}} = \begin{pmatrix} \dfrac{-2x_1x_2}{x_3} - \dfrac{u}{x_3} + \dfrac{x_4}{x_3} \\ 0 \\ x_2 \\ 0 \end{pmatrix} + \begin{pmatrix} 0 \\ 0 \\ 0 \\ 1 \end{pmatrix} w_T$$

$$y = (1 \quad 0 \quad 0 \quad 0)\mathbf{x} + v$$

where

$\mathbf{x} = (\dot{\sigma} \quad \dot{r} \quad r \quad a_T)^T$,
$u = a_M$,
v is the noise associated with the measurement of the line-of-sight rate of change.

Differentiating this state equation yields (Eqn (5.67))

$$\mathbf{A} = \begin{pmatrix} \dfrac{-2\hat{x}_2^+(n)}{\hat{x}_3^+(n)} & \dfrac{-2\hat{x}_1^+(n)}{\hat{x}_3^+(n)} & \dfrac{2x_1^+(n) - \hat{x}_4^+(n)}{(\hat{x}_3^+(n))^2} & \dfrac{1}{\hat{x}_3^+(n)} \\ 0 & 0 & 0 & 0 \\ 0 & 1 & 0 & 0 \\ 0 & 0 & 0 & 0 \end{pmatrix}$$

The sampling period $T = 1$ is short compared with the system dynamics, justifying the use of Eqns (5.68) and (5.69) with

$$\varphi = \mathbf{I} + \mathbf{A} \cdot 1$$

$$\mathbf{\Gamma} = (0 \quad 0 \quad 0 \quad 1)^T$$

$$\mathbf{Q}_d = \mathbf{Q} \cdot 1 = 500$$

To account for the rapidly increasing measurement noise as the missile approaches the target, set:

$$\mathbf{R}_d = \mathbf{R} = \frac{360}{r^2}$$

Substituting these in Eqns (5.50)–(5.54) yields the estimates shown in Figure 5.18 for a target manoeuvre consisting of a step in lateral acceleration of 50 m s^{-2}, $1 \leqslant t \leqslant 2$.

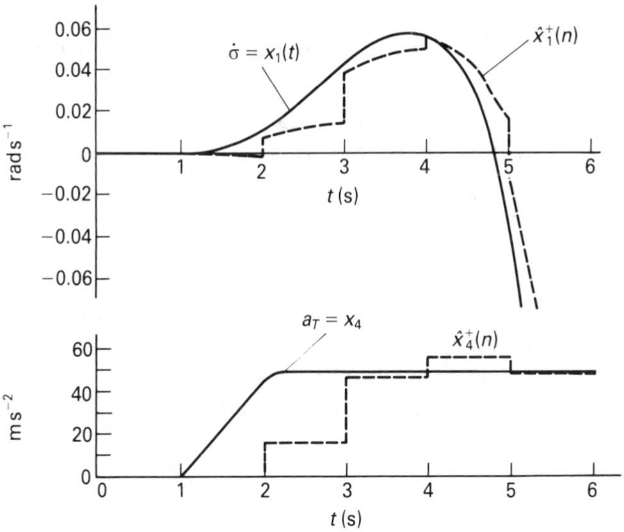

Figure 5.18 Extended Kalman filter behaviour.

The Kalman filter detects the manoeuvre at the sampling instant following it, and generates good estimates of the new situation after two or three further sampling periods. The improvements in the estimates calculated at the sampling instants are noteworthy.

5.9 Adaptive Kalman filters

As explained in Section 5.5, the value of the process noise strength, $\mathbf{Q}_d(n)$, used in the Kalman filter algorithm, can sometimes be modified, with advantage, empirically. Indeed, in many practical cases the true value of $\mathbf{Q}_d(n)$ is unknown and some guesswork and tuning of $\mathbf{Q}_d(n)$ (and, perhaps less commonly, $\mathbf{R}_d(n)$) is mandatory.

One type of adaptive Kalman filter relies on systematic methods of tuning $\mathbf{Q}_d(n)$ to achieve some desirable performance criterion, usually defined in terms of the filter 'residual', or difference between the system output, $\mathbf{y}(n)$, and the equivalent filter 'output', $\hat{\mathbf{y}}(n)$. In terms of the split filter of Section 5.7, the residual is

$$\begin{aligned}\tilde{\mathbf{r}}_d(n) &= \mathbf{y}(n) - \hat{\mathbf{y}}(n) \\ &= \mathbf{y}(n) - (\mathbf{C}_d(n)\hat{x}^-(n) + \mathbf{D}_d(n)u(n))\end{aligned} \tag{5.70}$$

It can be shown that $\tilde{\mathbf{r}}_d(n)$ is a vector of Gaussian white noise of strength:

$$\tilde{\mathbf{R}}_d(n) = \mathbf{C}_d(n)\mathbf{G}^-(n)\mathbf{C}_d^T(n) + \mathbf{R}_d(n)$$

Substituting from Eqn (5.54), to involve $\mathbf{Q}_d(n)$:

$$\tilde{\mathbf{R}}_d(n) = \mathbf{C}_d(n)[\boldsymbol{\varphi}(n)\mathbf{G}^+(n-1)\boldsymbol{\varphi}^T(n) + \boldsymbol{\Gamma}(n)\mathbf{Q}_d(n)\boldsymbol{\Gamma}^T(n)]\mathbf{C}_d^T(n) + \mathbf{R}_d(n) \tag{5.71}$$

If the system (and Kalman filter) states are stationary, or nearly so, over some reasonable number of samples, k, $\tilde{\mathbf{R}}_d(n)$ is readily calculated:

$$\tilde{\mathbf{R}}_d(n) = \frac{1}{k}\sum_{j=n-k+1}^{n} (\tilde{\mathbf{r}}_d(j)\tilde{\mathbf{r}}_d^T(j)) \tag{5.72}$$

Equating $\tilde{\mathbf{R}}_d(n)$ in Eqns (5.71) and (5.72) gives an indication of how well the filter is operating, and suggests the possibility of redefining $\mathbf{Q}_d(n)$ to give a filter performance which is improved in the sense that its residual is reduced.

One scheme, referred to as a covariance matching method [ref. 8], depends on calculating the 'correct' $\mathbf{Q}_d(n)$ by using $\tilde{\mathbf{R}}_d(n)$, calculated from Eqn (5.72), in Eqn (5.71) to give

$$\mathbf{C}_d(n)\boldsymbol{\Gamma}(n)\mathbf{Q}_d(n)\boldsymbol{\Gamma}^T(n)\mathbf{C}_d^T(n) = \tilde{\mathbf{R}}_d(n) - \mathbf{C}_d(n)\boldsymbol{\varphi}(n)\mathbf{G}^+(n-1)\boldsymbol{\varphi}^T(n)\mathbf{C}_d(n) - \mathbf{R}_d(n) \tag{5.73}$$

Eqn (5.73) does not yield a unique $\mathbf{Q}_d(n)$ if there are fewer outputs than states ($r < m$), but if $\mathbf{Q}_d(n)$ is taken as diagonal, and the model states are arranged so that the first r diagonal elements of $\mathbf{Q}_d(n)$ are chosen for adaptation, Eqn (5.73) is readily solved to yield these elements, which can then be updated in the Kalman filter algorithm.

A logical shortcoming of this scheme is that $\mathbf{G}^+(n-1)$ in Eqn (5.73) itself depends on the unadapted – and presumably incorrect – value of $\mathbf{Q}_d(n)$, and there is no real guarantee that the newly calculated $\mathbf{Q}_d(n)$ is a real improvement. Furthermore, the elements of $\mathbf{Q}_d(n)$ which can be adapted are often defined by $\mathbf{C}_d(n)$ and $\Gamma(n)$, and these elements may not correspond to the states directly affected by process noise. Nevertheless, this method is certainly effective in many cases, and in summary it operates as follows.

1. The filter is run over a number of sampling periods during which the states are more or less stationary ($\mathbf{u}(n)$ constant). $\tilde{\mathbf{R}}_d(n)$ (Eqn (5.72)) is calculated.
2. A decision is taken as to whether the current error convariance matrix, $\mathbf{G}^+(n-1)$, matches this satisfactorily, using Eqn (5.71). If not, $\mathbf{Q}_d(n)$ is calculated and updated using Eqn (5.73).
3. The sequence is repeated from (1).

Other, more complex, schemes [ref. 8] depend on adjusting $\mathbf{Q}_d(n)$, $\mathbf{R}_d(n)$ according to the calculated likelihood of the filter estimates (Bayesian estimation method); calculating the most likely values of $\mathbf{Q}_d(n)$, $\mathbf{R}_d(n)$ (maximum likelihood method); or calculating the Kalman filter gain from the autocorrelation of the system output (correlation method).

A second, quite distinct, type of adaptive filter relies simply on choosing between a number of different filters according to the current perceived conditions, notably the system inputs, $\mathbf{u}(n)$, and outputs, $\mathbf{y}(n)$. Such a strategy must, of course, avoid the possibility of repeated and rapid switching between filters, with the consequent transients, so some overlap, or hysteresis, in the switching criterion is usually necessary.

5.10 Practical Kalman filter algorithms

Kalman filters (which have found surprisingly diverse application [ref. 9]) can exhibit practical convergence and stability problems arising from the limited accuracy of the calculations. In particular, if the computed error covariance matrix, $\mathbf{G}(n)$ (Eqns (5.37) and (5.54)) becomes non-positive semi-definite – a theoretical impossibility – the algorithm can diverge.

Difficulty can also be encountered with the matrix inversion in the algorithms (Eqns (5.36) and (5.50)), especially where fairly large matrices are involved.

Two popular methods of avoiding these difficulties are described here; there are many elaborations in the literature [refs 2, 5, 6, 8].

5.10.1 The covariance square root filter

This depends on the fact that any (square, symmetric) positive semi-definite matrix can be factorized into two square triangular matrices, one the transpose of the other. In particular,

$$\mathbf{G}(n) = \mathbf{S}(n) \cdot \mathbf{S}^T(n) \tag{5.74}$$

Here $\mathbf{S}(n)$ is a lower triangular, and $\mathbf{S}^T(n)$ an upper triangular matrix.

This factorization can be realized efficiently by the Crout [ref. 7] or Cholesky [ref. 8] decomposition algorithms. The latter is as follows:

$$s_{ij} = \begin{cases} \dfrac{1}{s_{jj}}\left[g_{ij} - \sum\limits_{k=1}^{j-1} s_{ik}s_{jk}\right], & j = 1, 2, \ldots, (i-1) \\ \left(g_{ii} - \sum\limits_{k=1}^{i-1} (s_{ik})^2\right)^{1/2}, & j = i \\ 0, & j > \mathrm{i} \end{cases} \tag{5.75}$$

where

g_{ij} is the (i, j)th element of $\mathbf{G}(n)$,

s_{ij} is the (i, j)th element of $\mathbf{S}(n)$.

The algorithm starts by calculating $s_{11} = \sqrt{g_{11}}$, and proceeds to calculate the rows of $\mathbf{S}(n)$ (for which $s_{ij} = 0, j > i$) one after the other.

Example

Consider the positive semi-definite matrix:

$$\mathbf{G}(n) = \begin{pmatrix} 2 & -2 & 3 \\ -2 & 5 & 2 \\ 3 & 2 & 13 \end{pmatrix}$$

Then, from Eqn (5.75),

$$s_{11} = \sqrt{2} = 1.414, \quad s_{12} = 0, \quad s_{13} = 0$$

$$s_{21} = \frac{1}{\sqrt{2}}(-2-0) = -1.414, \quad s_{22} = (5-(2))^{1/2} = 1.732, \quad s_{23} = 0$$

$$s_{31} = \frac{1}{\sqrt{2}}(3-0) = 2.121, \quad s_{32} = \frac{1}{1.732}(2+(2.121)(1.414)) = 2.887$$

$$s_{33} = [13-(2.121)^2-(2.887)^2]^{1/2} = -0.408$$

Thus

$$\mathbf{S}(n) = \begin{pmatrix} 1.414 & 0 & 0 \\ -1.414 & 1.732 & 0 \\ 2.121 & 2.887 & 0.408 \end{pmatrix}$$

This satisfies Eqn (5.74).

The algorithm of Eqns (5.50)–(5.54) can be cast in terms of 'square roots' of this kind, automatically precluding the possibility of $\mathbf{G}(n)$ becoming non-positive semi-definite. The problem of matrix inversion (Eqn (5.50)) is tackled by regarding the system as single output, $r = 1$, $R_d(n)$ a scalar, and extending this by repetition to deal with multioutput cases.

For this 'single output' case ($r = 1$), Eqn (5.50) becomes

$$\mathbf{J}(n) = \mathbf{G}^-(n)\mathbf{C}_d^T(n) \cdot \frac{1}{\alpha(n)}$$

where

$$\alpha(n) = R_d(n) + \mathbf{C}_d(n)\mathbf{G}^-(n)\mathbf{C}_d^T(n)$$

Here, $\alpha(n)$ is a scalar and $\mathbf{C}_d(n)$ a row vector. Eqn (5.52), clearly the most likely source of aberration in the symmetrical positive semi-definite format of the error covariance matrix, is replaced with

$$\mathbf{S}^+(n)\mathbf{S}^{+T}(n) = \mathbf{S}^-(n)\left[\mathbf{I} - \frac{1}{\alpha(n)}\mathbf{S}^{-T}(n)\mathbf{C}_d^T(n)\mathbf{C}_d(n)\mathbf{S}^-(n)\right]\mathbf{S}^{-T}(n) \tag{5.76}$$

where $\mathbf{S}^-(n)$ is the lower triangular 'square root' of $\mathbf{G}^-(n)$, derived by the algorithm of Eqn (5.75).

The 'square root' of the bracketed term can be shown to be [ref. 2]

$$\mathbf{I} - \frac{1}{\alpha(n)}\gamma(n)\mathbf{S}^{-T}(n)\mathbf{C}_d^T(n)\mathbf{C}_d(n)\mathbf{S}^-(n) \tag{5.77}$$

where

$$\gamma(n) = \frac{1}{1 + \sqrt{\dfrac{R_d(n)}{\alpha(n)}}}$$

Substitution of Eqns (5.76) and (5.77) in Eqn (5.52) shows that the latter can be replaced with the 'square root' equation:

$$\mathbf{S}^+(n) = \mathbf{S}^-(n)\left[\mathbf{I} - \frac{\gamma(n)}{\alpha(n)}\mathbf{S}^{-T}(n)\mathbf{C}_d^T(n)\mathbf{C}_d(n)\mathbf{S}^-(n)\right] \tag{5.78}$$

This guarantees that $\mathbf{G}^+(n)$ is always symmetric positive semi-definite.

It remains to specify a time update definition of $\mathbf{S}^-(n+1)$. This is done simply by evaluating Eqn (5.54) – which will not lose its symmetric positive format – using $\mathbf{S}^+(n)$:

$$\mathbf{G}^-(n+1) = (\boldsymbol{\varphi}(n)\mathbf{S}^+(n))(\mathbf{S}^{+T}(n)\boldsymbol{\varphi}^T(n) + \boldsymbol{\Gamma}(n)\mathbf{Q}_d(n)\boldsymbol{\Gamma}^T(n) \tag{5.79}$$

$\mathbf{G}^-(n+1)$ is then factorized to generate $\mathbf{S}^-(n+1)$.

The algorithm thus consists of Eqns (5.50)–(5.54) as before, but with Eqn (5.54) replaced with Eqn (5.79) and Eqn (5.52) replaced with a Cholesky decomposition algorithm (Eqns (5.75) and (5.78)). The whole cycle is repeated r times at each sampling instant to deal with r system outputs.

5.10.2 The *U–D* factorized covariance filter

This algorithm, which is usually more efficient in computational and storage terms than the covariance square root algorithm, and matches it in accuracy, depends on the fact that any symmetric (square) positive semi-definite matrix can be factorized as follows:

$$\mathbf{G}(n) = \mathbf{U}(n)\mathbf{D}_g(n)\mathbf{U}^T(n) \tag{5.80}$$

where $\mathbf{U}(n)$ is an upper unit triangular matrix:

$$\mathbf{U}(n) = \begin{pmatrix} 1 & u_{12}(n) & u_{13}(n) & \dots & u_{1m}(n) \\ 0 & 1 & u_{23}(n) & & u_{2m}(n) \\ 0 & 0 & 1 & & u_{3m}(n) \\ \vdots & & & & \vdots \\ 0 & 0 & 0 & \cdots & 1 \end{pmatrix}$$

$$\mathbf{D}_g(n) = \begin{pmatrix} d_{11}(n) & 0 & 0 & \cdots & 0 \\ 0 & d_{22}(n) & 0 & & 0 \\ \vdots & & & & \\ 0 & 0 & 0 & & d_{mm}(n) \end{pmatrix}$$

An algorithm for factorizing $\mathbf{G}(n)$ in this way, using its elements, g_{ij}, is based on the Cholesky decomposition algorithm (Eqn (5.75)):

$$d_{mm}(n) = g_{mm}(n)$$

$$u_{im} = \begin{cases} 1, & i = m \\ \dfrac{g_{im}}{d_{mm}}, & i = 1, 2, \ldots, (m-1) \end{cases} \tag{5.81}$$

This generates the mth column of $\mathbf{U}(n)$.

For the remaining columns:

$$d_{jj}(n) = g_{jj}(n) - \sum_{k=j+1}^{m} d_{kk}u_{jk}^2, \quad j = (m-1), (m-2), \ldots, 1$$

$$u_{ij} = \begin{cases} 0, & i > j \\ 1, & i = j \\ g_{ij} - \sum\limits_{k=j+1}^{m} (d_{kk}u_{ik}u_{jk})\dfrac{1}{d_{jj}}, & i = (j-1), (j-2), \ldots, 1 \end{cases} \tag{5.82}$$

Example

Consider the positive semi-definite matrix:

$$\mathbf{G}(n) = \begin{pmatrix} 2 & -2 & 3 \\ -2 & 5 & 2 \\ 3 & 2 & 13 \end{pmatrix}$$

From Eqn (5.81):

$$d_{33} = 13$$

$$u_{13} = \frac{3}{13} = 0.231, \quad u_{23} = \frac{2}{13} = 0.154, \quad u_{33} = 1$$

From Eqn (5.82):

$$d_{22} = 5 - (13(0.154)^2) = 4.692$$

$$u_{12} = -2 - 13(0.231)(0.154)\cdot\frac{1}{4.692} = -0.525, \quad u_{22} = 1, \quad u_{32} = 0$$

$$d_{11} = 2 - [4.692(-0.525)^2 + 13(0.231)^2] = 0.013$$

$$u_{11} = 1, u_{21} = 0, u_{31} = 0$$

Thus

$$\mathbf{U}(n) = \begin{pmatrix} 1 & -0.525 & 0.231 \\ 0 & 1 & 0.154 \\ 0 & 0 & 1 \end{pmatrix}$$

and

$$\mathbf{D}_g(n) = \begin{pmatrix} 0.013 & 0 & 0 \\ 0 & 4.692 & 0 \\ 0 & 0 & 13 \end{pmatrix}$$

These satisfy Eqn (5.80).

As with square root filter, the matrix inversion problem is dealt with by considering a single output system, and repeating the process r times to deal with r outputs. Assuming $\mathbf{G}^-(n)$ to be factorized (Eqns (5.81) and (5.82)), Eqn (5.50) becomes

$$\mathbf{J}(n) = \mathbf{U}^-(n)\mathbf{D}_g^-(n)\mathbf{U}^{-T}(n)\mathbf{C}_d^T(n)\frac{1}{\alpha(n)} \tag{5.83}$$

where $\alpha(n)$ is the scalar:

$$\alpha(n) = R_d(n) + \mathbf{C}_d(n)\mathbf{U}^-(n)\mathbf{D}_g^-(n)\mathbf{U}^{-T}(n)\mathbf{C}_d^T(n)$$

Substituting in Eqn (5.52):

$$\mathbf{G}^+(n) = \mathbf{U}^-(n)\left[\mathbf{D}_g^-(n) - \frac{1}{\alpha(n)}\cdot\boldsymbol{\beta}(n)\cdot\boldsymbol{\beta}^T(n)\right]\mathbf{U}^{-T}(n) \tag{5.84}$$

where

$$\boldsymbol{\beta}(n) = \mathbf{D}_g^-(n)\mathbf{U}^{-T}(n)\mathbf{C}_d^T(n)$$

$\mathbf{G}^+(n)$ can now be factorized (Eqns (5.81) and (5.82)) to yield:

$$\mathbf{G}^+(n) = \mathbf{U}^+(n)\mathbf{D}_g^+(n)\mathbf{U}^{+T}(n) \tag{5.85}$$

The cycle is completed by expressing $\mathbf{U}^-(n+1)$, $\mathbf{D}_g^-(n+1)$ in terms of $\mathbf{U}^+(n)$, $\mathbf{D}_g^+(n)$, using Eqn (5.54).

Assuming for the moment that $\mathbf{Q}_d(n)$ is diagonal – which is often the case – Eqn (5.54) can be seen to be satisfied by

$$\tilde{\mathbf{U}}\tilde{\mathbf{D}}\tilde{\mathbf{U}}^T$$

where

$$\tilde{\mathbf{U}} = (\varphi(n)\mathbf{U}^+(n) \,\vdots\, \mathbf{\Gamma}(n)) \quad \text{(an } m \times (m+s) \text{ matrix)} \tag{5.86}$$

$$\tilde{\mathbf{D}} = \begin{pmatrix} \mathbf{D}_g^+(n) & \mathbf{0} \\ \text{---------} & \\ \mathbf{0} & \mathbf{Q}_d(n) \end{pmatrix} \quad \text{(an } (m+s)^2 \text{ diagonal matrix)} \tag{5.87}$$

$\mathbf{U}^-(n+1)$, $\mathbf{D}_g^-(n+1)$ are recovered from $\tilde{\mathbf{U}}$, $\tilde{\mathbf{D}}$, by the following algorithm. Let:

$$\tilde{\mathbf{U}}^T = (\mathbf{a}_1 \quad \mathbf{a}_2 \quad \mathbf{a}_3 \quad \ldots \quad \mathbf{a}_m)$$

$$\mathbf{b}_i = \tilde{\mathbf{D}}\mathbf{a}_i, \quad i = 1, 2, \ldots, m$$

Then:

$$d_{ii}(n+1) = \mathbf{a}_i^T \cdot \mathbf{b}_i, \quad i = 1, 2, \ldots, m \tag{5.88}$$

Let:

$$\mathbf{c}_i = \frac{\mathbf{b}_i}{d_{ii}(n+1)}$$

Then:

$$u_{ji}(n+1) = \mathbf{a}_j^T \cdot \mathbf{c}_i, \; j = 1, 2, \ldots (i-1) \tag{5.89}$$

$\mathbf{U}^-(n+1)$ and $\mathbf{D}_g^-(n+1)$ are constructed from these elements according to the pattern of Eqn (5.80).

If $\mathbf{Q}_d(n)$ is not diagonal, $(\mathbf{\Gamma}(n)\mathbf{Q}_d(n)\mathbf{\Gamma}^T(n))$ can be factorized (Eqns (5.81) and (5.82)) to yield new matrices $\bar{\mathbf{\Gamma}}(n)$, $\bar{\mathbf{Q}}_d(n)$, which suit the algorithm.

The Kalman filter algorithm consists of Eqns (5.50)–(5.54), with Eqn (5.50) replaced by (5.83), Eqn (5.52) replaced by (5.84) and (5.85), and Eqn (5.54) replaced by Eqns (5.86)–(5.89).

The whole process is repeated r times at each sampling instant to deal with r outputs.

EXERCISES

These exercises require CAD facilities for system simulation in continuous and discrete time.

1 Consider the deterministic dynamical system modelled by:

$$\dot{\mathbf{x}} = \begin{pmatrix} 0 & 1 \\ -101 & -2 \end{pmatrix}\mathbf{x} + \begin{pmatrix} 0 \\ 101 \end{pmatrix}u, \quad u = \begin{cases} 0, & t < 0 \\ 1, & t \geqslant 0 \end{cases}$$

$$y = (1 \quad 0)\mathbf{x}$$

Design an asymptotic state estimator for the system with eigenvalues at $s = -50 + j50$, and simulate the behaviour of the system and estimator for:

$$\mathbf{x}(0) = 0,\ \hat{\mathbf{x}}(0) = 0;\ \mathbf{x}(0) = 0,\ \hat{\mathbf{x}}(0) = \begin{pmatrix} -10 \\ 0 \end{pmatrix};\ \mathbf{x}(0) = 0,\ \hat{\mathbf{x}}(0) = \begin{pmatrix} -10 \\ +10 \end{pmatrix}$$

2 Design a discrete time state estimator for the system in (1) using a sampling period $T = 0.1$ and setting the estimator eigenvalues at $z = 0.1 \pm j0.1$. (The necessary continuous-to-discrete time model conversion can be done manually, or by CAD.) Simulate the behaviour of the estimator under the conditions in (1).

3 Design Luenberger observers in the continuous and discrete time domains for the system in (1) using the procedures described in Sections 5.2.2 and 5.4.2. Simulate and test the observers' behaviour under the conditions in (1), and compare their performances with the asymptotic estimators designed in (1) and (2).

4 Write programs (FORTRAN, PASCAL or 'C') to solve
(a) the matrix Riccati Eqn (5.14), $t \to \infty$, for a single-input, single-output second order time-invariant system to yield the gain (Eqn (5.15)) of a suboptimal filter.
(b) the Kalman filter Eqns (5.35)–(5.37), $n \to \infty$, for a single-input, single-output second order system, to yield the gain of a suboptimal filter.

5 Use the programs developed in (4) to design suboptimal Kalman–Bucy and Kalman filters for the system in (1), this time corrupted with process and measurement noise, $\mathbf{w}(t)$, $v(t)$, strengths $\mathbf{Q} = \begin{pmatrix} 0.1 & 0 \\ 0 & 0.1 \end{pmatrix}$, $R = 0.1$, respectively, and assuming a sampling period of $T = 0.2$.
Simulate the behaviour of the filters under the conditions in (1), for
(a) $\omega(t) = 0$, $v(t) = 0$,
(b) the design conditions, namely $\mathbf{Q} = \begin{pmatrix} 0.1 & 0 \\ 0 & 0.1 \end{pmatrix}$, $R = 0.1$,
(c) $\mathbf{Q} = \begin{pmatrix} 0.1 & 0 \\ 0 & 0.1 \end{pmatrix}$, $R = 0$.
Note that in the simulation, $\mathbf{w}(t)$, $v(t)$ must in fact be band-limited.

6 Develop the program of (4(b)) to cover the case where the process and measurement noise vectors are correlated, using the scheme described in Section 5.6.4.

7 Using the program developed in (6), redesign the Kalman filter of (5) to cover the case where the cross-correlation matrix (Eqn (5.49)) of $\mathbf{w}(n)$, $v(n)$, is:

$$\mathbf{R}_{dwv}(k) = (0.5 \quad 0.5)^T \delta(k)$$

Simulate and test the filter under the conditions outlined in (1), using:
(a) uncorrelated $\mathbf{w}(n)$, $v(n)$,

(b) correlated $\mathbf{w}(n)$, $v(n)$, $\mathbf{R}_{dwv}(k) = (0.5 \quad 0.5)^T\delta(k)$,
(c) correlated $\mathbf{w}(n)$, $v(n)$, $\mathbf{R}_{dwv}(k) = (1.0 \quad 1.0)^T\delta(k)$.

8 The dynamic behaviour of a pendulum is modelled by the non-linear heuristic equations:

$$\dot{\mathbf{x}} = \begin{pmatrix} x_2 \\ -0.01x_2 - 0.5\sin x_1 \end{pmatrix} + \begin{pmatrix} 0 \\ u \end{pmatrix} + \begin{pmatrix} 0 \\ w \end{pmatrix}$$

$$y = (1 \quad 0)\mathbf{x} + v$$

where

x_1 is the angular displacement from vertical (rad),
x_2 is the angular velocity (rad s^{-1}),
u is an input torque (rad s^{-1}),
w is a Gaussian white noise torque, representing wind disturbance (rad s^{-2}), variance 0.2 ($\mathrm{rad}^2\,\mathrm{s}^{-4}$),
v is a Gaussian white noise displacement measurement error (rad), variance 0.1 (rad^2).

Design an extended Kalman filter, $T = 0.25$, using the scheme described in Section 5.8, and simulate the system and filter behaviour, $\mathbf{x}(0) = 0$,
(a) under noise-free conditions,
(b) for $w(t)$, $v(t)$ white Gaussian noise, variances 0.2 ($\mathrm{rad}^2\,\mathrm{s}^{-4}$), 0.1 ($\mathrm{rad}^2$) respectively,
(c) for $w(t)$, $v(t)$ white Gaussian noise, variances 0.5 ($\mathrm{rad}^2\,\mathrm{s}^{-4}$), 0.2 ($\mathrm{rad}^2$) respectively.

9 (a) Design a covariance square root filter algorithm (Section 5.10.1) for the Kalman filter of (5).
(b) Design a U–D factorized filter algorithm (Section 5.10.2) for the Kalman filter of (5).
Simulate and test these filters under the conditions outlined in (5).

10 Design an adaptive Kalman filter for the system of (8) (Section 5.9), and realize it in simulation. Test its behaviour under the conditions in (8), and compare it with that of the non-adaptive filter.

REFERENCES

[1] Bozic, S. M., *Digital and Kalman Filtering*, Arnold, London, 1979.
[2] Maybeck, P. S., *Stochastic Models, Estimation and Control*, Vol 1, Academic Press, New York, 1979.
[3] Kalman, R. E. and Bucy, R. S., 'New results in linear filtering and prediction theory', *Trans ASME J. Basic Eng.*, **83**, 95–108, 1961.
[4] Kalman, R. E., 'A new approach to linear filtering and prediction problems', *Trans ASME J. Basic Eng.*, **83**, 35–45, 1960.

[5] Leondes, C. T., 'Advances in the techniques and technology of the application of nonlinear filters and Kalman filters', *AGARDograph*, no. 256, 1982.
[6] Leondes, C. T., 'Theory and application of Kalman filtering', *AGARDograph*, no. 139, 1970.
[7] Press, W. H., Flannery, B. P., Teukolsky, S. A. and Vetterling, W. T., *Numerical Recipes: The art of scientific computing*, CUP, Cambridge 1986.
[8] Mehra, R. K., 'Approaches to adaptive filtering', *IEEE Trans. on Automatic Control*, **AC-17**(5), 693–8 (1972).
[9] Sorenson, H., 'Special issue on applications of Kalman filtering', *IEEE Trans. on Automatic Control*, 1983.

6 Control of stochastic systems

6.1 Introduction

The design of control strategies for stochastic systems presents all the difficulties encountered with deterministic systems in addition to those associated with noise. The general problem is represented in block diagram form in Figure 6.1, where a dynamical system, S, corrupted by process noise, $\mathbf{w}(t)$, and measurement noise, $\mathbf{v}(t)$, is controlled by a strategy, C, to which an input demand, $\mathbf{r}(t)$ is fed.

The control strategy, C, which may entail output or state feedback (or both), is in general required to ensure that the system outputs $\mathbf{y}(t)$, or some (or even all) of the states, $\mathbf{x}(t)$, respond 'well' to the input demand $\mathbf{r}(t)$ while minimizing the adverse effects of $\mathbf{w}(t)$, $\mathbf{v}(t)$. Where C is a 'regulator', $\mathbf{r}(t)$ has a fixed value, and the task of C is purely to minimize the effects of $\mathbf{w}(t)$; where C is a 'tracking' controller, the task is to follow some predefined track, $\tilde{\mathbf{x}}(t)$, and $\mathbf{r}(t)$ has some precalculated set of values; where the system is a servomechanism, C must be such that $\mathbf{y}(t)$, or some of the states, $\mathbf{x}(t)$, 'follow' $\mathbf{r}(t)$ closely, for any $\mathbf{r}(t)$.

Most deterministic control systems have natural noise reducing properties resulting from the low-pass filter characteristics of the systems themselves and from the naturally corrective effects of feedback strategies. Where noise levels are low, therefore, deterministic system designs which simply ignore noise can be satisfactory. A very extensive literature on deterministic control design methods exists (e.g. classical control [refs. 1, 2]; modern control [refs. 1, 2, 3]; on-line computer techniques [refs. 4, 5]), and while these are not explored in detail here, where the control of stochastic systems is the topic of interest, their relevance to the reduction of noise effects is examined, using continuous and discrete time classical control ideas, in Sections 6.2 and 6.3. Section 6.4 concerns minimum variance strategies, which are applicable to

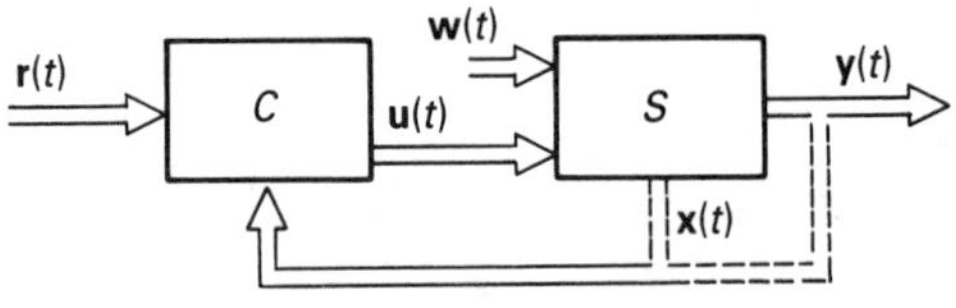

Figure 6.1 Control of a stochastic system.

some simple discrete systems. Sections 6.5 and 6.6 describe the design of deterministic regulators and tracking control strategies using linear quadratic optimal (LQO) control, and Section 6.7 deals with the application of these ideas to stochastic systems. In Section 6.8, the use of Kalman filters in control systems is considered, while Section 6.9 introduces the concept of H_∞ control.

6.2 Classical design – continuous time systems

In this section classical continuous time control systems, which are assumed to be linear time-invariant, single-input single-output, are considered from the point of view of their behaviour in the presence of noise. A typical classical control system subject to process noise is shown in Figure 6.2(a).

The linear time-invariant system, S, is subject to coloured (non-white) Gaussian process noise, $n(t)$. The system dynamics are represented by $G(s)$ and $G_N(s)$, which may contain common features; $n(t)$ is modelled by white noise, $w(t)$, processed by a filter, transfer function $G_F(s)$ (Section 2.11).

A classical deterministic control strategy, designed by one of the methods referred to in Section 6.1, typically consists of output feedback and a shaping network,

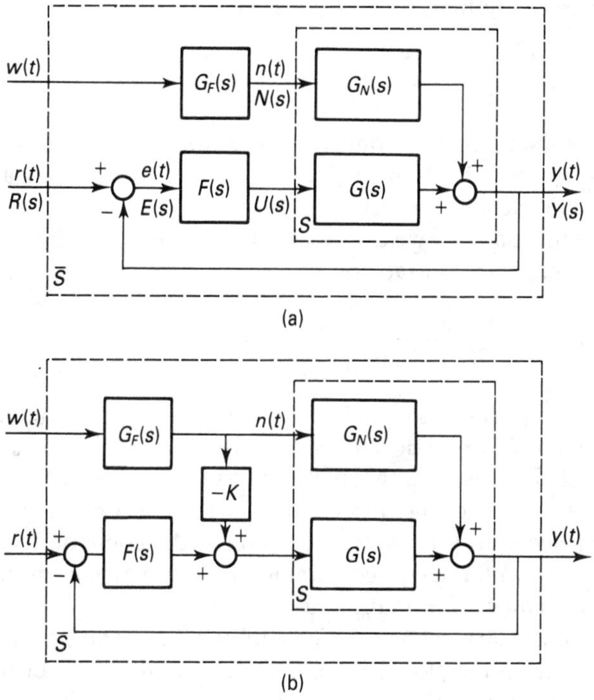

Figure 6.2 Classical system subject to process noise.

transfer function $F(s)$. If the closed loop system is a servomechanism, this strategy is designed so that the output, $y(t)$, follows the input, $r(t)$, well – with any error $(r(t) - y(t))$ removed, or at least reduced to negligible amplitude, quickly and stably. If the system is merely a regulator, $r(t)$ is constant and the objective of the control strategy is simply to reduce the effect of $n(t)$ as far as possible, while ensuring stable closed loop behaviour. In the case of a tracking controller, $r(t)$ is pre-defined as a required track, and the control strategy ensures that $y(t)$ follows this closely.

The s-domain model of the closed loop system, S, is

$$Y(s) = G(s)F(s)(R(s) - Y(s)) + G_N(s)N(s) \tag{6.1}$$

where

$Y(s)$ is the Laplace transform of the output, $y(t)$,
$R(s)$ is the Laplace transform of the input, $r(t)$,
$N(s)$ is the Laplace transform of the noise, $n(t)$.

Setting $N(s) = 0$ in Eqn (6.1) yields the transfer function between $R(s)$ and $Y(s)$:

$$\frac{Y(s)}{R(s)} = \bar{G}(s) = \frac{G(s)F(s)}{1 + G(s)F(s)} \tag{6.2}$$

Correspondingly, the closed loop frequency response is

$$\frac{Y(j\omega)}{R(j\omega)} = \bar{G}(j\omega) = \frac{G(j\omega)F(j\omega)}{1 + G(j\omega)F(j\omega)} \tag{6.3}$$

The ideal – but obviously unattainable – control system would have $\bar{G}(j\omega) = 1$ (i.e. $|\bar{G}(j\omega)| = 1$, $\underline{/\bar{G}(j\omega)} = 0$, $0 \leqslant \omega \leqslant \infty$), but in practice, characteristics such as those shown in Figure 6.3(a), are usually achieved.

Setting $R(s) = 0$ in Eqn (6.1) yields the transfer function between $N(s)$ and $Y(s)$. The corresponding frequency characteristic is

$$\frac{Y(j\omega)}{N(j\omega)} = \bar{G}_N(j\omega) = \frac{G_N(j\omega)}{1 + G(j\omega)F(j\omega)} \tag{6.4}$$

In the ideal – unattainable – system, this would be zero, but, in practice, characteristics of the kind shown in Figure 6.3(b), can be achieved using classical control design techniques. For servomechanism design, the dual objective is to choose $F(s)$ so that a good system response to $r(t)$ and high attenuation of the noise $n(t)$ are realized. These objectives are not necessarily incompatible, though the common features in Eqns (6.3) and (6.4) do represent a design constraint. Good servomechanism design can, in fact, yield good regulator performance automatically.

'Feed-forward' control strategies, in which additional demand signals are fed directly into some point within the closed loop, are a common feature of practical classical systems. The control strategy, in these cases, is left with the residual task of removing error in the system response to the feed-forward signal, and this can be

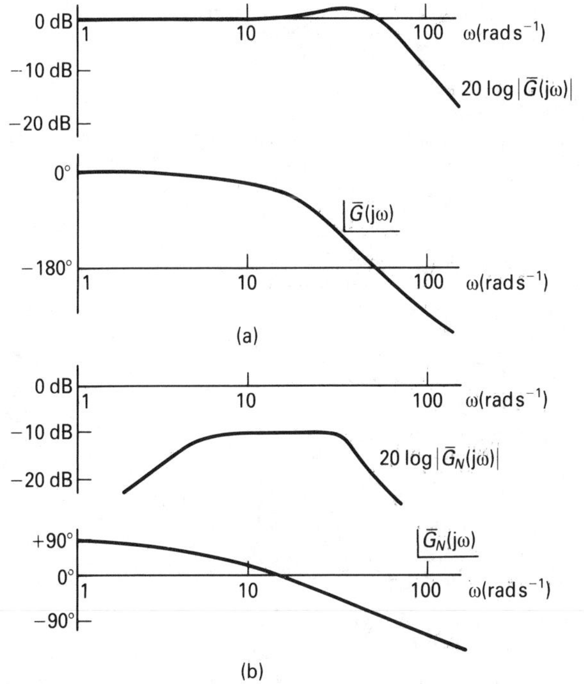

Figure 6.3 Typical closed loop system characteristics.

advantageous. If the process noise, $n(t)$, is measurable, or if it can be estimated, a feed-forward strategy using $n(t)$ can be very effective. This is illustrated in Figure 6.2(b), where the feed-forward gain, $-K$, would normally be chosen to nullify the steady state effects of $n(t)$ on $G(s)$:

$$K = \left[\frac{G_N(s)}{G(s)}\right]\bigg|_{s=0}$$

These points are illustrated by example.

Example

The volumetric flow rate of a liquid in an industrial process is monitored by a suitable transducer and controlled via an amplifier, actuator and valve, as illustrated in Figure 6.4(a). The liquid density varies randomly so that a given valve position does not correspond to a constant mass flow rate, as indicated by the sample function in Figure 6.5(b).

A block diagram model of the system, including a classical feedback control strategy, $F(s)$, is constructed, as shown in Figure 6.5(c); $F(s)$ is to be designed (cf.

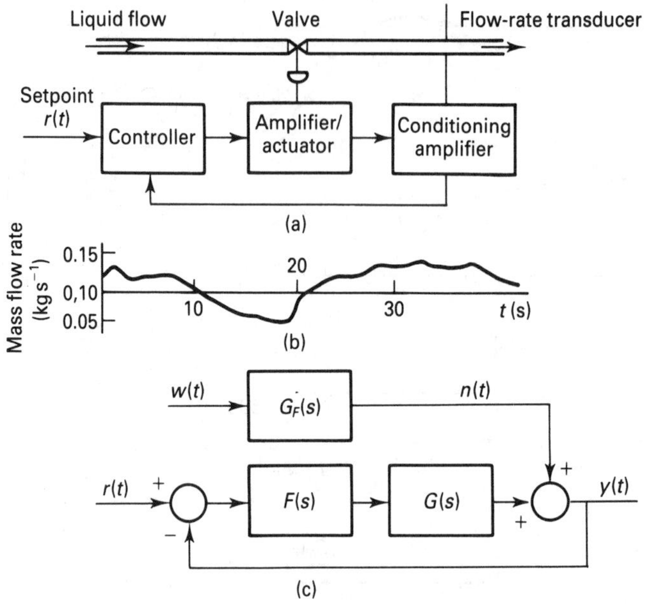

Figure 6.4 Stochastic system and classical control.

Figure 6.2). The dynamical behaviour of the amplifier, actuator, valve, liquid, and transducer is represented by the transfer function:

$$G(s) = \frac{6e^{-0.3s}}{(s+2)(s+3)}$$

The effect of liquid density variation is represented by white noise, $w(t)$, of unit strength, and the filter:

$$G_F(s) = \frac{2}{(s+0.2)}$$

A controller, incorporating integration to nullify steady state error, is designed, transfer function $F(s)$, by classical means (Bode plots, Nyquist diagrams, root locus plots, Nichols charts):

$$F(s) = \frac{10(s+1.5)}{s+15}$$

Closed loop frequency responses $G(j\omega)$, $G_N(j\omega)$ (Eqns (6.3) and (6.4)) are shown in Figure 6.5(a, b). These are both reasonably good in that $G(j\omega) \doteqdot 1$, at least for frequencies up to about 1 rad s^{-1}, while $\bar{G}_N(j\omega)$ is very small over the same range. A closed loop step function response, $w(t) = 0$, is shown in Figure 6.5(c), and sample functions of $n(t)$, $y(t)$, $r(t) = 0$, for the open and closed loop systems are shown in Figure 6.5(d): the improvement realized by the control strategy is obvious.

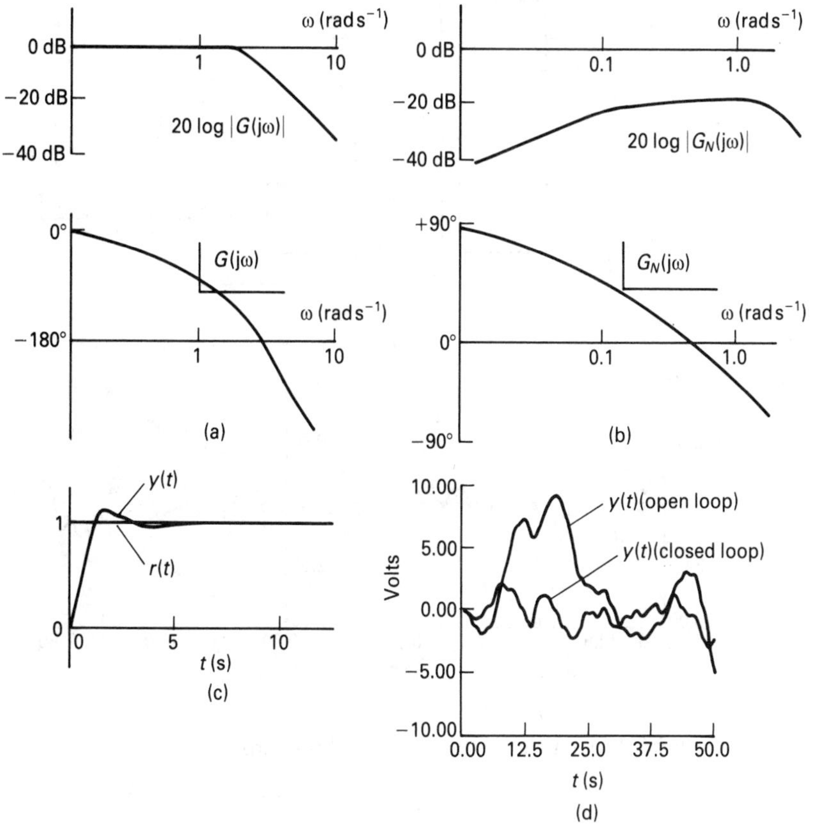

Figure 6.5 Stochastic system characteristics and responses.

If the liquid density can be measured by a suitable transducer mounted near the flow rate transducer – which may be difficult in practice – a feed-forward strategy, Figure 6.6(a), may be a possibility. K is chosen to nullify steady state noise effects in the open loop system (Eqn (6.5)):

$$K = \left(\frac{1(s+2)(s+3)}{6e^{-0.3s}} \right) \Bigg|_{s=0} = 1$$

A sample function of $y(t)$, $K = 1$, Figure 6.6(b), shows a worthwhile improvement over that in Figure 6.6(c) ($K = 0$) and illustrates the possible benefit of feeding forward the noise signal.

6.3 Classical design – discrete time systems

The shaping network $F(s)$, Figure 6.2, can be replaced with an on-line digital computer – usually a microprocessor – with appropriate analog-to-digital and digital-

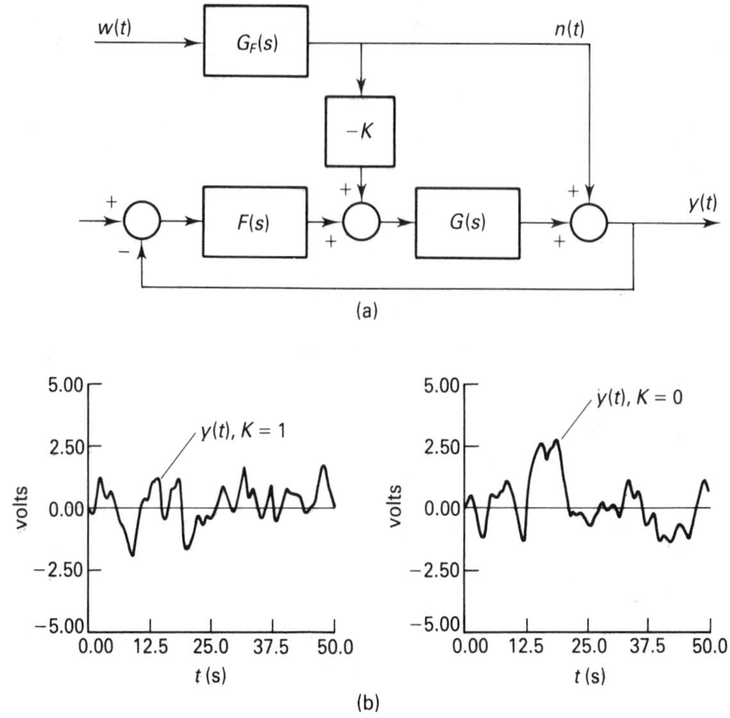

Figure 6.6 Stochastic system with feed forward.

to-analog converters (ADCs and DACs). An open loop sampled data system appropriate to this is shown in Figure 6.7(a) and the corresponding closed loop system in Figure 6.7(b). Figure 6.7(c) shows a closed loop system with a feed-forward control strategy, applicable where $n(t)$ can be measured.

The discrete time system Σ consists of the system S of Figure 6.2 with a DAC and ADC buffering its input, $u(t)$, and output, $y(t)$, respectively, as illustrated in Figure 6.7(a), allowing interfaces with a digital computer via signals $u(n)$, $y(n)$. The sampling interval is T. A typical closed loop system is shown in Figure 6.7(b); $\bar{G}(z)$ is the z domain transfer function of the DAC, $G(s)$, and ADC (Appendix A, Eqn (A1.19)):

$$\bar{G}(z) = Z\left\{\frac{1 - e^{-sT}}{s} \cdot G(s)\right\}$$

where the right-hand side is interpreted as the z transform 'equivalent' to the Laplace transform in the brackets (Appendix A, Figures A1.1, A1.3). $\bar{D}(z)$ represents the z transform of the control algorithm, $\bar{R}(z)$, $\bar{E}(z)$, $\bar{V}(z)$, $\bar{Y}(z)$ are the z transforms of the

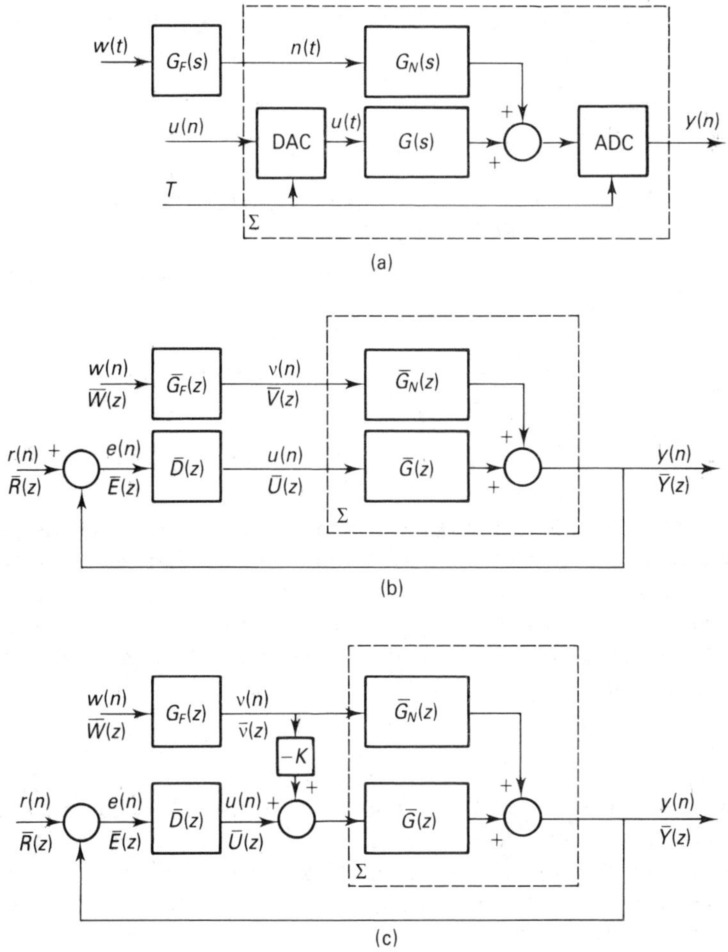

Figure 6.7 Classical sampled data system subject to process noise.

discrete signals $r(n)$, $e(n)$, $v(n)$, $y(n)$, as shown in the diagram, $v(n)$ (the system noise) is represented by white noise $w(n)$, filtered by a digital filter, transfer function $\bar{G}_F(z)$, and applied to Σ within which a transfer function $\bar{G}_N(z)$ models its effect, $w(n)$ and $\bar{G}_F(z)$ are defined (Section 3.9) in such a way that their effect on the system is indistinguishable, at sampling instants, from that of $w(t)$ and $G_F(s)$.

The control strategy is designed by one of the methods mentioned in Section 6.1 (discrete time domain Bode plot, Nyquist diagram, Nichols chart, roots locus [refs. 2, 4, 5] and realized by a microprocessor algorithm, transfer function $\bar{D}(z)$.

The closed loop z-domain model is (cf. Eqn (6.1))

$$\bar{Y}(z) = \bar{G}(z)\bar{D}(z)[\bar{R}(z) - \bar{Y}(z)] + \bar{G}_N(z)\bar{N}(z) \tag{6.5}$$

As with the continuous time system, the responses of the system output, $y(n)$, to $r(n)$ and $v(n)$ are found:

$$\frac{\bar{Y}(z)}{\bar{R}(z)} = \bar{\bar{G}}(z) = \frac{\bar{G}(z)\bar{D}(z)}{1 + \bar{G}(z)\bar{D}(z)} \tag{6.6}$$

$$\frac{\bar{Y}(z)}{\bar{v}(z)} = \bar{\bar{G}}_N(z) = \frac{\bar{G}_N(z)}{1 + \bar{G}(z)\bar{D}(z)} \tag{6.7}$$

The frequency responses, or Fourier transfer functions, are given by replacing z with $e^{j\beta}$ (Appendix A, Section A1.3).

$$\bar{\bar{G}}(e^{j\beta}) = \frac{\bar{G}(e^{j\beta})\bar{D}(e^{j\beta})}{1 + \bar{G}(e^{j\beta})\bar{D}(e^{j\beta})} \tag{6.8}$$

$$\bar{\bar{G}}_N(e^{j\beta}) = \frac{\bar{G}_N(e^{j\beta})}{1 + \bar{G}(e^{j\beta})\bar{D}(e^{j\beta})} \tag{6.9}$$

where β is the frequency (rad per sample), and $\beta = \omega T$. In the ideal case, which is of course unattainable, $\bar{\bar{G}}(e^{j\beta}) = 1$ and $\bar{\bar{G}}_N(e^{j\beta}) = 0, 0 \leqslant \beta \leqslant \pi$. In practice, characteristics equivalent to those of Figure 6.3, $\omega = \beta/T$ can usually be realized. The dual objective of good design is to choose $\bar{D}(z)$ so that good system response to $r(n)$, and high attenuation of $w(n)$, are achieved. As before, of course, if $r(n)$ can be measured, feed forward can be helpful.

Example

An automated bacon slicer delivers groups of five slices of bacon to a weighing machine where each group is weighed after a delay of three group-times. The group weight is fed to a control strategy which manipulates the slice thickness of the following group. Figure 6.8(a) shows the arrangement. The surface areas and densities of the slices vary considerably, and a control strategy is devised to minimize the group weight variance.

In Figure 6.8(b), the group weights are modelled as deviations from a datum weight, yielding a linear model with zero input demand – a linear regulator. The noise, $v(n)$, is modelled by white noise, $w(n)$, processed by a filter with transfer function:

$$\bar{G}_F(z) = \frac{0.05z}{z - 0.95}$$

Also,

$$\bar{G}_N(z) = 1$$

Assuming a well-behaved local servomechanism controlling the slice thickness (held constant during each slice group), the machine transfer function is:

$$\bar{G}(z) = \frac{1}{z^3}$$

This yields the open loop frequency response to noise:

$$\bar{G}_F(e^{j\beta}) = \frac{\bar{Y}(e^{j\beta})}{\bar{W}(e^{j\beta})} = \frac{0.05e^{j\beta}}{e^{j\beta} - 0.95}$$

This is illustrated in Figure 6.9(a).

A closed loop control strategy is designed using classical techniques in the discrete time domain to yield good behaviour, which in this case amounts to achieving the closed loop noise frequency response approximating that of Figure 6.9(b). The control strategy transfer function is

$$\bar{D}(z) = \frac{0.4(z+1)(z-0.3)^2}{(z-1)(z+0.2)^2}$$

This corresponds, in the time domain, to the algorithm (Appendix A, Eqns (A1.15) and (A1.17)):

$$u(n) = 0.4e(n) + 0.16e(n-1) - 0.20e(n-2) + 0.036e(n-3)$$
$$+ 0.60u(n-1) + 0.36u(n-2) + 0.04u(n-3)$$

The closed loop frequency response, Eqn (6.9), is illustrated in Figure 6.9(b), and sample function responses of the open and closed loop systems to (filtered) white noise, in Figure 6.9(d). Figure 6.9(b) also indicates the 'perfect' closed loop frequency response attainable in this case: a zero gain over the frequency range $0 \leqslant \beta < \pi/3$ and unity gain $\pi/3 \leqslant \beta \leqslant \pi$, corresponding to the delay of three samples in the open loop system (which disallows the control strategy from reacting in under three samples).

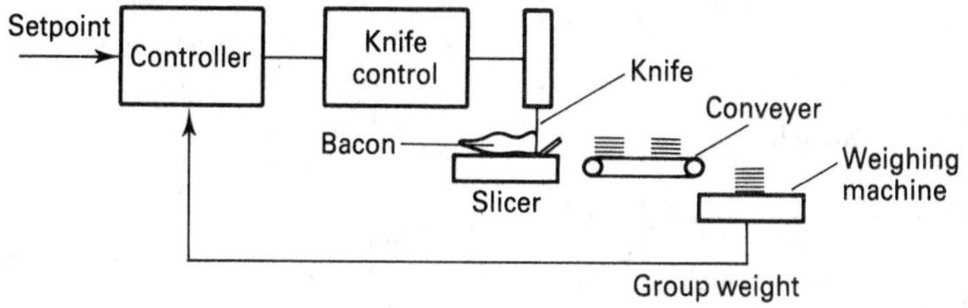

(a)

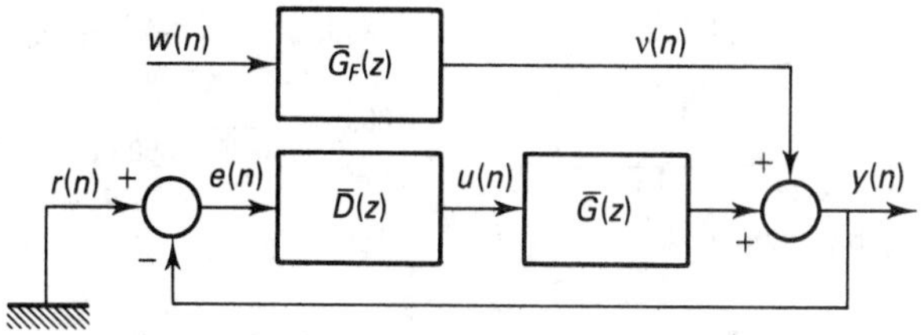

(b)

Figure 6.8 Bacon slicer and dynamical model.

The noise frequency response would no doubt be materially improved by a feed-forward strategy, which would entail measuring $v(n)$. However, this seems impractical.

Figure 6.9(d) shows the improvement of the closed loop behaviour over that of the open loop; as in the continuous time example, designing for good servomechanism behaviour automatically yields a good regulator design.

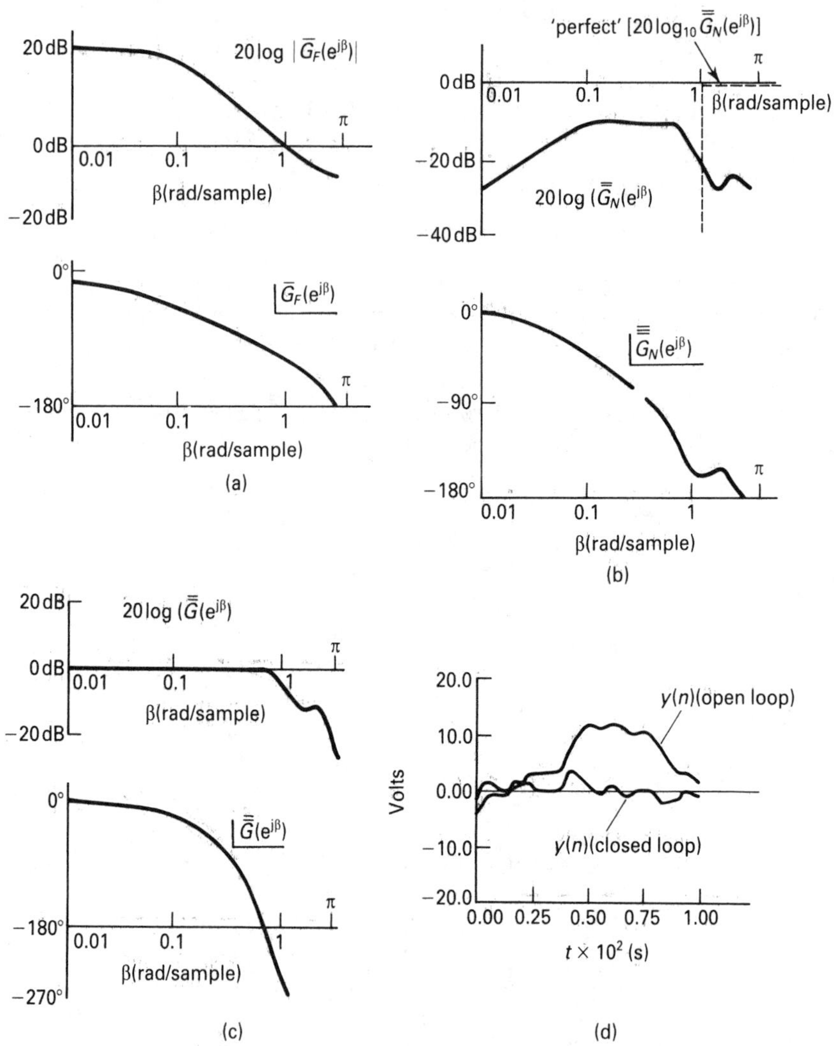

Figure 6.9 Bacon slicer behaviour.

6.4 Minimum variance regulator design

Where the open loop system is relatively simple, an effective regulator control algorithm can sometimes be designed algebraically with a view simply to reducing the variance of the system output, $y(n)$, to a minimum, regardless of stability or other considerations [ref. 6]. This can result in the cancellation of system poles with control strategy zeros, and so in an uncontrollable closed loop system (Appendix A, Section A3.3), but where this is tolerable, i.e. where the uncontrollable modes are stable and 'fast', useful results can be achieved, possibly with some empirical manipulation.

No feedback strategy can, of course, nullify system delay, and in the following analysis delay terms are modelled separately to yield system and noise transfer functions $z^{-k}\bar{G}(z)$, $\bar{G}_N(z)$, respectively, where $\bar{G}(z)$, $\bar{G}_N(z)$ are proper fractions in z.

The system model shown in Figure 6.10 is simply that of Figure 6.7(b), with $r(n) = 0$, and $\bar{G}(z)$ slightly redefined.

The open loop system behaviour is given by

$$\bar{Y}(z) = z^{-k}\bar{G}(z)\bar{U}(z) + \bar{G}_N(z)\bar{G}_F(z)\bar{W}(z) \tag{6.10}$$

where

$z^{-k}\bar{G}(z)$ is the open loop system transfer function,
$\bar{D}(z)$ is the regulator control strategy,
$\bar{G}_N(z)$, $\bar{G}_F(z)$, $\bar{W}(z)$ represent the stochastic behaviour of the system.

For the purpose of understanding this design method, which is based on causality, it is useful to recast Eqn (6.10):

$$\bar{Y}(z) = \frac{A^*(z^{-1})}{z^k B^*(z^{-1})} \cdot \bar{U}(z) + \frac{C^*(z^{-1})}{B^*(z^{-1})} \cdot \bar{W}(z) \tag{6.11}$$

where $A^*(z^{-1})$, $B^*(z^{-1})$, $C^*(z^{-1})$ are polynomials in z^{-1} and may be regarded as backward operators on the processes by which they are multiplied.

$B^*(z^{-1})$ is easily found from the lowest common denominator of $\bar{G}(z)$, $\bar{G}_N(z)$, $\bar{G}_F(z)$. $A^*(z^{-1})$, $C^*(z^{-1})$ follow accordingly.

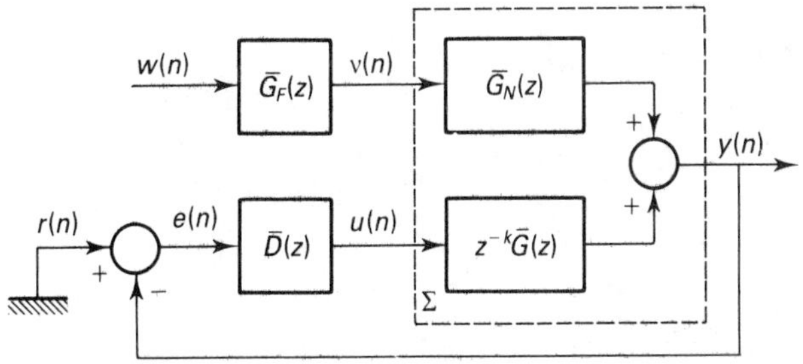

Figure 6.10 Minimum variance regulator.

At time n, $w(n)$ may be calculated using the information available, namely $u(n-1), u(n-2), \ldots, y(n), y(n-1) \ldots$. This is given by Eqn (6.11):

$$\bar{W}(z) = \frac{B^*}{C^*}\left[\bar{Y}(z) - \frac{A^*}{z^k B^*}\bar{U}(z)\right] \tag{6.12}$$

Also from Eqn (6.11):

$$z^k \bar{Y}(z) = \frac{A^*}{B^*}\bar{U}(z) + z^k \frac{C^*}{B^*}\bar{W}(z) \tag{6.13}$$

In the time domain, the terms in $(A^*/B^*)\bar{U}(z)$ represent terms occurring at n, $(n-1)$, $(n-2), \ldots$, all of which are known at time n. The terms in $z^k(C^*/B^*)\bar{W}(z)$ represent terms occurring at times $(n+1)$, $(n+2), \ldots, (n+k)$, in addition to others at n, $(n-1)$, $(n-2), \ldots$. Splitting it into the two parts corresponding to these:

$$z^k \frac{C^*}{B^*}\bar{W}(z) = z^k F^*(z^{-1})\bar{W}(z) + \frac{G^*}{B^*}\bar{W}(z) \tag{6.14}$$

where F^*, G^* are easily derived from $z^k(C^*/B^*)$, so that all the terms in $z^k F^* \bar{W}(z)$ contain positive powers of z and so represent terms occurring at $(n+1)$, $(n+2), \ldots, (n+k)$.

Substituting from Eqn (6.14) in Eqn (6.13) and then from Eqn (6.12) in the result gives

$$z^k \bar{Y}(z) = \left[\frac{A^*}{B^*}\bar{U}(z) + \frac{G^*}{C^*}\left(\bar{Y}(z) - \frac{A^*}{z^k B^*}\bar{U}(z)\right)\right] + z^k F^* \bar{W}(z) \tag{6.15}$$

Eqn (6.15) models the fact that, in the time domain, $y(n+k)$ is affected by $u(n)$, $u(n-1), \ldots, y(n), y(n-1) \ldots$ and by $w(n+1)$, $w(n+2), \ldots, w(n+k)$. Clearly, the variance of $y(n+k)$ is minimized if $u(n)$ is chosen to nullify the first of these functions, represented in the z domain by the terms in the brackets; the second function, involving terms in $w(n+1)$, $w(n+2), \ldots, w(n+k)$, is obviously unaffected by any choice of $u(n)$. Thus, the minimum variance strategy is such that

$$\frac{A^*}{B^*}\bar{U}(z) + \frac{G^*}{C^*}\left(\bar{Y}(z) - \frac{A^*}{z^k B^*}\bar{U}(z)\right) = 0$$

Thus, recalling that (Eqn (6.14))

$$\frac{z^k C^* - G^*}{B^*} = z^k F^*$$

$$\bar{U}(z) = -\bar{D}(z)\bar{Y}(z)$$

$$= -\frac{G^*(z^{-1})}{A^*(z^{-1})F^*(z^{-1})} \cdot \bar{Y}(z) \tag{6.16}$$

Eqn (6.16) is the minimum variance control strategy giving the algorithm $\bar{D}(z)$ (Figure

6.11); this is readily expressed in the time domain (Appendix A, Eqns (A1.15) and (A1.17)).

The stability of Eqn (6.16) depends on all the zeros of $A^*(z^{-1})$ lying within the z plane unit circle – otherwise the algorithm is unstable. If they do not lie within the unit circle – and there is nothing in the design method to ensure that they do – more complex strategies can be devised [ref. 6].

There is, of course, no guarantee that the behaviour of the closed loop system designed by this method is satisfactory in any respect other than the reduction of the output variance to a minimum; indeed, the closed loop system might be unstable. Its properties must be checked independently using the familiar techniques of classical control, and in cases where Eqn (6.16) cannot be used directly, it may at least suggest the structure of a satisfactory strategy. There is, of course, no reason why noise feed-forward should not be used in conjunction with such a strategy, if the noise can be measured.

Example

Consider the Example of Section 6.3 concerning a regulator design for a bacon slicer (Figure 6.7).

Recasting the system model in the form of Eqn (6.11):

$$\bar{Y}(z) = \frac{1 - 0.95z^{-1}}{z^3(1 - 0.95z^{-1})}\bar{U}(z) + \frac{0.05}{(1 - 0.95z^{-1})}\bar{W}(z)$$

From Eqn (6.14):

$$\frac{0.05z^3}{1 - 0.95z^{-1}}\bar{W}(z) = (0.05z^3 + 0.047z^2)\bar{W}(z) + \frac{0.045}{1 - 0.95z^{-1}}\bar{W}(z)$$

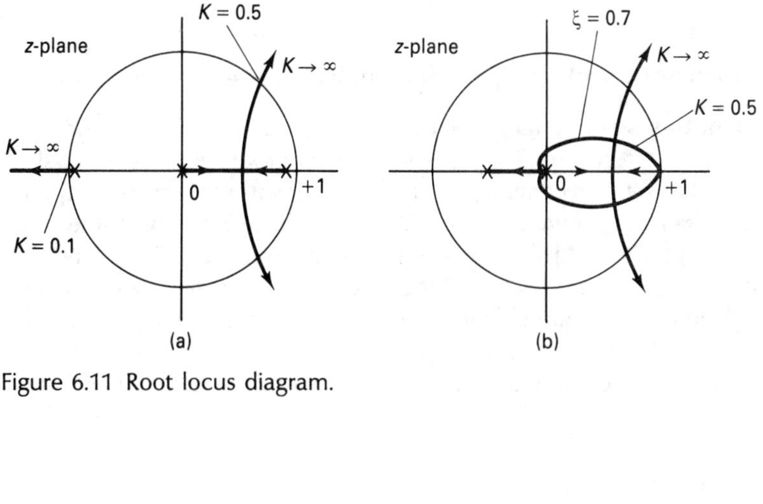

Figure 6.11 Root locus diagram.

Thus,

$$F^*(z^{-1}) = 0.05 + 0.0475z^{-1} = 0.05(1 + 0.95z^{-1})$$

$$G^*(z^{-1}) = 0.045$$

$$A^*(z^{-1}) = 1 - 0.95z^{-1}$$

The minimum variance control algorithm (Eqn (6.16)) is

$$\bar{U}(z) = \frac{-0.902}{(1 - 0.950z^{-1})(1 + 0.950z^{-1})}\bar{Y}(z)$$

or (Figure 6.10)

$$\bar{D}(z) = \frac{0.902z^2}{(z - 0.950)(z + 0.950)}$$

This algorithm, with its zeros at $z = 0$, in series with the bacon slicer with its three poles at $z = 0$, constitutes an uncontrollable system. Worse, a brief consideration of the root locus diagram for the system using this algorithm shows the closed loop system to be unstable (Figure 6.11(a)).

In spite of these problems, however, the structure of the algorithm looks promising, and some manipulation of its poles yields a much improved root locus diagram. This is shown in Figure 6.11(b). In this case:

$$\bar{D}(z) = \frac{0.5z^2}{(z - 1)(z + 0.5)} = \frac{0.5z^2}{z^2 - 0.5z - 0.5}$$

The corresponding algorithm (Appendix A, Eqns (A1.15) and (A.71)) is

$$u(n) = -0.5y(n) + 0.5u(n - 1) + 0.5u(n - 2)$$

This is not, of course, strictly a minimum variance strategy, but it is not too far removed from it, and it gives the closed loop system a performance almost indistinguishable from that shown in Figure 6.9(d).

6.5 Optimal control for linear deterministic systems – continuous time

Modern control system design, based on state space analysis (Appendix A, Section A2), depends on state feedback, as opposed to the output feedback of classical control systems. Of the many techniques available, that with the readiest application to stochastic systems is optimal control, where the objective is to drive the system states in such a way that some defined cost function is minimized. This turns out to have useful application in the design of regulators, where some steady state is to be maintained, and to tracking control strategies, where some predetermined state trajectory is to be followed. Optimal control does not apply readily to the design of servomechanisms, where any demand is to be followed – the fulfilment of a hitherto undefined requirement cannot be optimized, since a cost function cannot, in these circumstances, be specified.

Practical systems are very commonly corrupted with noise, and an attempt to apply optimal control strategies to stochastic cases, which are repeatedly driven away from some required target or trajectory by process noise, seems natural. This section outlines basic optimal control concepts and describes their application to linear deterministic continuous time systems, while Section 6.6 deals with the equivalent discrete time topics. Sections 6.7 and 6.8 describe the application of the ideas to linear stochastic systems.

Consider an mth order continuous time dynamical system, S (Eqns (2.67) and (2.68)),

$$\dot{\mathbf{x}} = \mathbf{f}(\mathbf{x}, \mathbf{u}, t) \tag{6.17}$$

$$\mathbf{y} = \mathbf{g}(\mathbf{x}, \mathbf{u}, t) \tag{6.18}$$

where $\mathbf{u}$, $\mathbf{x}$, $\mathbf{y}$, are the system inputs, states, and outputs respectively. The generalized objective of optimal control is to design an input $\mathbf{u}^*(t)$, normally some function of the states, $\mathbf{x}(t)$, which drives $\mathbf{x}(t)$ from its initial states, $\mathbf{x}(t_0)$, towards some target states, $\mathbf{x}_T$, while minimizing a scalar cost function which is practically useful.

A cost function structure which renders the problem mathematically tractable, and which is also useful, is

$$V = l(\mathbf{x}(t_f), \mathbf{u}(t_f), t_f) + \int_{t_0}^{t_f} L(\mathbf{x}(t), \mathbf{u}(t), t)\,\mathrm{d}t \tag{6.19}$$

The control process occurs over the time interval $t_0 \leqslant t \leqslant t_f$; $l(\mathbf{x}(t_f), \mathbf{u}(t_f), t_f)$ is a 'terminal cost' which represents a penalty quantifying the deviation of $\mathbf{x}(t_f)$ from the target states, $\mathbf{x}_T$; $\int_{t_0}^{t_f} L(\mathbf{x}(t), \mathbf{u}(t), t)\,\mathrm{d}t$ is a 'running cost' which accumulates during the control process; V can be cast to represent deviation from the target states, $\mathbf{x}_T$, providing a means of designing optimal regulators, or deviation from a predefined trajectory, providing a means of designing optimal tracking strategies.

The arrangement is shown in block form in Figure 6.12, in which the optimal strategy, designated C^*, generates the optimal forcing function, $\mathbf{u}^*(t)$, using the system states, designated $\mathbf{x}^*(t)$ for the optimally controlled case, and a closed loop system input, $\mathbf{r}^*(t)$, which may be required by the optimal strategy.

A Hamiltonian scalar function is defined:

$$H = L(\mathbf{x}, \mathbf{u}, \mathrm{t}) + \mathbf{f}^T(\mathbf{x}, \mathbf{u}, t)\,.\,\mathbf{p}(\mathbf{x}, \mathbf{u}, t) \tag{6.20}$$

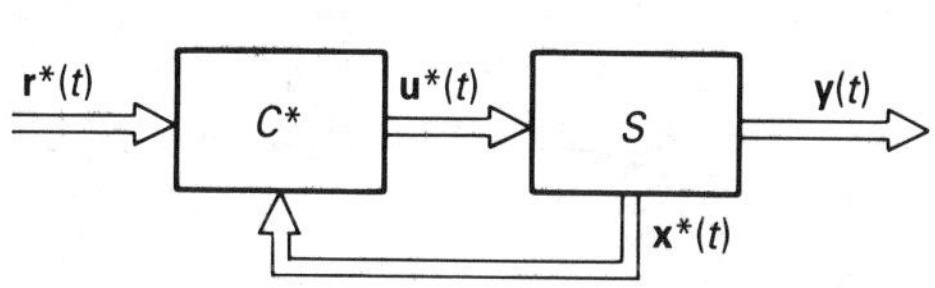

Figure 6.12 Optimal control system.

where $\mathbf{p}$ is a 'costate' vector governed by the costate equations

$$\dot{\mathbf{p}} = -\left(\frac{\partial H}{\partial \mathbf{x}}\right)^T \tag{6.21}$$

Pontryagin's principle [ref. 7] states that the control, $\mathbf{u}^*(t)$, which minimizes V, i.e. the optimal control, is that which minimizes H WRT $\mathbf{u}(t)$.

As a mathematical exercise, the design of $\mathbf{u}^*(t)$ takes the following sequence:

1. Define V (Eqn (6.19)), and, by implication, L and l.
2. Define H using L, Eqns (6.17) and (6.20), and the unknown costates, $\mathbf{p}(t)$. Hence, find $\mathbf{u}^*(t)$, the value of $\mathbf{u}(t)$ which minimizes H WRT $\mathbf{u}(t)$ in terms of $\mathbf{x}(t)$, $\mathbf{p}(t)$.
3. Write the costate equations (Eqn (6.21)) describing the behaviour of $\mathbf{p}(t)$.
4. Solve Eqns (6.17) and (6.21) with $\mathbf{u}(t) = \mathbf{u}^*(t)$, using the mixed boundary value conditions $\mathbf{x}(t_0)$, $\mathbf{x}_T$. Since these boundary values are not all initial values this is difficult, or even impossible in many cases, and various methods of overcoming or avoiding this difficulty exist. In principle, however, this yields the optimal state and costate, $\mathbf{x}^*(t)$ and $\mathbf{p}^*(t)$.
5. Substitute $\mathbf{x}^*(t)$, $\mathbf{p}^*(t)$ in the expression for $\mathbf{u}^*(t)$ (stage (2)) to yield the optimal control strategy $\mathbf{u}^*(t)$.

A simple example illustrates this procedure.

Example

Consider the R–C network of Figure 6.13(a). The behaviour of S is modelled by

$$\dot{x} = -x + u$$

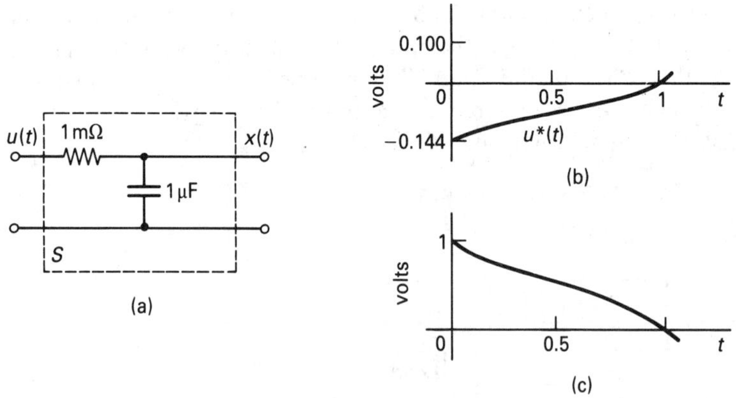

Figure 6.13 Optimal control system.

where

u represents the system input (volts),
x represents the single system state, and output (volts).

Suppose it is required to drive the state from $x(0) = 1$ to $x(1) = 0$, keeping $x(t)$ as small as possible throughout the process, meanwhile holding the amplitude of $u(t)$ within some reasonable limits:

1. A cost function corresponding to these requirements is specified (Eqn 6.19):

$$V = \int_0^1 (x^2 + u^2)\,dt$$

2. The Hamiltonian (Eqn (6.20)) is

$$H = x^2 + u^2 + (-x + u)p$$

H is minimized WRT u:

$$\frac{\partial H}{\partial u} = 2u + p = 0$$

The optimal control is therefore

$$u^*(t) = -0.5p^*(t)$$

3. The costate equation is (Eqn (6.21))

$$\dot{p} = -2x + p$$

4. Substituting $u^*(t)$ in the state and costate equations yields two differential equations describing the behaviour of the optimally controlled state and costate, $0 \leqslant t \leqslant 1$:

$$\dot{x}^* = -x^* - 0.5p^*$$

$$\dot{p} = -2x^* + p^*$$

The boundary conditions, which are mixed between initial and final times, are

$$x^*(0) = 1, \quad x^*(1) = x_T = 0$$

Because these are not both initial conditions, the differential equations are not soluble by Runge–Kutta or other similar techniques, and further progress is blocked. However, in this simple case the solution is known in closed form:

$$x^*(t) = 1.085e^{-\sqrt{2}t} - 0.085e^{+\sqrt{2}t}$$

$$p^*(t) = 0.439e^{-\sqrt{2}t} - 0.152e^{+\sqrt{2}t}$$

5. Hence, (Eqn (6.22)), the optimal control is

$$u^*(t) = -0.220e^{-\sqrt{2}t} + 0.076e^{+\sqrt{2}t}$$

The effect of this strategy (which is not in state feedback form) on the system behaviour is illustrated in Figure 6.13(b, c).

The ideas outlined so far do not amount to a design procedure, since the specification of $u^*(t)$ depends on solving an intractable mixed boundary value set of differential equations. If this could be avoided, and if $u^*(t)$ could be expressed as a function of $x^*(t)$, a practical state feedback control strategy would result.

If the system, S, is linear and the cost function is 'quadratic' in form, these problems can be solved to yield 'linear quadratic optimal' (LQO) control strategies for regulators and tracking strategies.

6.5.1 LQO regulator design

The objective of an optimal regulator is to generate a set of inputs, $\mathbf{u}^*(t)$, to drive the system from any initial states, $\mathbf{x}(t_0)$ to the target, $\mathbf{x}_T = \mathbf{0}$, while minimizing some weighted measure of the state amplitudes – and to maintain the states at $\mathbf{0}$ thereafter. A non-zero target state is accommodated by a suitable redefinition of the state variables.

Alternatively, it is possible to specify that a regulator controls the outputs, rather than the states, of a system. In the great majority of practical cases, however, the state variables can be defined directly to include the variables to be controlled; the regulation of the system state is therefore explored here.

Consider an mth order linear system with r inputs (Eqn (2.69):

$$\dot{\mathbf{x}} = \mathbf{A}(t)\mathbf{x} + \mathbf{B}(t)\mathbf{u} \tag{6.22}$$

A quadratic cost function (cf. Eqn (6.19)) is defined:

$$V = \int_{t_0}^{t_f} (\mathbf{x}^T\mathbf{E}\mathbf{x} + \mathbf{u}^T\mathbf{F}\mathbf{u})\,\mathrm{d}t \tag{6.23}$$

t_0, t_f are the initial and final times over which the control process is formally specified.

$\mathbf{E}$ is a positive semi-definite $(m \times m)$ constant matrix chosen to describe the quadratic function of the states to be minimized during the control process. If $\mathbf{E}$ is diagonal, it merely weights the relative importance of the different states to be minimized in the cost function. Otherwise, it can be used to specify state trajectories of a certain kind to yield a limited class of tracking strategy.

$\mathbf{F}$ is a positive definite $(r \times r)$ constant matrix chosen to constrain the amplitudes of the system input $\mathbf{u}^*(t)$ demanded by the control strategy. Essentially, the larger the elements of $\mathbf{F}$, the more constrained are the corresponding elements of $\mathbf{u}^*(t)$. Experimentation is used to select $\mathbf{F}$ so that $\mathbf{u}^*(t)$ never exceeds the control authority available in the system.

It can be shown [ref. 7] that the control which drives S from $\mathbf{x}(t_0)$ *to* $\mathbf{x}(t_f)$, while minimizing V, is

$$\mathbf{u}^*(t) = -\mathbf{K}(t)\mathbf{x}^*(t) \tag{6.24}$$

where

$$\mathbf{K}(t) = \mathbf{F}^{-1}\mathbf{B}^T(t)\mathbf{P}(t) \tag{6.25}$$

and $\mathbf{P}(t)$ is the $(m \times m)$ matrix solution of the 'Riccati' differential equation:

$$\dot{\mathbf{P}}(t) = -\mathbf{P}(t)\mathbf{A}(t) - \mathbf{A}^T(t)\mathbf{P}(t) + \mathbf{P}(t)\mathbf{B}(t)\mathbf{F}^{-1}\mathbf{B}^T(t)\mathbf{P}(t) - \mathbf{E} \tag{6.26}$$

with the end point boundary condition

$$\mathbf{P}(t_f) = \mathbf{0}$$

It can also be shown that the cost function corresponding to this optimal control is

$$V^* = \mathbf{x}^T(t_0)\mathbf{P}(t_0)\mathbf{x}(t_0) \tag{6.27}$$

Eqn (6.26) can readily be cast in reverse time, $\hat{t} = t_f - t$, and solved for $\mathbf{P}(\hat{t})$, given the initial condition $\mathbf{P}(\hat{t}) = \mathbf{0}$:

$$\frac{\mathrm{d}\mathbf{P}}{\mathrm{d}\hat{t}} = \mathbf{P}(\hat{t})\mathbf{A}^T(\hat{t}) + \mathbf{A}^T(\hat{t})\mathbf{P}(\hat{t}) - \mathbf{P}(\hat{t})\mathbf{B}(\hat{t})\mathbf{F}^{-1}\mathbf{B}^T(\hat{t})\mathbf{P}(\hat{t}) + \mathbf{E} \tag{6.28}$$

One possibility is therefore to solve Eqn (6.28) in reverse time, find $\mathbf{P}(t_0) = \mathbf{P}(\hat{t})|\hat{t} = t_f$, and use this as the initial condition to solve Eqn (6.26) in forward time, applying its solution as it evolves to Eqns (6.25) and (6.24) to yield $\mathbf{u}^*(t)$. An unfortunate property of the Riccati equation can, however, interfere with this. The solution typically undergoes little change over a long period, during which an effectively constant gain state feedback strategy is realized, and then undergoes rapid change as it approaches the target. This can lead to considerable inaccuracy in $\mathbf{P}(t)$, and so in $\mathbf{K}(t)$, as the target is approached. While this may not be particularly important, since the target is probably more or less attained before the inaccuracies appear, the feature is undesirable in that it leads to uncertainty about the exact system behaviour in the region of the target.

Another possibility is to solve Eqn (6.28) and keep a time record of $\mathbf{P}(\hat{t})$, which can then be used in reverse sequence, via Eqn (6.24), on the real-time process. This is, of course, somewhat clumsy to realize in practice.

The optimal strategy derived in these ways drives the system states to $\mathbf{0}$ as a single exercise, $t_0 \leqslant t \leqslant t_f$: it does nothing about maintaining the states at $\mathbf{0}$, $t > t_f$, a task frequently required of a regulator. This point is covered neatly, and the problem of solving Eqn (6.28) is also solved, if S is linear time-invariant, and if a suboptimal control, whose performance is usually almost indistinguishable from that of the optimal, is accepted.

If it is accepted that the system states approach the target asymptotically, rather than actually attain it, then t_f may be set to ∞ in Eqn (6.23), and Eqn (6.28) yields a steady state solution which satisfies the 'algebraic Riccati equation':

$$\frac{\mathrm{d}\mathbf{P}}{\mathrm{d}\hat{t}} = \mathbf{0} = \mathbf{PA} + \mathbf{A}^T\mathbf{P} - \mathbf{PBF}^{-1}\mathbf{B}^T\mathbf{P} + \mathbf{E} \tag{6.29}$$

This value of **P** is found and substituted in Eqn (6.25) to yield a constant state feedback matrix **K** and a suboptimal strategy (Eqn (6.24))

$$\mathbf{u}(t) = -\mathbf{K}\mathbf{x}(t) \tag{6.30}$$

This is usually very nearly as effective as the true optimal strategy for the single control exercise, $t_0 \leqslant t \leqslant t_f$. It is also much easier to realize, since **K** can be established off-line, independently of the strategy realization. Moreover, it completes the requirement that the regulator maintains the states at **0**, $t > t_f$, since it automatically applies continuously.

In summary, the procedure for the design of an LQO regulator is as follows:

1. Define V (Eqn (6.23)), choosing **E**, **F** to define the control objectives as described above.
2. Construct the relevant Riccati equation (Eqn (6.28) or (6.29)) and solve this to yield the optimal control strategy (Eqns (6.25) and (6.24)) or, much more usually, the suboptimal strategy (Eqn (6.30)).
3. Simulate the control system and check that $\mathbf{u}^*$ utilizes the control authority well but without exceeding its practical limits. Redefine **F** if necessary, and repeat from (2).
4. Examine the effects of redefining **E**, repeating from (1) until the best simulated performance has been achieved. This iterative process, which often involves considerable manipulation, can only be undertaken with proper computer aided design facilities.

Example

The vertical dynamics of an aircraft which is about to land are described by the equations:

$$\begin{pmatrix} \dot{U} \\ \dot{W} \\ \dot{q} \\ \dot{\theta} \\ \dot{h} \\ \dot{e} \end{pmatrix} = \begin{pmatrix} -0.058 & 0.065 & 0 & -0.171 & 0 & 1 \\ -0.303 & -0.685 & 1.109 & 0 & 0 & 0 \\ 0.072 & -0.658 & -0.947 & 0 & 0 & 0 \\ 0 & 0 & 1 & 0 & 0 & 0 \\ 0 & -1 & 0 & 1.133 & 0 & 0 \\ 0 & 0 & 0 & 0 & 0 & -0.571 \end{pmatrix} \begin{pmatrix} U \\ W \\ q \\ \theta \\ h \\ e \end{pmatrix}$$

$$+ \begin{pmatrix} 0 & 0 & -0.119 \\ -0.054 & 0 & 0.074 \\ -1.117 & 0 & 0.115 \\ 0 & 0 & 0 \\ 0 & 0 & 0 \\ 0 & 0.571 & 0 \end{pmatrix} \begin{pmatrix} \mu \\ \gamma \\ \delta \end{pmatrix} \tag{6.31}$$

where

U is the forward speed along the aircraft main body axis ($\mathrm{m\,s^{-1}}$),
W is the downward velocity at right-angles to the main body axis ($\mathrm{m\,s^{-1}}$),
q is the angular velocity (degrees $\mathrm{s^{-1}}$), is the pitch WRT the ground (degrees),
h is the height above the ground (m),
e is the forward acceleration due to throttle action ($\mathrm{m\,s^{-2}}$)
μ is the elevator angle (degrees),
γ is the throttle acceleration value ($\mathrm{m\,s^{-2}}$),
δ is the aileron angle (degrees),

The system is illustrated in Figure 6.14.

Consider the design of a state feedback strategy (Figure 6.12, $\mathbf{r}(t) = \mathbf{0}$), yielding $(\mu\ \gamma\ \delta)^T$, which brings the aircraft to land, from any reasonable initial state, while minimizing a simple measure of the amplitudes of pitch, θ, pitch velocity, q, and height, h:

$$V = \int_{t_0}^{t_f} (\theta^2 + q^2 + 0.04h^2)\,\mathrm{d}t$$

The target state is set at zero, the model recast appropriately – in this case yielding Eqn (6.31) without alteration – and cost function matrix corresponding to V is specified (Eqn (6.23):

$$\mathbf{E} = \begin{pmatrix} 0 & 0 & 0 & 0 & 0 & 0 \\ 0 & 0 & 0 & 0 & 0 & 0 \\ 0 & 0 & 1 & 0 & 0 & 0 \\ 0 & 0 & 0 & 1 & 0 & 0 \\ 0 & 0 & 0 & 0 & 0.04 & 0 \\ 0 & 0 & 0 & 0 & 0 & 0 \end{pmatrix}$$

$\mathbf{F}$ is chosen arbitrarily, and a control strategy is designed by solving Eqn (6.28),

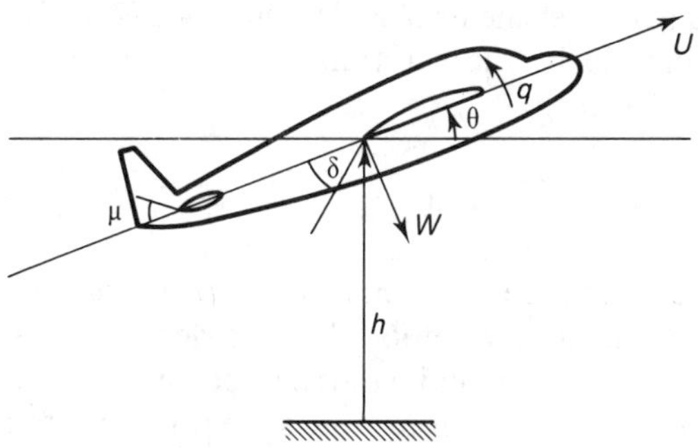

Figure 6.14 Aircraft landing.

$\mathbf{P}(\hat{0}) = \mathbf{0}$, and finding the steady state solution (which also satisfies Eqn (6.29)). The strategy is then simulated and **F** adjusted to limit $\mathbf{u}^*$ to the control authority available, or, alternatively, to utilize that authority more fully. The process is repeated several times, with **E** also being adjusted if considered desirable, until good performance is achieved.

After several iterations, **F** is established:

$$\mathbf{F} = \begin{pmatrix} 0.003 & 0 & 0 \\ 0 & 0.200 & 0 \\ 0 & 0 & 0.003 \end{pmatrix}$$

Eqns (6.28), (6.29) and (6.30) yield the strategy

$$\mathbf{K} = \begin{pmatrix} -0.761 & 1.914 & -18.298 & -20.524 & -1.115 & -0.458 \\ 0.116 & -0.063 & 0.007 & 0.114 & 0.043 & 0.157 \\ -2.429 & 3.966 & 1.484 & -5.099 & -3.459 & -2.068 \end{pmatrix}$$

A simulated landing flare and the corresponding behaviours of q, θ, μ, δ are shown in Figure 6.15, for the initial conditions $U(0) = 5$, $W(0) = -2.5$, $q(0) = -1$, $\theta(0) = -3$, $h(0) = 15$, $e(0) = 0.5$.

The aircraft is forced into the correct landing attitude by rapid initial manoeuvre of the control surfaces, and brought to land with θ, q held small. By altering the **E** matrix elements (and the **F** matrix elements correspondingly to prevent $\mathbf{u}(t)$ exceeding the control authority), different landing conditions can be specified.

As mentioned above, it is possible to use the suboptimal regulator design to design a tracking strategy which drives the system states from $\mathbf{x}(t_0)$ to $\mathbf{x}(t_f)$ along a specified trajectory, or track, of a particular type. In this case the symmetric matrix **E** (Eqn (6.23)) is selected with off-diagonal elements so chosen that

$$\mathbf{x}^T\mathbf{E}\mathbf{x} = (\sqrt{e_{11}}x_1 + \sqrt{e_{22}}x_2 + \cdots + \sqrt{e_{mm}}x_m)^2$$

where $e_{11}, e_{22}, \ldots, e_{mm}$ are the diagonal elements of **E**. This has the effect of forcing the state trajectory – as far as is allowed by the constraint (represented by **F**) on the control authority – **u**, to follow the track

$$\sqrt{e_{11}}x_1 + \sqrt{e_{22}}x_2 + \cdots + \sqrt{e_{mm}}x_m = 0 \qquad (6.32)$$

Such a track is, of course, rather specialized: it is not explicitly time dependent, and amounts to a track of the kind which is asymptotically approached with a constant gain state feedback strategy. In fact, the suboptimal design procedure simply selects the constant state feedback gain matrix which forces the state trajectory to approach that track asymptotically.

More general track specifications, which are naturally of interest, are pursued in Section 6.5.2.

Example

Consider the design of a state feedback strategy (Figure 6.12, $\mathbf{r}(t) = \mathbf{0}$) yielding $(\mu\ \gamma\ \delta)^T$ such that aircraft of the preceding example lands along a track defined by a time-independent equation of the structure of Eqn (6.32). A flare path $h = h_0 e^{-0.2t}$, for example, can be shown to be such a trajectory (in spite of its evident time dependence), by regarding it as the solution to the equations:

$$\dot{h} = -0.2h, \quad h(t_0) = h_0$$

Substituting for $\dot{h}$ from Eqn (6.31) yields (cf. Eqn (6.32)) the (explicitly) time-independent trajectory specification:

$$-W + 1.133\theta + 0.2h = 0$$

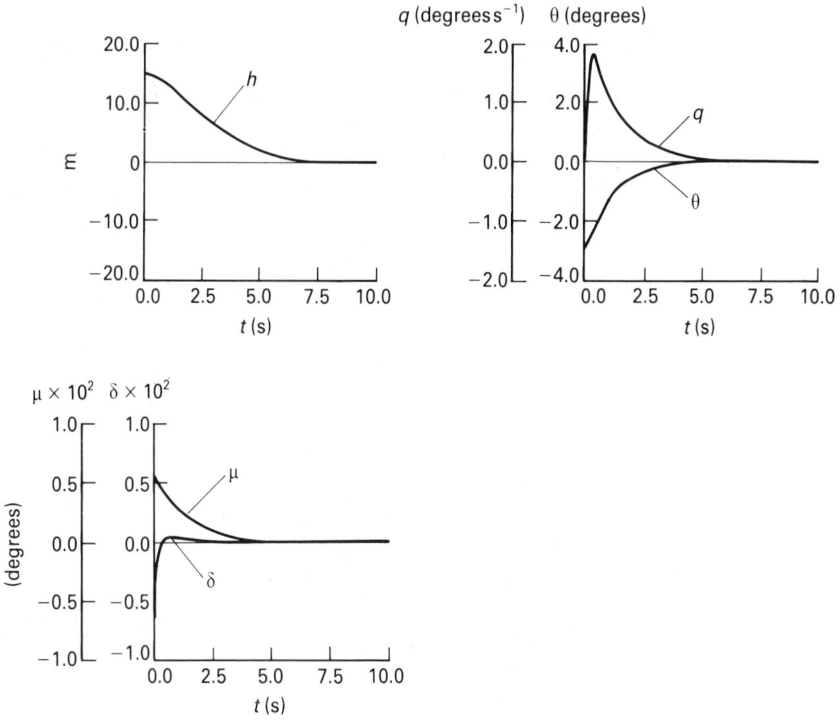

Figure 6.15 Aircraft landing flare.

It is easy to show that this corresponds to

$$\mathbf{E} = \begin{pmatrix} 0 & 0 & 0 & 0 & 0 & 0 \\ 0 & 1.000 & 0 & -1.133 & -0.200 & 0 \\ 0 & 0 & 0 & 0 & 0 & 0 \\ 0 & -1.133 & 0 & 1.284 & 0.227 & 0 \\ 0 & -0.200 & 0 & 0.227 & 0.040 & 0 \\ 0 & 0 & 0 & 0 & 0 & 0 \end{pmatrix}$$

Some experimentation with modelling the system and checking the excursions of μ, γ, δ so that they are fully used, but do not exceed their limits, yields

$$\mathbf{F} = \begin{pmatrix} 0.0032 & 0 & 0 \\ 0 & 0.179 & 0 \\ 0 & 0 & 0.0008 \end{pmatrix}$$

and the suboptimal constant feedback matrix (Eqn (6.30))

$$\mathbf{K} = \begin{pmatrix} 1.731 & 2.138 & -2.179 & -6.486 & -0.890 & -0.659 \\ 0.020 & -0.009 & 0.006 & 0.018 & 0.002 & 0.029 \\ -3.099 & 30.483 & -1.883 & -41.788 & -6.929 & -0.919 \end{pmatrix}$$

A simulated suboptimal landing flare and corresponding control inputs are shown in Figures 6.16(a, b), for the initial conditions $U(0) = 5$, $W(0) = 2.5$, $q(0) = -1$, $\theta(0) = -3$, $h(0) = 15$, $e(0) = 0.5$. The aircraft is forced into the correct attitude by rapid manoeuvre of the control surfaces, and is then held close to the specified landing flare throughout its descent. By altering the **E** matrix, different landing flares can be specified.

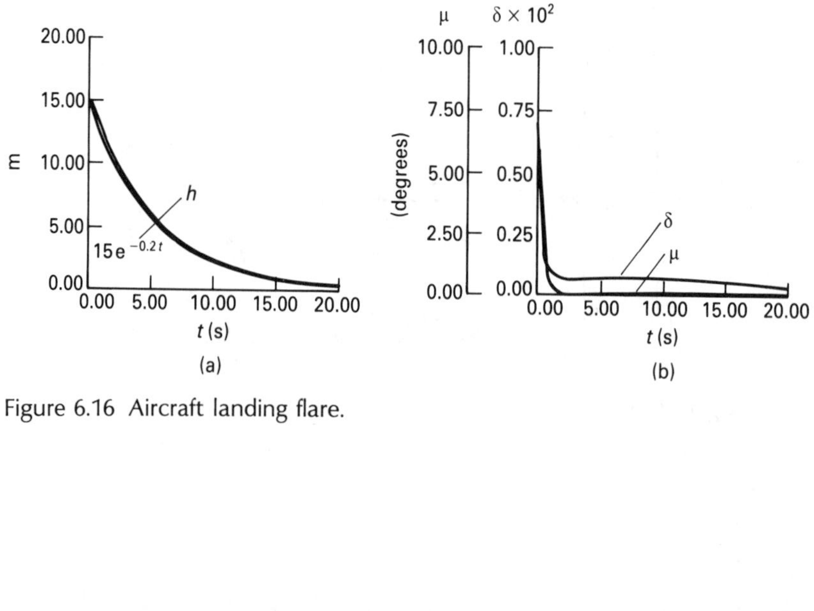

Figure 6.16 Aircraft landing flare.

6.5.2 LQO tracking strategy design

The objective of an LQO tracking strategy is to generate an optimal input, $\mathbf{u}^*(t)$, to drive the system states onto and along a predefined track in the state space, $\tilde{\mathbf{x}}$t). There is no mathematical restriction on $\tilde{\mathbf{x}}(t)$, but the control strategy generates $\mathbf{u}^*(t)$, which, while optimal in the sense that it minimizes a cost function, is not necessarily effective in any worthwhile practical sense. For good results, an 'achievable' $\tilde{\mathbf{x}}(t)$, which does not markedly infringe the natural dynamics of the closed loop system, must be defined.

As in the case of the regulator, the desired track can be defined in terms of the system outputs, rather than states, and the control strategy developed in terms of output feedback. However, since the states can almost invariably be recast to describe such a track directly, to yield the identical strategy expressed in terms of state feedback, this topic is not pursued here. Consider an mth order linear system (Eqn (2.69)):

$$\dot{\mathbf{x}} = \mathbf{A}(t)\mathbf{x} + \mathbf{B}(t)\mathbf{u} \tag{6.33}$$

A desired trajectory, $\tilde{\mathbf{x}}(t)$, is defined, $t_0 \leqslant t \leqslant t_f$ and a corresponding cost function synthesized (cf. Eqn (6.23)):

$$V = \int_{t_0}^{t_f} ((\mathbf{x} - \tilde{\mathbf{x}})^T\mathbf{E}(\mathbf{x} - \tilde{\mathbf{x}}) + \mathbf{u}^T\mathbf{F}\mathbf{u})\,\mathrm{d}t \tag{6.34}$$

$\mathbf{E}$ is a positive semi-definite $(m \times m)$ constant matrix chosen to decribe a quadratic function of the difference between the optimal trajectory $\mathbf{x}^*(t)$ and the demanded trajectory $\tilde{\mathbf{x}}(t)$ which is to be minimized over the duration of the control process. Normally, $\mathbf{E}$ is a diagonal matrix describing a sum of weighted squared differences:

$$e_{11}(x_1^* - \tilde{x}_1)^2 + e_{22}(x_2^* - \tilde{x}_2)^2 + \cdots + e_{mm}(x_m^* - \tilde{x}_m)^2$$

where $e_{11}, e_{22}, \ldots, e_{mm}$ are diagonal elements of $\mathbf{E}$. $\mathbf{F}$ is a positive definite $(r \times r)$ constant matrix, usually diagonal, chosen to constrain the amplitudes of $\mathbf{u}^*(t)$, as in regulator design.

It can be shown [ref. 7] that the control which drives S from $\mathbf{x}(t_0)$ to $\mathbf{x}(t_f)$, while minimizing V, is (cf. Eqn (6.24))

$$\mathbf{u}^*(t) = -\mathbf{K}(t)\mathbf{x}^*(t) + \mathbf{F}^{-1}\mathbf{B}^T(t)\mathbf{r}^*(t) \tag{6.35}$$

where (cf. Eqn (6.25))

$$\mathbf{K}(t) = \mathbf{F}^{-1}\mathbf{B}^T(t)\mathbf{P}(t) \tag{6.36}$$

$\mathbf{P}(t)$ is the solution of Eqn (6.26), $\mathbf{P}(t_f) = \mathbf{0}$ and $\mathbf{r}^*(t)$ is the solution of

$$\dot{\mathbf{r}}^* = (\mathbf{A}(t) - \mathbf{B}(t)\mathbf{F}^{-1}\mathbf{B}^T(t)\mathbf{P}(t))^T\mathbf{r}^* - \mathbf{E}\tilde{\mathbf{x}}(t) \tag{6.37}$$

with the end point boundary condition:

$$\mathbf{r}^*(t_f) = \mathbf{0}$$

As in the regulator case, Eqns (6.26) and (6.37) can be recast in reverse time and solved using Runge–Kutta or similar techniques. In reverse time, Eqn (6.37) becomes

$$\frac{\mathbf{dr}^*}{\mathrm{d}\hat{t}} = -(\mathbf{A}(\hat{t}) - \mathbf{B}(\hat{t})\mathbf{F}^{-1}\mathbf{B}^T(\hat{t})\mathbf{P}(\hat{t}))^T\mathbf{r}^* + \mathbf{E}\tilde{\mathbf{x}}(t) \tag{6.38}$$

where $\hat{t} = t_f - t$ and $\mathbf{r}^*(\hat{0}) = \mathbf{0}$.

As with the regulator, if S is linear time invariant, and a suboptimal solution is acceptable, t_f can be set to ∞ and steady state algebraic solutions of Eqns (6.28) and (6.38) yield a constant feedback matrix, $\mathbf{K}$ (Eqn (6.30)) and a simplified input function $\mathbf{r}^*$, found by setting $\dot{\mathbf{r}}^* = \mathbf{0}$ in Eqn (6.37):

$$\mathbf{u}(t) = -\mathbf{K}\mathbf{x}(t) + \mathbf{F}^{-1}\mathbf{B}^T[(\mathbf{A} - \mathbf{B}\mathbf{F}^{-1}\mathbf{B}^T\mathbf{P})^T]^{-1}\mathbf{E}\tilde{\mathbf{x}}(t) \tag{6.39}$$

where $\mathbf{P}$ is the steady state value of $\mathbf{P}(\hat{t})$, $\hat{t} \to \infty$, and $\mathbf{A}$, $\mathbf{B}$ are constant matrices. How satisfactory such a suboptimal solution turns out to be depends strongly on the track $\tilde{\mathbf{x}}(t)$ and the dynamics of the system (as does the optimal solution). Simulation of the system behaviour is an essential component in arriving at an effective strategy.

In summary, the procedure for the design of an LQO tracking strategy is as follows:

1. Define V (Eqn (6.34)), choosing $\mathbf{E}$, $\mathbf{F}$, $\tilde{\mathbf{x}}(t)$ to define the control objectives as described above.
2. Construct the relevant reverse-time Ricatti equation (Eqn (6.28) and input equation (Eqn 6.38)) and solve to give the optimal (Eqn 6.35)) or suboptimal (Eqn (6.39)) strategy.
3. Simulate the control system and check that $\mathbf{u}^*(t)$ utilizes the control authority well, but without exceeding its limits. Redefine $\mathbf{F}$ if necessary and repeat from (2).
4. Consider the effects of redefining $\mathbf{E}$, repeating from (1), until a satisfactory simulated performance is achieved.

As with regulator design, proper computer-aided design tools are essential if all this is to be undertaken effectively.

Example

Referring to the aircraft of Figure 6.14, consider the design of a state feedback track strategy (Figure 6.12), yielding $(\mu\ \gamma\ \delta)^T$ such that the aircraft landing flare is a straight-line trajectory at constant forward velocity:

$$h(t) = 15 - 1.5t, \quad h(10) = 0$$

This requirement is used to specify a track, and consequently an $\mathbf{E}$ matrix, such that all the states are left 'free' except h:

$$\mathbf{E} = \begin{pmatrix} 0 & 0 & 0 & 0 & 0 & 0 \\ 0 & 0 & 0 & 0 & 0 & 0 \\ 0 & 0 & 0 & 0 & 0 & 0 \\ 0 & 0 & 0 & 0 & 0 & 0 \\ 0 & 0 & 0 & 0 & 1 & 0 \\ 0 & 0 & 0 & 0 & 0 & 0 \end{pmatrix}$$

Selecting $\mathbf{F}$ as in the preceding example, the suboptimal constant feedback matrix is developed from Eqns (6.28) and (6.25):

$$\mathbf{K} = \begin{pmatrix} -1.775 & 5.388 & -2.132 & -10.219 & -3.010 & -0.664 \\ 0.002 & -0.014 & 0.067 & 0.026 & 0.008 & 0.028 \\ -3.262 & 21.028 & -2.323 & -31.536 & -34.839 & -1.040 \end{pmatrix}$$

Eqn (6.39) yields the suboptimal strategy

$$\begin{pmatrix} \mu \\ \gamma \\ \delta \end{pmatrix} = -\mathbf{K} \begin{pmatrix} U \\ W \\ q \\ \theta \\ h \\ e \end{pmatrix} + \begin{pmatrix} -3.010 \\ 0.008 \\ -34.839 \end{pmatrix} (15 - 1.5t)$$

A simulated suboptimal landing flare and its control inputs, μ, δ, are shown in Figure 6.17. The aircraft is dragged into the right attitude and almost onto the required landing flare, and held there throughout its descent.

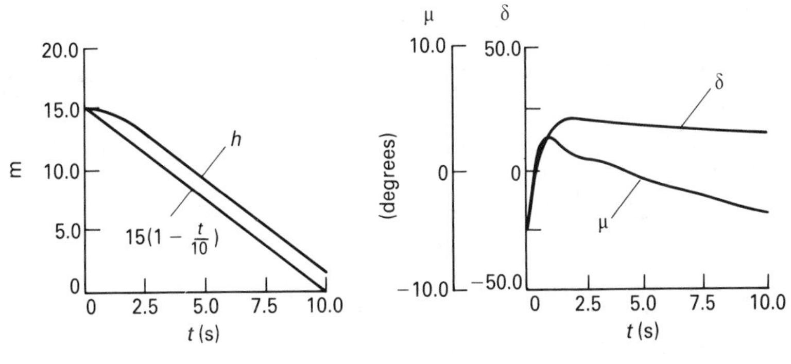

Figure 6.17 Aircraft landing flare.

6.6 Optimal control for linear deterministic systems – discrete time

Practical realization of the control strategies of Section 6.5 usually involves on-line computer control. The design of algorithms for these is considered in this section.

The design methods follow those described in Section 6.5, this time in the discrete time domain, and involve casting the linear system model in discrete time form (Sections 3.8 and 3.9), setting up cost functions, and solving difference equations in reverse discrete time to yield optimal, or suboptimal, regulator and tracking strategy control algorithms.

The general arrangement is shown in block form in Figure 6.18 (cf. Figure 6.12).

In Figure 6.18, Σ represents the discrete time linear system to be controlled, C^* the optimal control strategy and $\mathbf{r}^*(n)$ the optimal control input. The design process to establish C^* and $\mathbf{r}^*(n)$ can be expressed neatly in the form of recursive reverse time algorithms, which can be realized without undue difficulty. The designs of regulators and tracking strategies using this approach are considered here.

6.6.1 LQO discrete time regulator design

Consider a linear discrete time system (Eqns (3.32) and (3.33)):

$$\mathbf{x}(n+1) = \boldsymbol{\varphi}(n)\mathbf{x}(n) + \boldsymbol{\psi}(n)\mathbf{u}(n)$$

A linear quadratic cost function is defined (cf. Eqn (6.23)):

$$V = \sum_{n=n_0}^{n_0+N} (\mathbf{x}^T(n)\mathbf{E}\mathbf{x}(n) + \mathbf{u}^T(n)\mathbf{F}\mathbf{u}(n)) \tag{6.40}$$

where

n_0, $n_0 + N$ are the initial and final times over which the control process is specified,
E is a positive semi-definite $(m \times m)$ constant matrix,
F is a positive definite $(q \times q)$ constant matrix.

E, F are defined to achieve the system design objectives in the same way as **E, F** in Section 6.5.1.

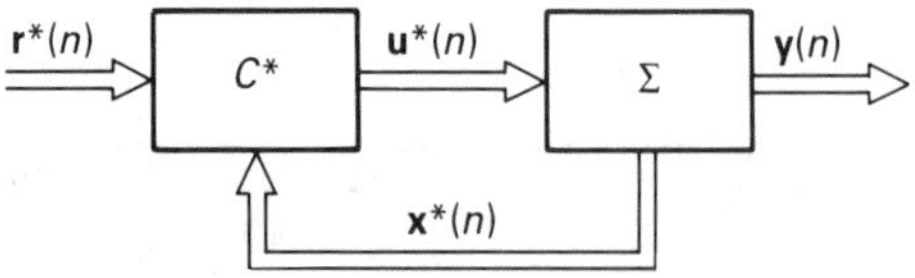

Figure 6.18 Discrete time optimal control system.

As with the continuous time regulator, it is assumed that the target states are zero, $\mathbf{x}_T = \mathbf{0}$. More elaborate algorithms [refs 4, 7] are of course available to cater for other cases, but generally it is possible to recast the state equations so that the desired target is at $\mathbf{x}_T = \mathbf{0}$.

The design problem is to devise an input, $\mathbf{u}^*(n)$, which drives the system from its initial state, $\mathbf{x}(n_0)$ to $\mathbf{0}$, in N steps, while minimizing V, and, usually, to maintain the states at $\mathbf{0}$ thereafter.

It can be shown that the following time-varying state feedback strategy achieves the first of these objectives:

$$\left.\begin{aligned} \mathbf{u}^*(n_0) &= -\mathbf{K}(n_0)\mathbf{x}(n_0) \\ \mathbf{u}^*(n_0+1) &= -\mathbf{K}(n_0+1)x^*(n_0+1) \\ &\vdots \\ \mathbf{u}^*(n) &= -\mathbf{K}(n)\mathbf{x}^*(n) \\ &\vdots \\ \mathbf{u}^*(n_0+N) &= -\mathbf{K}(n_0+N)\mathbf{x}^*(n_0+N) \end{aligned}\right\} \tag{6.41}$$

where

$\mathbf{u}^*(n)$ is the optimal input vector,
$\mathbf{x}^*(n)$ is the optimal state vector,
$\mathbf{K}(n)$ is found from the reverse-time algorithm

$$\left.\begin{aligned} \mathbf{K}(n-1) &= (\mathbf{F} + \psi^T(n)\mathbf{P}(n)\psi(n))^{-1}\psi^T(n)\mathbf{P}(n)\varphi(n) \\ \mathbf{M}(n) &= \mathbf{P}(n) - \mathbf{P}(n)\psi(n)(\mathbf{F} + \psi^T(n)\mathbf{P}(n)\psi(n))^{-1}\psi^T(n)\mathbf{P}(n) \\ \mathbf{P}(n-1) &= \varphi^T(n)\mathbf{M}(n)\varphi(n) + \mathbf{E} \end{aligned}\right\} \tag{6.42}$$

$$\mathbf{P}(n_0+N) = \mathbf{E}, \quad \mathbf{K}(n_0+N) = \mathbf{0}$$

Eqns (6.42) are realized, off-line, in reverse-time order to generate the control constants required in Eqns (6.41). The cost incurred over the N steps of the optimal process can be shown to be (cf. Eqn (6.27))

$$V^* = \mathbf{x}^T(n_0)\mathbf{P}(n_0)\mathbf{x}(n_0) \tag{6.43}$$

A noteworthy feature of the optimal control gains, $\mathbf{K}(n)$ (Eqns (6.42)) is that they are independent of N, i.e. the optimal strategy for a short sequence is identical to the latter part of the optimal strategy in a longer sequence. Clearly, if a particular sequence length is necessary for good behaviour, a shorter sequence will be unsatisfactory, and this points up the fact that the system behaviour is merely optimal (in the sense that the cost function is minimized) – not necesarily good. Evidently, it is necessary to select a reasonable sequence length, N, for the system behaviour to be both optimal and good. Too large a choice of N, of course, leads to time wasting.

As in the continuous time case, if Σ is linear time invariant, and if a suboptimal control is acceptable – which it generally is – N can be set to ∞ in Eqns (6.42) to yield a steady state value **K**, which can then be applied as a straightforward constant gain state feedback strategy (cf. Eqn (6.30)):

$$\mathbf{u}(n) = -\mathbf{K}\mathbf{x}(n) \tag{6.44}$$

Example

The vertical dynamics of a submarine travelling at a constant forward velocity are modelled by the equations (cf. Eqns (5.56 and 5.57)):

$$\begin{pmatrix} \dot{W} \\ \dot{q} \\ \dot{\theta} \\ \dot{h} \end{pmatrix} = \begin{pmatrix} -0.064 & 1 & 0 & 0 \\ 0.0008 & -0.308 & -0.008 & -0.0014 \\ 0 & 1 & 0 & 0 \\ 0 & 0 & -4.000 & 0 \end{pmatrix} \begin{pmatrix} W \\ q \\ \theta \\ h \end{pmatrix} + \begin{pmatrix} -0.053 & -0.104 \\ 0.056 & -0.016 \\ 0 & 0 \\ 0 & 0 \end{pmatrix} \begin{pmatrix} \delta b \\ \delta s \end{pmatrix} \tag{6.45}$$

where

W is the downward velocity at right-angles to the body axis ($\mathrm{m\,s^{-1}}$),
q is the pitch rate ($\mathrm{rad\,s^{-1}}$),
θ is the pitch (rad),
h is the depth (m) WRT a datum depth, h_d, below the mean sea level (m),
δb is the bow-plane control surface angle (rad),
δs is the stern-plane control surface angle (rad).

The system is illustrated in Figure 6.19.

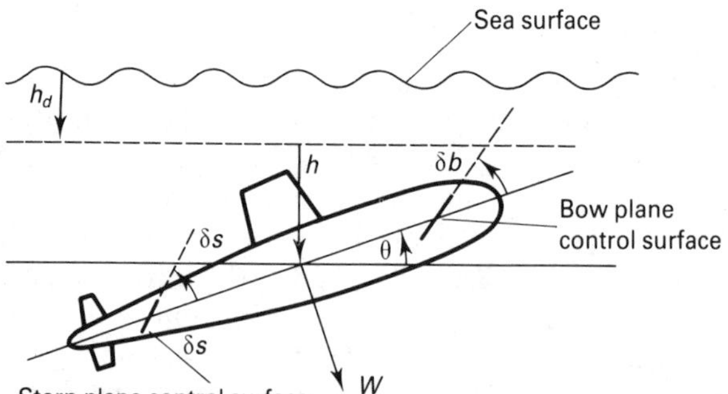

Figure 6.19 Submarine vertical dynamics.

Consider the design of an autopilot whose control strategies are to be realized in an on-line microprocessor. One control algorithm is required to change submarine depth with minimum pitch change.

A sampling period, $T = 2.0$ s, is chosen, and the dynamical model discretized (Appendix A, Eqns (A3.18) and (A3.19)):

$$\begin{pmatrix} W(n+1) \\ q(n+1) \\ \theta(n+1) \\ h(n+1) \end{pmatrix} = \begin{pmatrix} 0.880 & 1.388 & -0.006 & -0.002 \\ -0.001 & 0.535 & -0.003 & -0.002 \\ 0 & 1.488 & 0.993 & -0.002 \\ 1.877 & -4.999 & -7.982 & 1.005 \end{pmatrix} \begin{pmatrix} W(n) \\ q(n) \\ \theta(n) \\ h(n) \end{pmatrix}$$

$$+ \begin{pmatrix} -0.011 & -0.220 \\ 0.083 & -0.024 \\ 0.092 & -0.026 \\ -0.296 & -0.144 \end{pmatrix} \begin{pmatrix} \delta b(n) \\ \delta s(n) \end{pmatrix} \tag{6.46}$$

A cost function (Eqn (6.40)) is constructed which minimizes pitch while allowing the other variables to move 'freely', and **F** is chosen to utilize, but not exceed, the control authority provided by the control surface, about 0.5 rad in each case. After some experimentation and simulation runs:

$$\mathbf{E} = \begin{pmatrix} 0 & 0 & 0 & 0 \\ 0 & 0 & 0 & 0 \\ 0 & 0 & 10 & 0 \\ 0 & 0 & 0 & 0 \end{pmatrix}, \quad \mathbf{F} = \begin{pmatrix} 1 & 0 \\ 0 & 1 \end{pmatrix}$$

The strategy of Eqn (6.41) is realized, and some simulation results, $h(0) = 1.0$, are shown in Figure 6.20.

The submarine is forced into an ascending attitude by a violent initial manoeuvre of the control surfaces, which then behave in such a way that θ is kept small. This differs from traditional submarine control surface behaviour, which holds $\delta b = -\delta s$.

A suboptimal strategy is also easily designed for this linear time-invariant case by setting N large in Eqns (6.42) and finding the steady state feedback matrix, **K**, in reverse time:

$$\mathbf{K} = \begin{pmatrix} -0.180 & 5.200 & 2.474 & -0.036 \\ -0.133 & -1.115 & -0.339 & -0.005 \end{pmatrix}$$

The behaviour of this system, $h(0) = 10$, is almost indistinguishable from that of Figure 6.20.

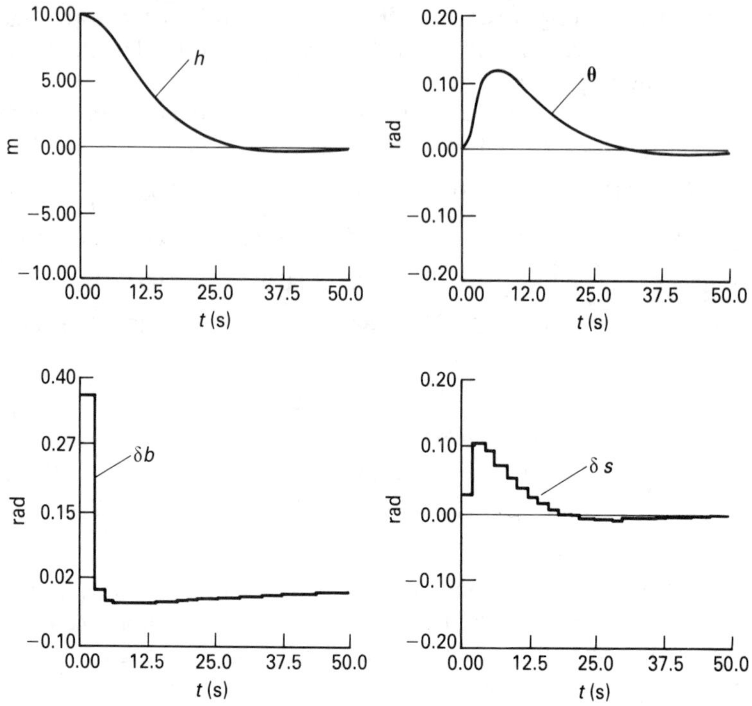

Figure 6.20 Submarine depth change.

6.6.2 LQO discrete time tracking strategy design

LQO discrete time tracking strategy design, like that for continuous time, results in the state feedback established for the optimal regulator and an additional input $\mathbf{r}^*(n)$ (Figure 6.18).

As before, the linear system model is

$$\mathbf{x}(n+1) = \varphi(n)\mathbf{x}(n) + \varphi(n)\mathbf{u}(n)$$

The objective in this case is to force the system state to move onto and along a defined track, $\tilde{\mathbf{x}}(n)$ from any initial state $\mathbf{x}(n_0)$. A suitable cost function is defined (cf. Eqn (6.34)):

$$V = \sum_{n=n_0}^{n_0+N} [(\mathbf{x}(n) - \tilde{\mathbf{x}}(n))^T \mathbf{E}(\mathbf{x}(n) - \tilde{\mathbf{x}}(n)) + \mathbf{u}^T(n)\mathbf{F}\mathbf{u}(n)] \tag{6.47}$$

In this case, it turns out (cf. Eqn (6.35)) that

$$\mathbf{u}^*(n) = -\mathbf{K}(n)\mathbf{x}^*(n) + \bar{\mathbf{K}}(n)\mathbf{r}^*(n) \tag{6.48}$$

Where $\mathbf{K}(n)$ is developed in reverse time according to Eqns (6.42), and, using the same algorithm but with:

$$\bar{\mathbf{K}}(n-1) = (\mathbf{F} + \boldsymbol{\psi}^T(n)\mathbf{P}(n)\boldsymbol{\psi}(n))^{-1}\boldsymbol{\psi}^T(\mathbf{n}) \tag{6.49}$$

$\mathbf{r}^*(n)$ is developed by the reverse-time formula (cf. Eqn (6.38)):

$$\left.\begin{aligned} &\mathbf{r}^*(n) = (\boldsymbol{\varphi}^T(n) + \mathbf{K}^T(n)\boldsymbol{\psi}^T(n))\mathbf{r}^*(n+1) - \mathbf{E}\tilde{\mathbf{x}}(n) \\ &\mathbf{r}^*(n_0 + N) = 0 \end{aligned}\right\} \tag{6.50}$$

As with the regulator, if Σ is linear time invariant, and a suboptimal strategy is acceptable – which depends on the system dynamics and the demanded track – N can be set to ∞ to yield the suboptimal strategy (cf. Eqn (6.39))

$$\mathbf{u}(n) = -\mathbf{K}\mathbf{x}(n) + \bar{\mathbf{K}}(\boldsymbol{\varphi}^T + \mathbf{K}^T\boldsymbol{\psi}^T - \mathbf{I})^{-1}\mathbf{E}\tilde{\mathbf{x}}(n) \tag{6.51}$$

where $\mathbf{K}$ and $\bar{\mathbf{K}}$ are developed from Eqns (6.42) and (6.49), solved in reverse time, N large, and $\mathbf{r}^*(n)$ is replaced by the solution of Eqn (6.50), $\mathbf{r}^*(n) = \mathbf{r}^*(n+1)$.

Example

Consider the design of a control algorithm for the autopilot for the submarine modelled by Eqn (6.46), which specifies that the track, on changing depth, $h(0) = 10$, is a straight line:

$$h = 10\left(1 - \frac{n}{25}\right)$$

Setting the states, other than h, 'free', $\mathbf{E}$ is chosen:

$$\mathbf{E} = \begin{pmatrix} 0 & 0 & 0 & 0 \\ 0 & 0 & 0 & 0 \\ 0 & 0 & 0 & 0 \\ 0 & 0 & 0 & 1 \end{pmatrix}$$

After some experimentation, $\mathbf{F}$ is chosen:

$$\mathbf{F} = \begin{pmatrix} 1 & 0 \\ 0 & 1 \end{pmatrix}$$

The design process outlined above yields the strategy (Eqn (6.51))

$$\mathbf{u}(n) = \begin{pmatrix} -1.581 & 8.791 & 7.251 & -0.412 \\ -0.587 & 1.427 & 2.436 & -0.246 \end{pmatrix} \begin{pmatrix} W(n) \\ q(n) \\ \theta(n) \\ h(n) \end{pmatrix} + \begin{pmatrix} 0.035 \\ -0.018 \end{pmatrix} 10\left(1 - \frac{n}{25}\right)$$

Some simulation results are given in Figure 6.21.

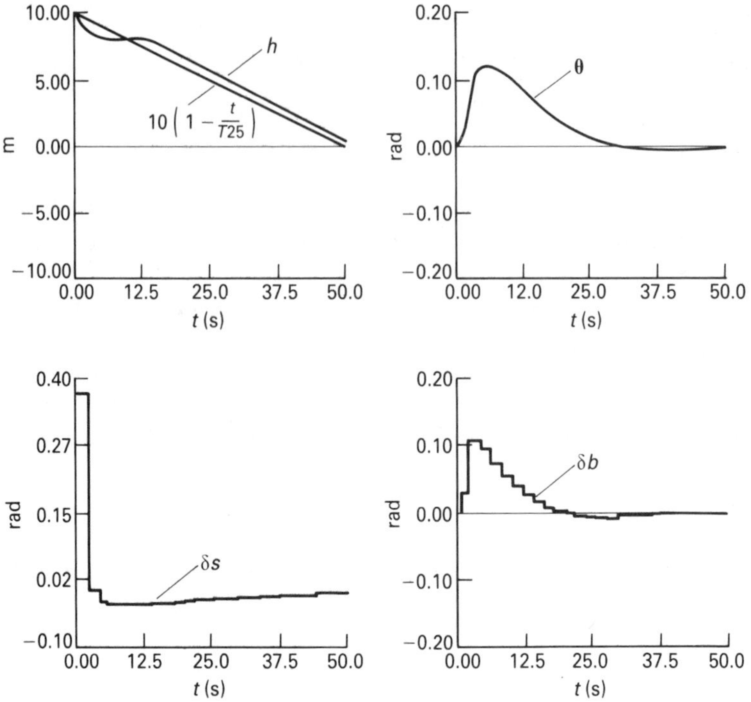

Figure 6.21 Submarine suboptimal depth change.

The submarine is forced into the correct ascending attitude and then held reasonably accurately on the linear design path throughout its ascent. In this case the suboptimal control is virtually as good as the considerably more complex optimal control.

6.7 Optimal control for linear stochastic systems

If the process noise corruption of a linear stochastic system (Section 2.10) is Gaussian, the optimal regulator and tracking strategy designs described in Sections 6.5 and 6.6 apply almost directly with appropriate changes in the cost function definitions. The controlled stochastic system, shown in block form in Figure 6.22(a) (cf. Figure 6.12) is referred to as a linear (system) quadratic (cost function) Gaussian (noise), or LQG system. The discrete time equivalent (cf. Figure 6.18) is shown in Figure 6.22(b).

In Figure 6.22(a) the continuous time stochastic system, S, subject to Gaussian process noise disturbance, $\mathbf{w}(t)$, is optimally controlled by the strategy C^*, which processes an optimal input, $\mathbf{r}^*(t)$, and the system states, $\mathbf{x}^*(t)$, to generate an optimal forcing function, $\mathbf{u}^*(t)$. As with deterministic systems, C^* can be an optimal or suboptimal regulator strategy, or, alternatively, a tracking strategy. As explained in Section 6.5, servomechanism strategies cannot readily be designed using optimal control criteria. The discrete time equivalent is shown in Figure 6.22(b).

In this section, the design objectives for such systems are set out, the design methods are described, mainly by reference to Sections 6.5 and 6.6, and some examples are used to illustrate the results which can be achieved. Both continuous and discrete time systems are considered.

6.7.1 Continuous time LQG systems (cf. Section 6.5)

Consider the heuristic model of a linear mth order stochastic system with r inputs, subject to Gaussian white noise disturbance (Eqn (2.75)):

$$\dot{\mathbf{x}}(t) = \mathbf{A}(t)\mathbf{x}(t) + \mathbf{B}(t)\mathbf{u}(t) + \mathbf{H}(t)\mathbf{w}(t) \tag{6.52}$$

where $\mathbf{w}(t)$ is a Gaussian (extended) white noise vector, strength $\mathbf{Q}(t)$.

As explained in Section 2.10, the detailed proof [refs. 7, 8] of LQG optimal design methods must of course depend on the rigorous system model (Eqn (2.77)), not on that of Eqn (6.52); here the design methods – as opposed to mathematical proofs – are of interest, so the heuristic model is adequate.

Quadratic cost functions are constructed using the expected values of scalar integral functions (Eqn (2.29)) of the stochastic processes $\mathbf{x}(t)$, $\mathbf{u}(t)$, considered, as always with stochastic systems, across an ensemble of controlled systems. For regulator design (cf. Eqn (6.23)):

$$V = \varepsilon\left[\int_{t_0}^{t_f} (\mathbf{x}^T\mathbf{E}\mathbf{x} + \mathbf{u}^T\mathbf{F}\mathbf{u})\,\mathrm{d}t\right] \tag{6.53}$$

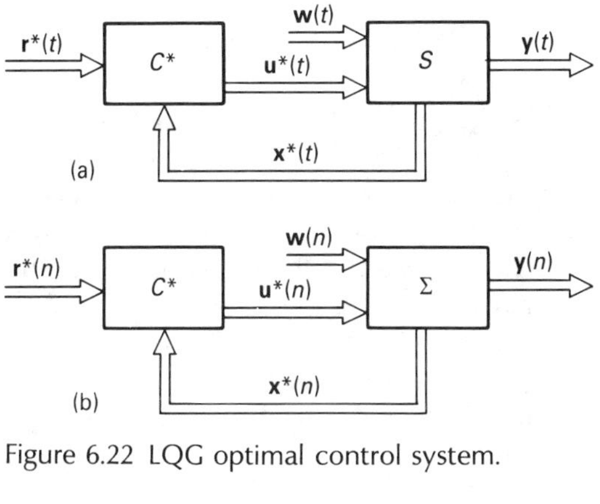

Figure 6.22 LQG optimal control system.

For tracking strategy design (cf. Eqn (6.34)):

$$V = \varepsilon\left[\int_{t_0}^{t_f} ((\mathbf{x}-\tilde{\mathbf{x}})^T\mathbf{E}(\mathbf{x}-\tilde{\mathbf{x}}) + \mathbf{u}^T\mathbf{F}\mathbf{u})\,\mathrm{d}t\right] \tag{6.54}$$

where $\mathbf{E}$, $\mathbf{F}$, t_0, t_f, $\tilde{\mathbf{x}}$ are all as defined in Section 6.5. The design objective of an LQG regulator is to generate an input $\mathbf{u}^*(t)$, $t_0 \leqslant t \leqslant t_f$ which drives the system states towards the target states $\mathbf{x}_T$, meanwhile minimizing the cost function of Eqn (6.53) and, usually, maintain the target states in spite of disturbance, $t > t_f$. The design objective of an LQG tracking strategy is to drive $\mathbf{x}(t)$ onto and along $\tilde{\mathbf{x}}(t)$, by minimizing the cost function of Eqn (6.54).

The design methods are simply those of Section 6.5, which is not surprising, since these strategies drive the states onto the optimal – or suboptimal – trajectories from points not on them, the very task being (repeatedly) set by the effect of process noise. The strength of that noise, $\mathbf{Q}(t)$, does not affect the strategies, which are optimal (though not necessarily satisfactory), but it naturally affects the results achieved.

Two elaborations of the optimal control strategies are worth mentioning.

The first concerns systems which do not inherently include some integrating feature such as a motor drive. State feedback, optimal or not, is insufficient in such cases to guarantee zero steady state error between the input(s) and the controlled state(s). Where the process noise has a steady state content – and there is no reason in practice why it should not – the absence of such an integrating feature in fact guarantees a steady state error, and this can be unacceptable. To deal with this, it is usually a simple matter to incorporate a suitable integrator in the control strategy which integrates the relevant state 'error' to generate an additional state which is also, of course, available to the optimal feedback strategy. An arrangement of this kind is shown in Figure 6.23(a).

The mth order system S has no inherent integrating feature, and it is required to design C^* so that no steady state error in one of the states, $x_i(t)$, is allowed to exist. In this case $x_i(t)$ is simply integrated to yield a new state, $x_{m+1}(t)$, the model is augmented to include $\dot{x}_{m+1} = -x_i(t)$, and a suboptimal control strategy is devised for this in the usual way to yield a feedback matrix

$$\mathbf{K} = (k_1 \quad k_2 \quad k_3 \quad \dots \quad k_m \quad k_{m+1})$$

The effect is that no steady state error in $x_i(t)$ is allowed to exist between.

The second point concerns the possibility of using feedforward where the process noise can be measured or, perhaps, estimated. This can be just as effective as with classical control strategies (Section 6.2) and is fairly easily realized within the optimal strategy if the measured noise can reasonably be modelled as filtered white noise, and the system state equation augmented to accommodate this.

An arrangement of this kind is shown in Figure 6.23(b), where the process noise is modelled as s white noise processes filtered by an array of s first order filters to produce $\mathbf{x}_s(t)$, an additional s states, which are, of course, uncontrollable (Appendix A, Section A2.3) from the input, $\mathbf{u}(t)$. If these can be measured and combined with the other states in the optimal control strategy, significant benefit can result.

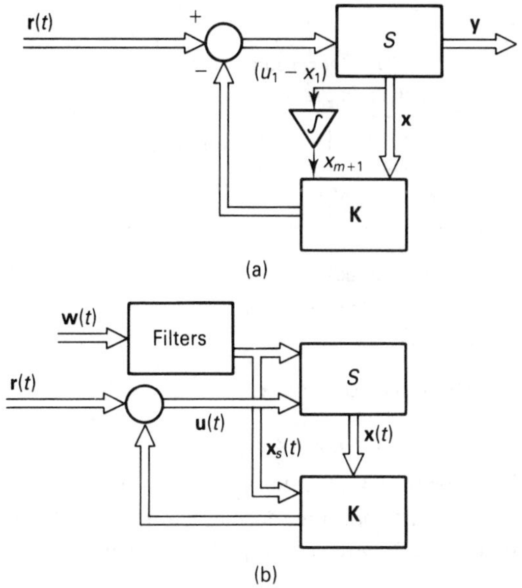

Figure 6.23 Augmented state feedback strategies.

Being uncontrollable the noise states cannot, of course, be influenced by the control strategy, and the corresponding elements in the matrix **E** should be set to 0. Failure to do this in fact has no effect on the optimal strategy.

These two points are perhaps best explained by example.

Examples

(a) An LQG regulator. Referring to the example of Figure 2.24, consider the design of an optimal suspension (i.e. regulator) system consisting of a control network and hydraulic ram, as shown in Figure 6.24(a), where all the states are available to the control strategy, *C*.

Assuming a well-behaved hydraulic actuator, the (heuristic) state equations of the open loop system are:

$$\dot{\mathbf{x}} = \begin{pmatrix} 0 & 1 & 0 & 0 \\ -80 & -40 & 80 & 200 \\ 0 & 0 & 0 & 50 \\ 20 & 1 & -170 & -50 \end{pmatrix} \mathbf{x} + \begin{pmatrix} 0 \\ 1 \\ 0 \\ -1 \end{pmatrix} u + \begin{pmatrix} 0 \\ 0 \\ 0 \\ 150 \end{pmatrix} w$$

where u is the (scaled) input to the hydraulic actuator ($\mathrm{m\,s^{-2}}$). A reasonable suboptimal regulator design objective is to minimize a combination of the variances

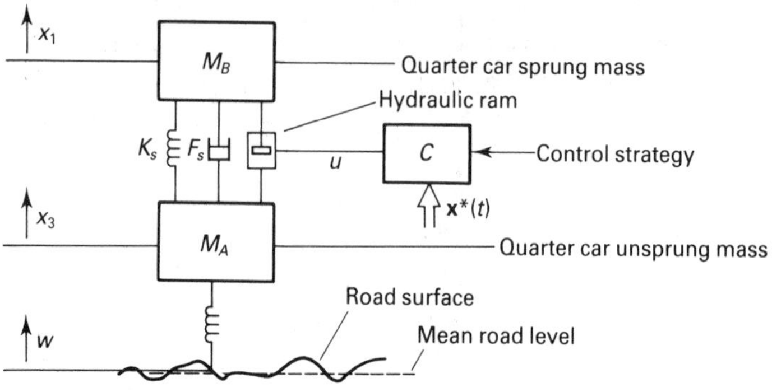

Figure 6.24 Quarter car with active suspension.

of the sprung mass vertical displacement and velocity, $\varepsilon(x_1^2 + x_2^2)$; minimizing the vertical acceleration explicitly is not an option, since acceleration is not a state variable.

The cost function matrices, **E**, **F**, are developed by the iterative procedure described in Section 6.5:

$$\mathbf{E} = \begin{pmatrix} 1 & 0 & 0 & 0 \\ 0 & 1 & 0 & 0 \\ 0 & 0 & 0 & 0 \\ 0 & 0 & 0 & 0 \end{pmatrix}$$

$$\mathbf{F} = (0.1)$$

Such a strategy does, of course, assume that all the states are available to it, which is not easily achieved in this case, since measurement of displacement from the mean road level is not practical. (State estimates based on a Kalman filter (Section 6.8) may of course be substituted, with a corresponding, but hopefully small, deterioration in performance.)

Some simulated sample functions for the uncontrolled open loop system and for the suboptimal regulator strategy designed using $Q = 0.025$ ($\mathrm{m}^2\,\mathrm{s}^{-4}$), namely,

$$u(t) = -(-2.060 \quad -2.253 \quad +0.714 \quad -3.238)\mathbf{x}$$

are shown in Figure 6.25(a, b) respectively. The graphs show that the noise displacement and velocity of the sprung mass are reduced by factors of about 10 by the suboptimal regulator strategy.

A further improvement could be obtained if a transducer, TDR, were available to measure – or at least assess – the road surface excursion from the mean level. This is illustrated in Figure 6.26(a). In this case the measured road noise, x_5, is modelled as white noise seen through a filter of transfer function (cf. Eqn (2.89)):

$$\mathbf{F}(s) = \frac{\lambda}{s + \lambda}$$

where λ is the speed of the vehicle (m s^{-1}).

Consider the case $\lambda = 20$ (m s^{-1}). The state equations are augmented accordingly, and the suboptimal strategy design is applied to the fifth order system:

$$\dot{\mathbf{x}} = \begin{pmatrix} 0 & 1 & 0 & 0 & 0 \\ -80 & -4 & 80 & 200 & 0 \\ 0 & 0 & 0 & 50 & 0 \\ 20 & 1 & -170 & -50 & 150 \\ 0 & 0 & 0 & 0 & -20 \end{pmatrix} \mathbf{x} + \begin{pmatrix} 0 \\ 1 \\ 0 \\ -50 \\ 0 \end{pmatrix} u + \begin{pmatrix} 0 \\ 0 \\ 0 \\ 0 \\ 20 \end{pmatrix} w$$

$$\mathbf{E} = \begin{pmatrix} 1 & 0 & 0 & 0 & 0 \\ 0 & 1 & 0 & 0 & 0 \\ 0 & 0 & 0 & 0 & 0 \\ 0 & 0 & 0 & 0 & 0 \\ 0 & 0 & 0 & 0 & 0 \end{pmatrix}$$

$\mathbf{F} = 0.1.$

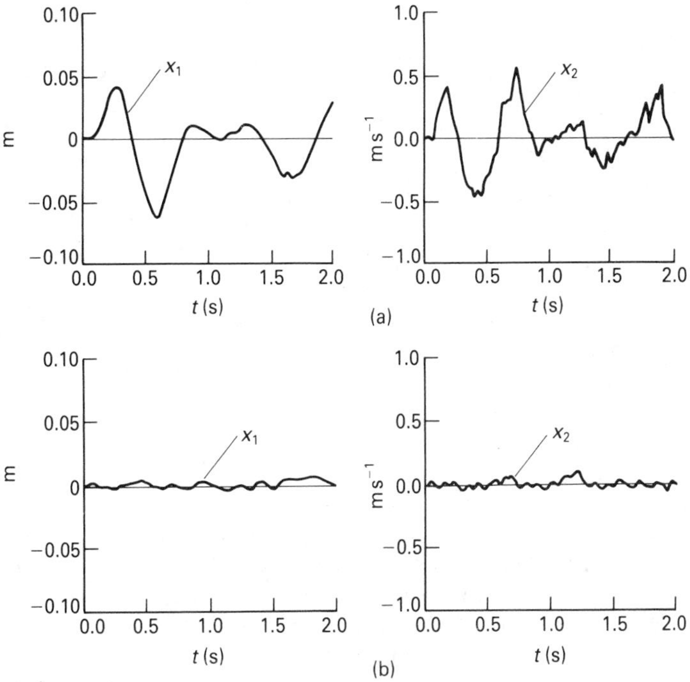

Figure 6.25 Active suspension system.

This yields the suboptimal strategy, for $Q = 0.025$ ($m^2 s^{-4}$):

$$u(t) = -(-2.502 \quad -3.083 \quad 1.139 \quad -3.886 \quad -2.557)\mathbf{x}(t)$$

No useful attempt can be made to adjust element (5, 5) of **E**: x_5 is uncontrollable and the strategy is unaffected by the value of this element.

The sample function results shown in Figure 6.26(b) show a worthwhile improvement over those given in Figure 6.25(b). The use of integration in the control strategy would be inappropriate in this case, since it is clearly desirable to reduce only the non-steady-state vertical displacement and velocity of the quarter car.

(b) A LQG tracking strategy. Consider the design of an LQG tracking strategy for the autoland system of Figure 6.14, which in this case is required to deal with air turbulence affecting the aircraft behaviour.

The open loop system is described by Eqn (6.31), augmented (rather crudely) by a noise vector and gain matrix (Eqn (6.52)):

$$\mathbf{Hw} = \begin{pmatrix} 0 & 0 \\ 1 & 0 \\ 0 & 1 \\ 0 & 0 \\ 0 & 0 \\ 0 & 0 \end{pmatrix} \mathbf{w}$$

where **w** represents Gaussian white noise, strength

$$\mathbf{Q} = \begin{pmatrix} 1 & 0 \\ 0 & 1 \end{pmatrix}$$

The suboptimal tracking strategy of Section 6.5.2 is designed:

$$\begin{pmatrix} \mu \\ \gamma \\ \delta \end{pmatrix} = -\mathbf{K} \begin{pmatrix} U \\ W \\ q \\ \theta \\ h \\ e \end{pmatrix} + \begin{pmatrix} -3.010 \\ 0.008 \\ -34.839 \end{pmatrix} (15 - 1.5t)$$

Some simulation results for this tracking strategy are given in Figure 6.27.

The aircraft is dragged into the correct attitude and (almost) onto the specified descent path; it is repeatedly disturbed but is dragged back towards that path by the control strategy. The control surface activity required to achieve this bears interesting comparison with that of Figure 6.17.

The process noise in this case is certainly unmeasureable, so there is no

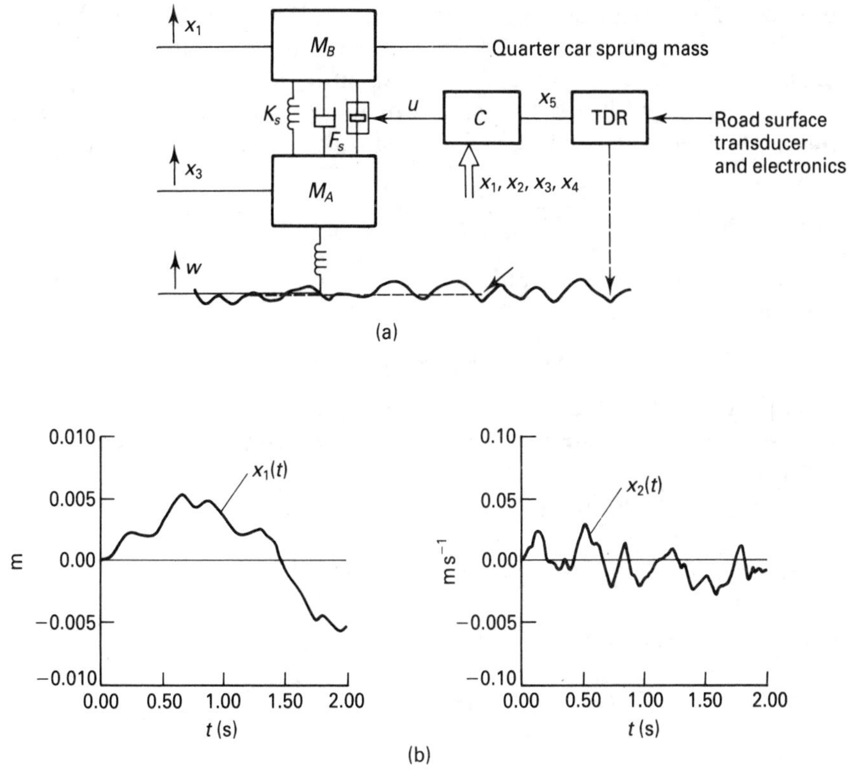

Figure 6.26 Active suspension with noise measurement.

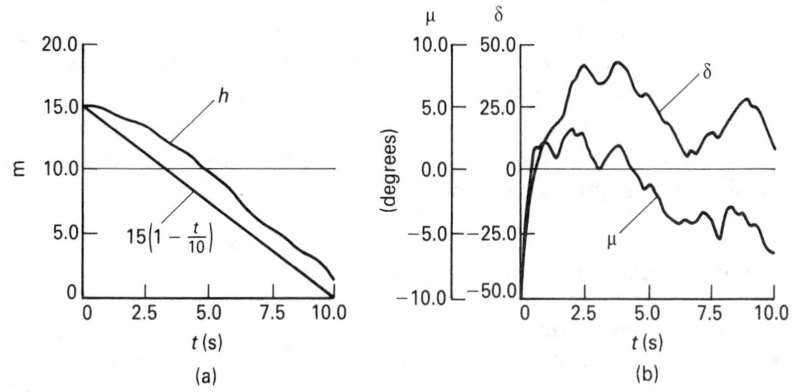

Figure 6.27 LQG controlled autoland facility.

prospect of realizing a feed-forward strategy, at least directly. It might, of course, be possible to construct an estimate of the noise using a state estimator or Kalman filter, in which case this could be included in the strategy.

The use of integration of one or more of the states here is inappropriate, since the dynamic model of the aircraft already includes an integrating feature.

6.7.2 Discrete time LQG systems (cf. Section 6.6)

Consider the mth order linear stochastic system corrupted by Gaussian (extended) white noise (Eqn (3.36)):

$$\mathbf{x}(n+1) = \varphi(n)\mathbf{x}(n) + \psi(n)\mathbf{u}(n) + \mathbf{\Gamma}(n)\mathbf{w}(n) \tag{6.55}$$

As in the continuous time case, quadratic cost functions are constructed as expected values of the appropriate functions of $\mathbf{x}(n)$, $\mathbf{u}(n)$, considered across an ensemble of systems.

For regulator design (cf. Eqn (6.40)):

$$V = \varepsilon\left[\sum_{n=n_0}^{n_0+N} (\mathbf{x}^T(n)\mathbf{E}\mathbf{x}(n) + \mathbf{u}^T(n)\mathbf{F}\mathbf{u}(n))\right] \tag{6.56}$$

For tracking strategy design (cf. Eqn (6.47)):

$$V = \varepsilon\left[\sum_{n=n_0}^{n_0+N} (\mathbf{x}(n) - \tilde{\mathbf{x}}(n))\mathbf{E}(\mathbf{x}(n) - \tilde{\mathbf{x}}(n)) + \mathbf{u}^T(n)\mathbf{F}\mathbf{u}(n)\right] \tag{6.57}$$

The design objective of the LQG discrete time regulator is to generate a system input $\mathbf{u}^*(n)$, $n_0 \leqslant n \leqslant n_0 + N$, which drives the system states to the target states $\mathbf{x}_T = \mathbf{0}$ in N steps, while minimizing the cost function of Eqn (6.56), and usually to maintain the target states thereafter. That of a tracking strategy is to drive $\mathbf{x}(n)$ onto and along $\tilde{\mathbf{x}}(n)$, which is achieved by minimizing the cost function of Eqn (6.57).

As in the continuous time case, the design methods are precisely those for LQO deterministic systems (Section 6.6.2), the algorithms (but not their effects) being unaffected by the strength of the process noise, $\mathbf{w}(n)$. In addition, the benefits of incorporating state integration and noise feedforward, based on noise measurement or estimation, are as important as in the continuous time case.

The examples of Section 6.7.1 are readily discretized and the methods of Section 6.6 are used to design discrete time control strategies. Results very similar to those shown in Figures 6.26 and 6.27 are achieved without undue difficulty.

6.8 Kalman filters and control strategies

The control strategies described in Sections 6.5–6.7, which all involve state feedback, depend on the states of the controlled system being available ('accessible') to the control strategy. In practice this is frequently not so, and the states must be estimated – almost invariably by a state estimator or Kalman filter of one of the kinds described in Chapter 5. The whole strategy thus appears as shown in Figure 6.28 (cf. Figure 6.22(b)). The discrete time case is illustrated since it is more realizable practically than its continuous time equivalent.

In Figure 6.28, Σ represents the stochastic system to be controlled, C^*, $\mathbf{r}^*(n)$, an optimal, or suboptimal control strategy, and KF a Kalman filter generating the state estimates, $\hat{\mathbf{x}}(n)$. Where the process noise can reasonably be represented by filtered white noise (Figure 6.23(b)), with a correspondingly augmented state equation model, it is natural to design the Kalman filter to include estimates of the noise states. This is, of course, advantageous if an (estimated) noise feed-forward strategy is to be realized.

In view of the importance of LQG optimal control strategies, the use of $\hat{\mathbf{x}}(n)$ naturally raises the question of how to design C^*, $\mathbf{r}^*(n)$ so that the control is optimal under these new conditions. In fact it can be shown that for linear systems, Gaussian noise and an optimal state estimator (as the Kalman filter is), the optimal control strategies, using the cost functions of Eqns (6.56) and (6.57), are identical to those described in Section 6.7.2. The same is true, using the cost functions of Eqns (6.53) and (6.54), for continuous time control systems incorporating Kalman–Bucy filters, but these are comparatively rare.

This is known as the 'Certainty Equivalence Principle', a special case of the 'Separation Theorem' [ref. 8], and means that the Kalman filter can be designed while ignoring the control problem, while the LQG control strategy can be designed as though the system states, $\mathbf{x}(n)$, were available, but in fact using the estimated states, $\hat{\mathbf{x}}(n)$, in the knowledge that the strategy is optimal in the same sense. The performance is not, of course, the same as in the direct state feedback case – the Kalman filter state estimates are themselves corrupted with noise – but it is optimal in the revised circumstances.

The Separation Theorem, and thus the Certainty Equivalence Principle, do not in fact hold for non-linear systems, or for non-Gaussian noise. This does not, of course, preclude the use of linearized Kalman filters in 'LQG' strategies for (linearized) non-linear systems; it merely means that the results are not optimal.

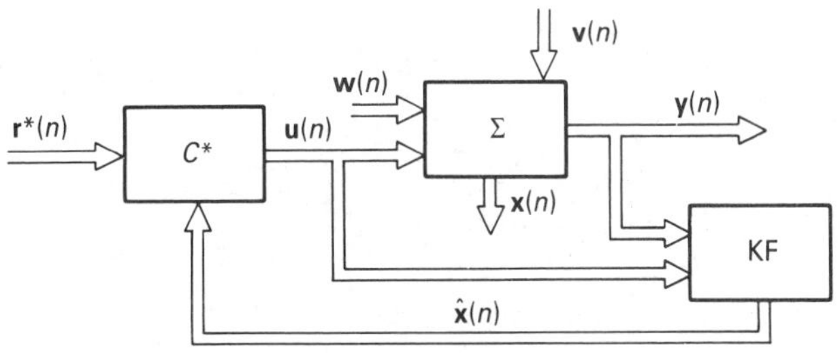

Figure 6.28 LQG control strategies using Kalman filters.

6.9 H_∞ optimal control strategies

An interesting, though complicated, development of recent years has been an attempt to extend the ideas of classical control, using output feedback and a shaping network (Figure 6.2(a)) to linear time-invariant multi-input, multi-output systems, and at the same time to optimize that shaping network, in some useful sense, over the whole class of possible linear shaping networks. This, of course, is considerably more elaborate than merely optimizing state feedback as LQO and LQG control designs seek to do. Such strategies are termed H_∞ optimal control strategies, after the infinite dimension 'Hardy' space of all possible stable transfer function matrices within which the closed loop transfer function matrix of any useful closed loop linear system must exist. The H_∞ optimal control problem, first formulated in a paper by Zames [Ref. 9], can be structured to allow the design of regulators, tracking strategies and even servomechanisms. The resulting design methods are elaborate and realizable only with the help of complex CAD algorithms which are beyond the scope of this book, but a brief introduction to the topic and references for further study are appropriate here. The main ideas are explained by considering their application to the design of an optimal regulator, the simplest case.

Consider a linear time-invariant regulator control system, in which the outputs are to be held at zero in spite of the presence of process noise. The system is shown, in continuous time transfer function matrix model form, in Figure 6.29, where

$\mathbf{G}(s)$ is the $(q \times q)$ transfer function matrix of the controlled system (assumed square for simplicity),
$\mathbf{Y}(s)$ is the q-vector system output,
$\mathbf{N}(s)$ is the s-vector coloured process noise,
$\mathbf{F}(s)$ is the $(q \times q)$ transfer function matrix of the regulator strategy, to be designed,
$\mathbf{U}(s)$ is the q-vector system input

Clearly,

$$\mathbf{Y}(s) = \mathbf{G}(s)\begin{pmatrix}\mathbf{N}(s)\\ \mathbf{U}(s)\end{pmatrix} = \mathbf{G}_1(s)\mathbf{N}(s) + \mathbf{G}_2(s)\mathbf{U}(s)$$

where $\mathbf{G}(s)$ is partitioned to

$$\mathbf{G}(s) = (\mathbf{G}_1(s) \vdots \mathbf{G}_2(s))$$

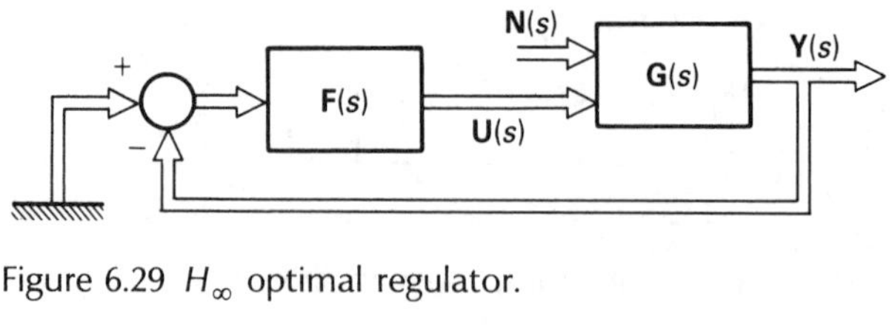

Figure 6.29 H_∞ optimal regulator.

Since $\mathbf{U}(s) = -\mathbf{F}(s)\mathbf{Y}(s)$, $\mathbf{U}(s)$ can be eliminated to give the relationship between $\mathbf{N}(s)$ and $\mathbf{Y}(s)$:

$$\mathbf{Y}(s) = [\mathbf{I} + \mathbf{G}_2(s)\mathbf{F}(s)]^{-1}\mathbf{G}_1(s)\mathbf{N}(s)$$
$$= \bar{\mathbf{G}}(\mathbf{G}(s), \mathbf{F}(s))\mathbf{N}(s) \tag{6.58}$$

where

$$\bar{\mathbf{G}}(\mathbf{G}(s), \mathbf{F}(s)) = [\mathbf{I} + \mathbf{G}_2(s)\mathbf{F}(s)]^{-1}\mathbf{G}_1(s) \tag{6.59}$$

Clearly, it is desirably to design $\mathbf{F}(s)$ so that $\bar{\mathbf{G}}(\mathbf{G}(s), \mathbf{F}(s))$ is 'minimum' in some useful sense over all frequencies, at the same time ensuring that the closed loop system is stable. A measure of $\bar{\mathbf{G}}(\mathbf{G}(s), \mathbf{F}(s))$ which is useful and at the same time mathematically tractable, if complicated, is the 'H_∞ norm':

$$\|\bar{\mathbf{G}}(\mathbf{G}(\mathrm{j}\omega), \mathbf{F}(\mathrm{j}\omega))\|$$

This is defined as the highest value, $0 \leqslant \omega \leqslant \infty$, of the maximum eigenvalue of the real square matrix:

$$\bar{\mathbf{G}}(\mathbf{G}(\mathrm{j}\omega), \mathbf{F}(\mathrm{j}\omega)) \cdot \bar{\mathbf{G}}(\mathbf{G}(-\mathrm{j}\omega), \mathbf{F}(-\mathrm{j}\omega))^T$$

The H_∞ optimal control design task is thus to find $\mathbf{F}^*(\mathrm{j}\omega)$, and so $\mathbf{F}^*(s)$, from the infinite class of such matrices of any order, which minimizes $\|\bar{\mathbf{G}}(\mathbf{G}(\mathrm{j}\omega), \mathbf{F}(\mathrm{j}\omega))\|$. This is the regulator shaping network transfer function matrix.

The task of finding $\mathbf{F}^*(s)$ is very considerable: several algorithms [refs. 10, 11] have been suggested, but the development of elegant algorithms, and indeed the successful application of the method itself, is the subject of current research. For a detailed description of the method, see ref. 12.

EXERCISES

These exercises require CAD facilities for:

(a) classical control system design in continuous and discrete time,
(b) system simulation in continuous and discrete time,
(c) solution of algebraic matrix (up to 3×3) Riccati equations.

1 The azimuth drive dynamics of an antenna dish are modelled by the transfer function (cf. Figure 6.2):

$$Y(s) = \frac{2}{s(s+2)}U(s)$$

where

$y(t)$ is the azimuth angle (potentiometer volts),
$u(t)$ is the drive motor amplifier input (volts), $|u(t)| \leqslant 5$,
$Y(s)$, $U(s)$ are the Laplace transforms of $y(t)$, $u(t)$.

The response of the system to wind disturbance is modelled by the transfer function:

$$\frac{Y(s)}{N(s)} = \frac{0.5}{s + 0.5}$$

where

$n(t)$ is the wind velocity (anemometer volts),
$N(s)$ is the Laplace transform of $n(t)$.

Using classical techniques, design continuous time regulators both with and without noise feed forward, and test the behaviour of the closed loop in simulation
(a) for white noise wind disturbance, variance 1 (volt^2),
(b) for coloured noise wind disturbance, represented by white noise seen through a filter, transfer function:

$$G_F(s) = \frac{0.1}{s + 0.1}$$

Note that in simulation, white noise must in fact be band limited.

2 Design discrete time regulators (Figure 6.7) for the system in (1), using sampling periods (a) $T = 0.2$, (b) $T = 0.4$. Test the behaviour of the regulators in simulation under the conditions in (1).

3 Cast the model of the system in (1) in state equation form (Appendix A, Section A2.4), the states x_1, x_2 representing azimuth angular displacement and velocity, respectively, and design a linear quadratic suboptimal regulator which
(a) minimizes the variance of the azimuth,
(b) minimizes the cost function:

$$\mathcal{E}\left[\int_0^\infty (x_1^2 + x_2^2)\,dt\right]$$

(c) forces the dish on to and along the state track:

$$x_1 + x_2 = 0$$

from any initial state, $|\mathbf{x}(0)| \leqslant 3$ to $\mathbf{x}(t_f)$, $|\mathbf{x}(t_f)| \leqslant 0.1$, in less than 1 unit of time.
Design a linear quadratic suboptimal tracking strategy which forces the system on to and along the track:

$$x_1(t) = 2 - t$$

Simulate the closed loop systems and test their behaviour.

4 Design discrete time control strategies for the systems in (3), first casting the state equations in discrete time form, using sampling periods (a) $T = 0.2$, (b) $T = 0.4$.
Test the behaviour of the closed loop systems in simulation under the conditions in (3).

5 The behaviour of the roll angle of a ship fitted with stabilizers is modelled, in greatly simplified form, by:

$$\dot{\mathbf{x}} = \begin{pmatrix} 0 & 1 \\ -0.01 & -0.04 \end{pmatrix} \mathbf{x} + \begin{pmatrix} 0 \\ 0.1 \end{pmatrix} u + \begin{pmatrix} 0 \\ 0.1 \end{pmatrix} w$$

where

x_1 is the roll angle (rad),
x_2 is the roll velocity (rad s^{-1}),
u is the torque generated by the stabilizers (rad s^{-2}), $|u| \leqslant 1$,
w is the disturbance torque of the sea (rad s^{-2})

Design a linear quadratic suboptimal regulator which minimizes
(a) roll angle,
(b) a combination of roll and roll velocity

$$\varepsilon\left[\int_0^\infty (x_1^2 + x_2^2)\,\mathrm{d}t\right]$$

Simulate and test the behaviour of the stabilized ship in both cases, assuming (band-limited) white Gaussian disturbance torque of variance 0.01 ($\text{rad}^2\ \text{s}^{-4}$).

6 Design discrete time control strategies for the tasks in (5), first casting the state equations in discrete time form, using sampling periods of (a) $T = 0.5$ s, (b) $T = 1$ s. Simulate and test the closed loop systems under the conditions in (5).

7 Laboratory equipment consisting of a flywheel driven via a torsion spring by a local position servo, and disturbed via a belt drive by a local torque servo, as shown in Figures 5.2 and 5.7, is modelled by Eqns (5.19) and (5.20).
Design a linear quadratic suboptimal regulator which minimizes the variance of the flywheel angular displacement:
(a) using an integral state feedback strategy (Section 6.7.1),
(b) using an integral state feedback and noise feedforward strategy (Section 6.7.1),
(c) using the strategy of (b) based on Kalman–Bucy estimates of the states (including the noise state).
Simulate and test the closed loop systems for white noise disturbance torque of variance 0.5 (volt^2).

8 Design discrete time suboptimal control strategies for the laboratory system in (7) for cases (a), (b) and (c) (for which a Kalman filter must be designed),

given that the discrete time model equations ($T = 0.05$) are:

$$\mathbf{x}(n+1) = \begin{pmatrix} 0.902 & 0.476 & 0.099 & 0.008 \\ -0.100 & 0.862 & 0.348 & 0 \\ -0.351 & -0.521 & 0.410 & 0.151 \\ 0 & 0 & 0 & 0.779 \end{pmatrix} \mathbf{x}(n) + \begin{pmatrix} 0.018 \\ 0.100 \\ 0.351 \\ 0 \end{pmatrix} u + \begin{pmatrix} 0.009 \\ 0.054 \\ 0.196 \\ 0.221 \end{pmatrix} w$$

Simulate and test the closed loop systems under the conditions in (7).

REFERENCES

[1] Banks, S. P., *Control Systems Engineering*, Prentice Hall, London, 1989.
[2] Borrie, J. A., *Modern Control Systems: A manual of design methods*, Prentice Hall International, London, 1986.
[3] Chen, C. T., *Linear System Theory and Design*, Holt, Rinehart and Winston, New York, 1984.
[4] Franklin, G. F., Powell, J. D., and Workman, M. L., *Digital Control of Dynamic Systems*, Addison Wesley, Reading, Mass., 1990.
[5] Katz, P., *Digital Control Using Control Microprocessors*, Prentice Hall International, London, 1981.
[6] Astrom, K. J. and Wittenmark, B., *Computer Controlled Systems: Theory and design*, Prentice Hall, Englewood Cliffs, NJ, 1984.
[7] Anderson, D. O. and Moore, J. B., *Optimal Control: Linear quadratic methods*, Prentice Hall, Englewood Cliffs, NJ, 1989.
[8] Maybeck, P. S., *Stochastic Models, Estimation and Control*, Academic Press, New York, 1979.
[9] Zames, G., 'Feedback and optimal sensitivity: Model reference transformation, multiplicative semi-norms and appropriate inverses', *IEEE Trans on Automatic Control*, **AC-26**, 1981, 301–20.
[10] Doyle, J., Glover, K., Khargonekar, P. and Francis, B., 'State space solutions to standard H_2 and H_∞ problems', *Proc. American Control Conf.*, Atlanta GA, 1988, pp. 1691–6.
[11] McFarlane, D. and Glover, K., 'A design procedure using robust stabilization of normalized coprime factors', *Proc. 27th IEEE Conf. on Decision and Control*, Austin TX, 1988, pp. 1343–8.
[12] Maciejowski, J. M., *Multivariate Feedback Design*, Addison Wesley, Wokingham, 1989.

APPENDIX A

Linear dynamical systems

The structure and main properties of standard-form models of the behaviour of deterministic linear dynamical systems are outlined briefly in this appendix. The treatment, which is necessarily brief, concentrates on the topics required to support the main text.

Laplace, z and Fourier transforms are covered in Section A1, continuous time systems in Section A2, and discrete time systems in Section A3. References for further study are given with each topic.

A1 Laplace, z and Fourier transforms

A1.1 Laplace transforms [refs 1, 2]

The Laplace transform of a function of time, $f(t)$, is defined:

$$\mathscr{L}(f(t)) = F(s) = \int_0^\infty f(t)\mathrm{e}^{-st}\,\mathrm{d}t \tag{A1.1}$$

Example

If

$$f(t) = \begin{cases} 0, & t < 0 \\ \mathrm{e}^{-at}, & t \geqslant 0 \end{cases}$$

$$\mathscr{L}(f(t)) = F(s) = \int_0^\infty \mathrm{e}^{-at}\mathrm{e}^{-st}\,\mathrm{d}t = \frac{1}{s+a}$$

s is a complex variable: $s = \sigma + \mathrm{j}\omega$.

The inverse Laplace transform is given by the line integral:

$$\mathscr{L}^{-1}(F(s)) = f(t) = \frac{1}{2\pi\mathrm{j}}\oint_C F(s)\mathrm{e}^{st}\,\mathrm{d}s \tag{A1.2}$$

where C represents a counter-clockwise closed contour in the s-plane Argand diagram enclosing all the singular points of $(F(s)\mathrm{e}^{st})$. For practical purposes most transforms and their inverses are found using a table of common Laplace transforms (Figure A1.1).

Example

If

$$F(s) = \frac{1}{s^2 + 3s + 2}$$

$$f(t) = \mathscr{L}^{-1}\left(\frac{1}{s^2 + 3s + 2}\right) = \mathscr{L}^{-1}\left(\frac{1}{s+1} - \frac{1}{s+2}\right)$$

$$= e^{-t} - e^{-2t}$$

(Figure A1.1, row 8).

	Function, $f(t)$, $t \geq 0$	Laplace transform, $F(s)$
1.	$f(t)$	$\int_0^\infty f(t)\, e^{-st} dt$
2.	$\frac{1}{2\pi j} \oint F(s)\, e^{st} ds$	$F(s)$
3.	$f(t - \tau)$, $t \geq \tau$	$e^{-s\tau} F(s)$
4.	$u^*(t)$	$\frac{1}{s}$
5.	t	$\frac{1}{s^2}$
6.	t^2	$\frac{2!}{s^3}$
7.	t^{n-1}	$\frac{(n-1)!}{s^n}$
8.	e^{-at}	$\frac{1}{s+a}$
9.	$\frac{1}{a}(u^*(t) - e^{-at})$	$\frac{1}{s(s+a)}$
10.	$t\, e^{-at}$	$\frac{1}{(s+a)^2}$
11.	$\sin(at)$	$\frac{a}{s^2 + a^2}$
12.	$\frac{1}{b} e^{-at} \sin(bt)$	$\frac{1}{(s+a)^2 + b^2}$
13.	$\cos(at)$	$\frac{s}{s^2 + a^2}$
14.	$e^{-at} \cos(bt)$	$\frac{s+a}{(s+a)^2 + b^2}$

Note: $u^*(t)$ is the unit step function: $u^*(t) = \begin{cases} 0, & t < 0 \\ 1, & t \geq 0 \end{cases}$

Figure A1.1 Laplace transforms.

One important property of Laplace transforms is that (Figure A1.1, row 3)

$$\mathscr{L}(f(t-\tau)) = e^{-s\tau}F(s) \tag{A1.3}$$

This gives the transform of $f(t)$ delayed by τ.

The most important property, however, concerns the transforms of derivatives of $f(t)$ WRT time:

$$\mathscr{L}\left(\frac{df(t)}{dt}\right) = sF(s) - f(0)$$

In general:

$$\mathscr{L}\left(\frac{d^m f(t)}{dt^m}\right) = s^m F(s) - s^{m-1}f(0) - s^{m-2}f^{(1)}(0) - \cdots - f^{(m-1)}(0) \tag{A1.4}$$

where

$$f^{(i)}(0) = \frac{d^i(f(t))}{dt^i}$$

This allows linear differential equations with constant coefficients to be solved, and corresponding dynamical systems to be analyzed.

Consider the equation:

$$\frac{d^m y}{dt^m} + b_{m-1}\frac{d^{m-1}y}{dt^{m-1}} + \cdots + b_0 y = a_{m-1}\frac{d^{m-1}u}{dt^{m-1}} + \cdots + a_0 u \tag{A1.5}$$

Suppose Eqn (A1.5) represents the behaviour of a dynamical system, S (Figure A1.2), where

$u(t)$ is the system input,
$y(t)$ is the system output.

S is 'linear time invariant' since it can be shown that the solution of Eqn (A1.5), $y(t)$, has a linear relationship with the system input, $u(t)$ (i.e. doubling $u(t)$ results in $y(t)$ being doubled), and since the coefficient terms are constant, or time invariant.

If Laplace transforms of Eqn (A1.5) are taken, assuming zero initial conditions (Eqn A1.4),

$$y(0) = y^{(1)}(0) = y^{(2)}(0) = \cdots = y^{(m-1)}(0) = 0$$

it is easy to show that:

$$Y(s) = G(s)U(s) \tag{A1.6}$$

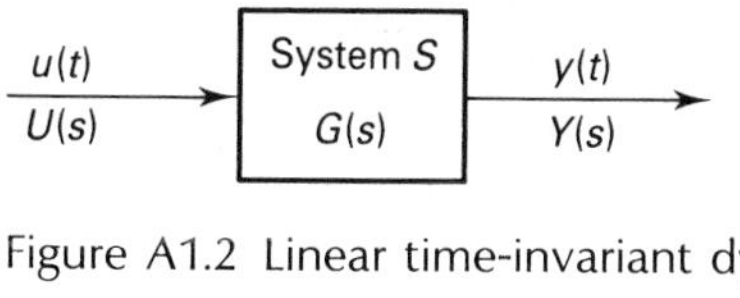

Figure A1.2 Linear time-invariant dynamical system.

where

$$G(s) = \frac{a_{m-1}s^{m-1} + a_{m-2}s^{m-2} + \cdots + a_0}{s^m + b_{m-1}s^{m-1} + \cdots + b_0} \tag{A1.7}$$

$G(s)$ is the 'transfer function' of the system S (Figure A1.2).

If an input $u(t)$ is specified, its transform, $U(s)$, can be found, $Y(s)$ calculated from Eqn (A1.6), and $y(t)$ then found by inverse Laplace transformation.

Example

Consider a linear time-invariant system whose behaviour is modelled by the third order differential equation:

$$\frac{d^3y}{dt^3} + 9\frac{d^2y}{dt^2} + 26\frac{dy}{dt} + 24y = 24u, \quad y(0) = y^{(1)}(0) = y^{(2)}(0) = 0 \tag{A1.8}$$

The system transfer function is (Eqn (A1.6), (A1.7)):

$$G(s) = \frac{Y(s)}{U(s)} = \frac{24}{s^3 + 9s^2 + 26s + 24}$$

Consider a step function input:

$$u(t) = \begin{cases} 0, & t < 0 \\ 1, & t \geqslant 0 \end{cases}$$

Then (Eqn (A1.6)):

$$Y(s) = \frac{24}{s^3 + 9s^2 + 26s + 24} \cdot \frac{1}{s}$$

By partial fractions:

$$Y(s) = \frac{1}{s} - \frac{6}{s+2} + \frac{8}{s+3} - \frac{3}{s+4}$$

From Figure (A1.1):

$$y(t) = u^*(t) - 6e^{-2t} + 8e^{-3t} - 3e^{-4t}, \quad t \geqslant 0$$

A1.2 *z transforms [refs 1, 4]*

The z transforms of a function of discrete time, $f(n)$, $n = 0, 1, 2, \ldots \infty$, is defined:

$$Z(f(n)) = \bar{F}(z) = \sum_{n=0}^{\infty} f(n)z^{-n} \tag{A1.9}$$

Example

If

$$f(n) = \begin{cases} 0, & n < 0 \\ e^{-an}, & n \geqslant 0 \end{cases}$$

$$Z(f(n)) = \bar{F}(z)$$

$$= \sum_{n=0}^{\infty} e^{-an} z^{-n} = \frac{1}{1 - e^{-a} z^{-1}} \qquad \text{(A1.10)}$$

z is a complex variable, conveniently represented in polar coordinate form:

$$z = r e^{j\beta} \qquad \text{(A1.11)}$$

	Function, $f(n)$, $n \geqslant 0$	z transform, $F(z)$
1.	$f(n)$	$\sum_{n=0}^{\infty} f(n) z^{-n}$
2.	$\frac{1}{2\pi j} \oint \bar{F}(z)\, dz$	$\bar{F}(z)$
3.	$f(n-k)$	$z^{-k}\bar{F}(z)$
4.	$u^*(n)$	$\frac{1}{1-z^{-1}}$
5.	n	$\frac{z^{-1}}{(1-z^{-1})^2}$
6.	n^2	$\frac{z^{-1}(1+z^{-1})}{(1-z^{-1})^3}$
7.	n^{m-1}	$\lim_{a \to 0} (-1)^{m-1} \left[\frac{\partial^{m-1}}{\partial a^{m-1}} \left(\frac{1}{1-e^{-a}z^{-1}} \right) \right]$
8.	e^{-an}	$\frac{1}{1-e^{-a}z^{-1}}$
9.	$\frac{1}{a}(u^*(n) - e^{-an})$	$\frac{(1-e^{-a})z^{-1}}{(1-z^{-1})(1-e^{-a}z^{-1})}$
10.	$n\, e^{-an}$	$\frac{e^{-a}z^{-1}}{(1-e^{-a}z^{-1})^2}$
11.	$\sin(an)$	$\frac{z^{-1}\sin(a)}{1-2z^{-1}\cos(a)+z^{-2}}$
12.	$\frac{1}{b} e^{-an}\sin(bn)$	$\frac{z^{-1}e^{-a}\sin(b)}{1-2z^{-1}e^{-a}\cos(b)+e^{-2a}z^{-2}}$
13.	$\cos(an)$	$\frac{1-z^{-1}\cos(a)}{1-2z^{-1}\cos(a)+z^{-2}}$
14.	$e^{-an}\cos(bn)$	$\frac{1-z^{-1}e^{-a}\cos b}{1-2z^{-1}e^{-a}\cos b+e^{-2a}z^{-2}}$

Note: $u^*(n)$ is the unit step function: $u^*(n = \begin{cases} 0, n < 0 \\ 1, n \geqslant 0 \end{cases}$

Figure A1.3 z transforms.

A table of common z transforms is given in Figure A1.3 (see previous page). Figures A1.1 and A1.3 are arranged to correspond, line by line.

Example

Consider $f(t) = \mathrm{e}^{-at}$, sampled at intervals T to yield the discrete function:

$$f(n) = \mathrm{e}^{-anT}$$

Then (Figure A1.3)

$$Z(f(n) = \frac{1}{1 - \mathrm{e}^{-aT}z^{-n}}$$

This agrees with Eqn (A1.10), $T = 1$.

The inverse z transform is given by the line integral (cf. Eqn (A1.2)):

$$Z^{-1}(\bar{F}(z)) = f(n) = \frac{1}{2\pi \mathrm{j}} \oint_C \bar{F}(z) z^{n-1}\, \mathrm{d}z \tag{A1.12}$$

where C represents a counter-clockwise closed contour in the z-plane Argand diagram enclosing all the singular points of $(\bar{F}(z)z^{n-1})$.

In practice, z transforms and their inverses are found by table look-up (Figure A1.3).

Example

Consider

$$\bar{F}(z) = \frac{z(1 - \mathrm{e}^{-1})}{(z-1)(z - \mathrm{e}^{-1})}$$

$$f(n) = Z^{-1}\left(\frac{z}{z-1} - \frac{z}{z - \mathrm{e}^{-1}}\right)$$

$$= u^*(n) - \mathrm{e}^{-n}$$

(Figure A1.3).

The most important property of the z transform concerns the transform of $f(n)$ advanced and delayed, respectively, in discrete time:

$$Z(f(n+k)) = z^k \bar{F}(z) - \sum_{n=0}^{k-1} z^{-n} f(n) \tag{A1.13}$$

$$Z(f(n-k)) = z^{-k}\bar{F}(z) \tag{A1.14}$$

This allows linear difference equations with constant coefficients to be solved, and corresponding dynamical systems to be analyzed.

Consider the behaviour of a discrete time dynamical system (Figure A1.4) described by the difference equation:

$$y(n) = a_0u(n) + a_1u(n-1) + \cdots + a_mu(n-m)$$
$$+ b_1y(n-1) + \cdots + b_my(n-m) \qquad \text{(A1.15)}$$

where

$u(n)$ is the system input,
$y(n)$ is the system output.

Eqn (A1.15) is an 'autoregressive moving average' (ARMA) model of the behaviour of Σ. Because its solution, $y(n)$, can be shown to hold a linear relationship with $u(n)$, and because its coefficients are constant, Eqn (A1.15) is a 'linear time-invariant' difference equation, and Σ is a linear time-invariant system.

Assuming zero 'initial' conditions,

$$y(-1) = y(-2) = \cdots = y(-m) = 0$$

taking z transforms of Eqn (A1.15) yields:

$$(1 - b_1z^{-1} - b_2z^{-2} - \cdots - b_mz^{-m})\bar{Y}(z) = (a_0 + a_1z^{-1} + \cdots + a_mz^{-m})\bar{U}(z)$$

where

$$\bar{Y}(z) = Z(y(n))$$
$$\bar{U}(z) = Z(u(n))$$

Hence,

$$\bar{Y}(z) = \bar{G}(z)\bar{U}(z) \qquad \text{(A1.16)}$$

where

$$\bar{G}(z) = \frac{a_0 + a_1z^{-1} + \cdots + a_mz^{-m}}{1 - b_1z^{-1} - \cdots - b_mz^{-m}} \qquad \text{(A1.17)}$$

is the 'transfer function' of the system.

The ARMA model, Eqn (A1.15), and the transfer function model, Eqn (A1.17), are equivalent; one can readily be constructed from the other by substitution of the appropriate coefficients. The system output, $y(n)$, corresponding to the input, $u(n)$, is most easily calculated by finding $\bar{U}(z)$, then $\bar{Y}(z)$ using Eqn (A1.16) and finally $y(n) = Z^{-1}(\bar{Y}(z))$ (Figure A1.3).

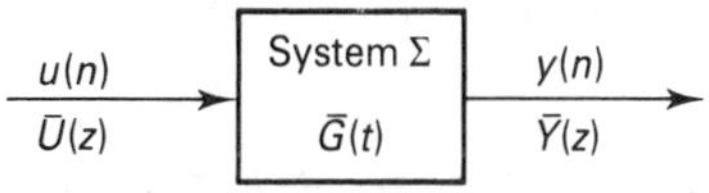

Figure A1.4 Linear time-invariant system.

Example

Consider the linear time-invariant system (Figure A1.4) whose behaviour is described by the ARMA model (Eqn (A1.15)):

$$y(n) = u(n) - 1.1y(n-1) + 0.3y(n-2), \quad y(-1) = y(-2) = 0 \qquad \text{(A1.18)}$$

The system transfer function is (Eqn (A1.16)):

$$\bar{G}(z) = \frac{\bar{Y}(z)}{\bar{U}(z)}$$

Substituting the appropriate coefficients in Eqn (A1.17):

$$\bar{G}(z) = \frac{1}{1 + 1.1z^{-1} - 0.3z^{-2}} = \frac{z^2}{z^2 + 1.1z - 0.3}$$

Consider a step function input:

$$u(n) = u^*(n) = \begin{cases} 0, & n < 0 \\ 1, & n \geqslant 0 \end{cases}$$

Then (Figure A1.3):

$$\bar{U}(z) = \frac{1}{1 - z^{-1}} = \frac{z}{z-1}$$

and (Eqn (A1.16)):

$$\bar{Y}(z) = \frac{z^2}{z^2 + 1.1z - 0.3} \cdot \frac{z}{z-1}$$

$$= \frac{5}{z-1} - \frac{10}{z-0.5} - \frac{15}{z-0.6}$$

Taking inverse z transforms (Figure A1. 3):

$$y(n) = 5u^*(n-1) + 10(0.5)^{n-1} - 15(0.6)^{n-1}, \quad n \geqslant 0$$

This is the solution of (A1.18).

It remains to consider the transfer function of a sampled data system Σ consisting of a continuous time LTI system buffered by DAC and ADC, sampling time T, as shown in Figure A1.5. If the Laplace transfer function of S, $G(s)$, and T, are known, then the transfer function of Σ, $\bar{G}(z)$, is:

$$\bar{G}(z) = Z\left\{\frac{1 - e^{-sT}}{s} \cdot G(s)\right\} \qquad \text{(A1.19)}$$

where the RHS represents the z transform (Figure A1.3) corresponding to the Laplace transform within the brackets (Figure A1.1, corresponding row).

Example

Consider

$$G(s) = \frac{1}{s+1}, \quad T = 0.5$$

Then:

$$\bar{G}(z) = Z\left\{\frac{1 - e^{-s \cdot 0.5}}{s} \cdot \frac{1}{s+1}\right\}$$

$$= Z\left\{(1 - e^{-s \cdot 0.5})\frac{1}{s(s+1)}\right\}$$

Noting that $e^{-s \cdot 0.5}$ denotes a delay of 0.5 (Eqn (A1.3)), equivalent to z^{-1} since $T = 0.5$ (Eqn (A1.14)):

$$\bar{G}(z) = (1 - z^{-1})Z\left\{\frac{1}{s(s+1)}\right\}$$

$$= (1 - z^{-1})\frac{(1 - e^{-0.5})z^{-1}}{(1 - z^{-1})(1 - e^{-0.5}z^{-1})}$$

(Figures A1.1 and A1.3, rows 9)

$$= \frac{0.393z^{-1}}{1 - 0.606z^{-1}}$$

A1.3 Fourier transforms

A1.3.1 Continuous time Fourier transforms [refs 1, 2]

The frequency characteristics, or content, of a function of time, $f(t)$, can be studied using its Fourier transform:

$$\mathcal{F}(f(t)) = F(j\omega) = \int_{-\infty}^{+\infty} f(t)e^{-j\omega t}\,dt \qquad \text{(A1.20)}$$

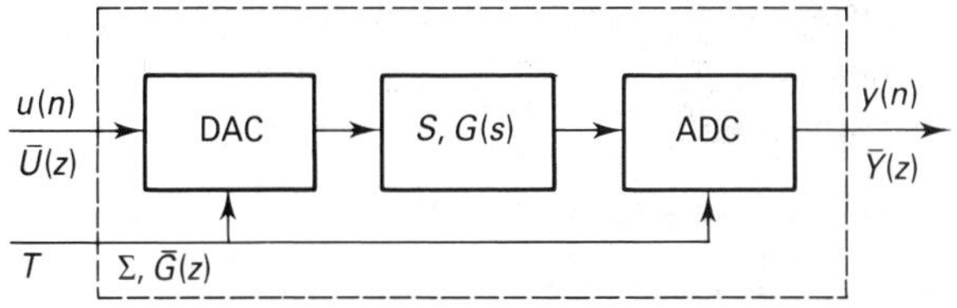

Figure A1.5 Sample data system.

where ω represents frequency in radians per unit time.

The inverse transform is

$$\mathscr{F}^{-1}(F(j\omega)) = f(t) = \frac{1}{2\pi}\int_{-\infty}^{+\infty} F(j\omega)e^{+j\omega t}\,d\omega \tag{A1.21}$$

Fourier transforms and their inverses are generally quite difficult to evaluate in analytic form: they are usually evaluated numerically using computer algorithms (Section 4.3.2).

Example

(a) Consider

$$f(t) = \begin{cases} 0, & t < 0 \\ e^{-at}, & t \geqslant 0 \end{cases}$$

$$F(j\omega) = \int_{-\infty}^{+\infty} e^{-at}e^{-j\omega t}\,dt = \frac{1}{a + j\omega}$$

(b) Consider:

$$f(t) = \cos(\omega_0 t)$$

$$F(j\omega) = \int_{-\infty}^{+\infty} \cos(\omega_0 t)e^{-j\omega t}\,dt = \pi\delta(\omega - \omega_0) + \pi\delta(\omega + \omega_0)$$

where $\delta(\omega)$ is a Dirac delta function (Section D1). This integral is evaluated only with some difficulty.

A table of common Fourier transforms is given in Figure A1.6, which shows that $F(j\omega)$ is generally complex. This may be seen by expressing it:

$$\begin{aligned} F(j\omega) &= \int_{-\infty}^{+\infty} f(t)(\cos(\omega t) - j\sin(\omega t))\,dt \\ &= F_R(\omega) - jF_I(\omega) \end{aligned}$$

where

$$F_R(\omega) = \int_{-\infty}^{+\infty} f(t)\cos(\omega t)\,dt$$

$$F_I(\omega) = \int_{-\infty}^{+\infty} f(t)\sin(\omega t)\,dt$$

It follows from these relationships that:

$$\left.\begin{array}{l} \text{if } f(t) \text{ is real, } F_R(\omega) \text{ is even, } F_I(\omega) \text{ is odd} \\ \text{if } f(t) \text{ is real and even, } F_I(\omega) = 0 \text{, and } F(j\omega) \text{ is even and real} \\ \text{if } f(t) \text{ is real and odd, } F_R(\omega) = 0 \text{, and } F(j\omega) \text{ is odd and imaginary} \end{array}\right\} \tag{A1.22}$$

The amplitude and phase characteristics of $f(t)$ can be expressed in terms of $F(\mathrm{j}\omega)$, $F_R(\omega)$, $F_I(\omega)$, as follows: $f(t)$ can be regarded as the sum of an infinite number of components of amplitude (Eqn (A1.21)):

$$\frac{1}{2\pi}\left|F(\mathrm{j}\omega)\right|\mathrm{d}\omega = \frac{1}{2\pi}\sqrt{F_R^2(\omega)+F_I^2(\omega)} \tag{A1.23}$$

and phase lag:

$$-\underline{|F(\mathrm{j}\omega)} = \tan^{-1}\left(\frac{F_I(\omega)}{F_R(\omega)}\right) \tag{A1.24}$$

Differential equations can be solved using Fourier transforms, though this is not their main use. However, it is useful to define Fourier transfer functions of LTI systems

	Function, $f(t)$	Fourier transform, $F(\mathrm{j}\omega)$
1.	$f(t)$	$\int_{-\infty}^{+\infty} f(t)\,\mathrm{e}^{-\mathrm{j}\omega t}\mathrm{d}t$
2.	$\frac{1}{2\pi}\int_{-\infty}^{+\infty} F(\mathrm{j}\omega)\,\mathrm{e}^{+\mathrm{j}\omega t}\mathrm{d}\omega$	$F(\mathrm{j}\omega)$
3.	$f(t+\tau)$	$\mathrm{e}^{-\mathrm{j}\omega t}F(\mathrm{j}\omega)$
4.	1	$2\pi\delta(\omega)$
5.	$\mathrm{e}^{-at}, t \geqslant 0, a > 0$	$\frac{1}{a+\mathrm{j}\omega}$
6.	$t\,\mathrm{e}^{-at}, t \geqslant 0, a > 0$	$\frac{1}{(a+\mathrm{j}\omega)^2}$
7.	$t^n\,\mathrm{e}^{-at}, t \geqslant 0, a > 0$	$\frac{n!}{(a+\mathrm{j}\omega)^{n+1}}$
8.	$\frac{a}{a^2+t^2}, t \geqslant 0, a > 0$	$\pi\,\mathrm{e}^{-a\lvert\omega\rvert}$
9.	$\cos(at)$	$\pi\delta(\omega - a) + \pi\delta(\omega + a)$
10.	$\sin(at)$	$-\pi \mathrm{j}\delta(\omega - a) + \pi \mathrm{j}d(\omega + a)$
11.	$\mathrm{e}^{-a\lvert t\rvert}, a > 0$	$\frac{2a}{a^2+\omega^2}$
12.	$u^*(t)$	$\pi\delta(\omega) + \frac{1}{\mathrm{j}\omega}$
13.	$\mathrm{e}^{jat}f(t)$	$F(j(\omega - a))$
14.	$a\delta t$	a

Note: $u^*(t)$ is the unit step function: $u^*(t) = \begin{cases} 0, t < 0 \\ 1, t \geqslant 0 \end{cases}$

Figure A1.6 Fourier transforms.

(cf. Eqns (A1.6) and (A1.7)):

$$G(\mathrm{j}\omega) = \frac{Y(\mathrm{j}\omega)}{U(\mathrm{j}\omega)} \tag{A1.25}$$

where

$Y(\mathrm{j}\omega)$ is the Fourier transform of the system output,
$U(\mathrm{j}\omega)$ is the Fourier transform of the system input,
$G(\mathrm{j}\omega)$ is the Fourier transfer function of the system.

A1.3.2 Discrete time Fourier transforms [refs 1, 2]

The frequency characteristics, or content, of a function of discrete time, $f(n)$, can be studied using its Fourier transform:

$$\begin{aligned}\mathscr{F}(f(n)) &= \bar{F}(\mathrm{e}^{\mathrm{j}\beta}) \\ &= \sum_{n=-\infty}^{+\infty} f(n)\mathrm{e}^{-\mathrm{j}\beta n}\end{aligned} \tag{A1.26}$$

where β represents frequency in radians per sample. $\bar{F}(\mathrm{e}^{\mathrm{j}\beta})$ is a complex function of continuous frequency, $-\infty \leqslant \beta \leqslant +\infty$ and is periodic with period 2π. Any interval of 2π is sufficient, therefore, to describe $\bar{F}(\mathrm{e}^{\mathrm{j}\beta})$; generally $-\pi < \beta \leqslant +\pi$ is used.

The corresponding inverse transform, usually difficult to evaluate analytically, is therefore:

$$\begin{aligned}\mathscr{F}^{-1}(\bar{F}(\mathrm{e}^{\mathrm{j}\beta})) &= f(n) \\ &= \frac{1}{2\pi}\int_{-\pi}^{+\pi} \bar{F}(\mathrm{e}^{\mathrm{j}\beta})\mathrm{e}^{\mathrm{j}\beta n}\,\mathrm{d}\beta\end{aligned} \tag{A1.27}$$

Where it is desirable to express the frequency in radians per unit time, the sampling interval T is involved, $\beta = \omega T$, and the relevant relationships are:

$$\begin{aligned}\mathscr{F}(f(nT)) &= \bar{F}(\mathrm{e}^{\mathrm{j}\omega T}) \\ &= \sum_{n=-\infty}^{+\infty} f(nT)\mathrm{e}^{-\mathrm{j}\omega nT}\end{aligned} \tag{A1.28}$$

where ω is frequency in radians per unit time, and T is the sampling period. $\bar{F}(\mathrm{e}^{\mathrm{j}\omega T})$ is periodic with period $2\pi/T$.

The inverse relationship is:

$$\begin{aligned}\mathscr{F}^{-1}(\bar{F}(\mathrm{e}^{\mathrm{j}\omega T})) &= f(nT) \\ &= \frac{T}{2\pi}\int_{-\pi/T}^{+\pi/T} \bar{F}(\mathrm{e}^{\mathrm{j}\omega T})\mathrm{e}^{\mathrm{j}\omega nT}\,\mathrm{d}\omega\end{aligned} \tag{A1.29}$$

As with continuous time transforms (cf. Eqn (A1.22)):

$$\left.\begin{array}{l}\text{if } f(n) \text{ is real and even, } \bar{F}(e^{j\beta}) \text{ is even and real} \\ \text{if } f(n) \text{ is real and odd, } \bar{F}(e^{j\beta}) \text{ is odd and imaginary}\end{array}\right\} \quad (A1.30)$$

It is also useful to develop transfer functions of LTI systems (cf. Eqns (A1.16) and (A1.17)):

$$\bar{G}(e^{j\beta}) = \frac{\bar{Y}(e^{j\beta})}{\bar{U}(e^{j\beta})} \quad (A1.31)$$

where

$\bar{Y}(e^{j\beta}) = \mathscr{F}(y(n))$,
$\bar{U}(e^{j\beta}) = \mathscr{F}(u(n))$,
$y(n)$ is the system output,
$u(n)$ is the system input.

A2 Continuous time linear systems [refs 1, 3]

A2.1 Model equations

The dynamical behaviour of a linear continuous time system, S, is modelled by the standard-form equations:

$$\dot{\mathbf{x}}(t) = \mathbf{A}(t)\mathbf{x}(t) + \mathbf{B}(t)\mathbf{u}(t) \quad (A2.1)$$

$$\mathbf{y}(t) = \mathbf{C}(t)\mathbf{x}(t) + \mathbf{D}(t)\mathbf{u}(t) \quad (A2.2)$$

where

$\mathbf{x}(t)$ is the state vector $(x_1 \quad x_2 \quad \dots \quad x_m)^T$, a real variable,
$\mathbf{u}(t)$ is the input vector $(u_1 \quad u_2 \quad \dots \quad u_q)^T$, a real variable,
$\mathbf{y}(t)$ is the output vector $(y_1 \quad y_2 \quad \dots \quad y_r)^T$, a real variable,
$\mathbf{A}(t)$, $\mathbf{B}(t)$, $\mathbf{C}(t)$, $\mathbf{D}(t)$ are time-dependent matrices of dimension $(m \times m)$, $(m \times q)$, $(r \times m)$, $(r \times q)$, respectively,
$\mathbf{A}(t)$ is the 'state matrix'.

The solution of Eqn (A2.1), given the initial condition, $\mathbf{x}(t_0)$, is:

$$\mathbf{x}(t) = \varphi(t, t_0)\cdot\mathbf{x}(t_0) + \int_{t_0}^{t} \varphi(\lambda, t_0)\mathbf{B}(\lambda)\mathbf{u}(\lambda)\,d\lambda \quad (A2.3)$$

where $\varphi(t, t_0)$ is the 'state transition matrix' which satisfies the differential equation:

$$\frac{d}{dt}(\varphi(t, t_0)) = \mathbf{A}(t)\,\varphi(t, t_0) \quad (A2.4)$$

In Eqn (A2.3), $\varphi(t, t_0)\mathbf{x}(t_0)$ represents the 'zero input response' of S – the behaviour of $\mathbf{x}(t)$ when $\mathbf{u}(t) = 0$, $\mathbf{x}(t_0) \neq \mathbf{0}$.

$\int_{t_0}^{t} \varphi(\lambda, t_0)\mathbf{B}(\lambda)\mathbf{u}(\lambda)\,\mathrm{d}\lambda$ represents the 'zero initial state response' – the behaviour of $\mathbf{x}(t)$ when $\mathbf{x}(t_0) = \mathbf{0}$, $\mathbf{u}(t) \neq \mathbf{0}$.

A large and useful subset of linear systems is 'linear time invariant' (LTI), modelled by the standard-form constant parameter equations:

$$\dot{\mathbf{x}}(t) = \mathbf{A}\mathbf{x}(t) + \mathbf{B}\mathbf{u}(t) \tag{A2.5}$$

$$\mathbf{y}(t) = \mathbf{C}\mathbf{x}(t) + \mathbf{D}\mathbf{u}(t) \tag{A2.6}$$

where **A, B, C, D** are constant matrices of appropriate dimension. The solution of Eqn (A2.4), $\mathbf{A}(t) = \mathbf{A}$, is:

$$\begin{aligned}\varphi(t, t_0) &= \mathbf{I} + \mathbf{A}(t - t_0) + \frac{1}{2!}\mathbf{A}^2(t - t_0)^2 + \cdots \\ &= \mathrm{e}^{\mathbf{A}(t - t_0)}\end{aligned} \tag{A2.7}$$

This yields the solution of Eqn (A2.5), given $\mathbf{x}(t_0)$:

$$\mathbf{x}(t) = \mathrm{e}^{\mathbf{A}(t-t_0)}\mathbf{x}(t_0) + \int_{t_0}^{t} \mathrm{e}^{\mathbf{A}\lambda}\mathbf{B}\mathbf{u}(\lambda)\,\mathrm{d}\lambda \tag{A2.8}$$

Example

Consider the system:

$$\dot{\mathbf{x}} = \begin{pmatrix} -2 & 1 \\ -101 & 0 \end{pmatrix}\mathbf{x} + \begin{pmatrix} 0 \\ 101 \end{pmatrix}u \tag{A2.9}$$

$$y = (1 \quad 0)\mathbf{x} \tag{A2.10}$$

For $\mathbf{x}(0) = (2 \quad 0)^T$, $u(t) = 0$, this yields the zero input response (Eqn (A2.8)):

$$\mathbf{x}(t) = \exp\begin{pmatrix} -2t & t \\ -101t & 0 \end{pmatrix}$$

$\mathbf{x}(t)$ is shown in Figure A2.1

For $\mathbf{x}(0) = \mathbf{0}$, $u(t) = 1$, $t \geqslant 0$, Eqn (A2.8) yields a zero initial state response:

$$\mathbf{x}(t) = \int_0^t \exp\begin{pmatrix} -2\lambda & \lambda \\ -101\lambda & 0 \end{pmatrix}\begin{pmatrix} 0 \\ 101 \end{pmatrix}\mathrm{d}\lambda$$

A2.2 Transfer functions, eigenvalues and stability (LTI systems)

For the LTI case, the transfer function of S is readily found by taking Laplace transforms of Eqns (A2.5) and (A2.6), for $\mathbf{x}(0) = \mathbf{0}$:

$$s\mathbf{X}(s) = \mathbf{AX}(s) + \mathbf{BU}(s)$$
$$\mathbf{Y}(s) = \mathbf{CX}(s) + \mathbf{DU}(s)$$

where
$$\mathbf{X}(s) = \mathscr{L}(\mathbf{x}(t))$$
$$\mathbf{U}(s) = \mathscr{L}(\mathbf{u}(t))$$
$$\mathbf{Y}(s) = \mathscr{L}(\mathbf{y}(t))$$

Eliminating $\mathbf{X}(s)$ from these equations:

$$\mathbf{Y}(s) = (\mathbf{C}(s\mathbf{I} - \mathbf{A})^{-1}\mathbf{B} + \mathbf{D})\mathbf{U}(s)$$
$$= \mathbf{G}(s)\mathbf{U}(s)$$

where $\mathbf{G}(s)$ is the $(r \times q)$ 'transfer function matrix' of S,

$$\mathbf{G}(s) = \mathbf{C}(s\mathbf{I} - \mathbf{A})^{-1}\mathbf{B} + \mathbf{D} \tag{A2.11}$$

Since

$$(s\mathbf{I} - \mathbf{A})^{-1} = \text{adjoint } (s\mathbf{I} - \mathbf{A}) \cdot \frac{1}{\det(s\mathbf{I} - \mathbf{A})}$$

Eqn (A2.11) yields the 'characteristic equation' of S, describing values of s for which $\mathbf{G}(s)$ has infinite elements:

$$\det(s\mathbf{I} - \mathbf{A}) = 0 \tag{A2.12}$$

Solutions of Eqn (A2.12) give the 'poles' of S.

The 'characteristic equation' of the matrix A, from which the eigenvalues are found, is:

$$\det(\lambda\mathbf{I} - \mathbf{A}) = 0 \tag{A2.13}$$

Eqns (A2.12) and (A2.13) show that the poles of S lie at the eigenvalues of $\mathbf{A}$.

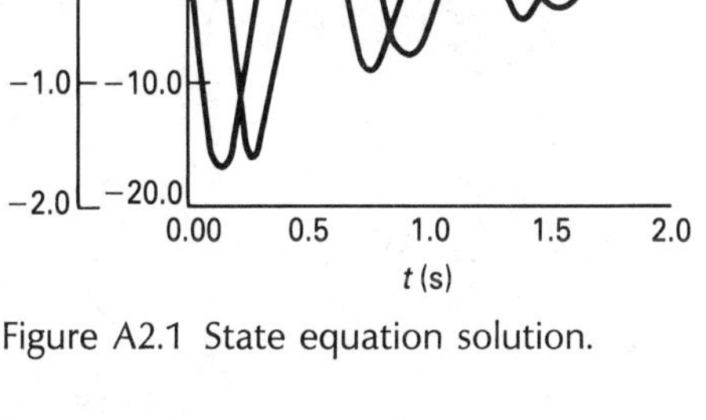

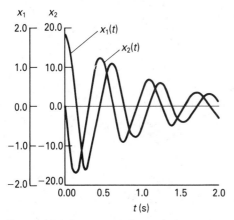

Figure A2.1 State equation solution.

The eigenvalues of **A** give an indication of the natural behaviour or natural modes of S:

A complementary pair of eigenvalues at $\lambda = \sigma \pm j\omega$ correspond to time domain behaviour of the 'shape':

$$e^{(\sigma + j\omega)t} + e^{(\sigma - j\omega)t} = 2e^{\sigma t}\cos(\omega t) \tag{A2.14}$$

This is an exponentially increasing (or decreasing) sinusoidal process.

A single real eigenvalue, $\lambda = \sigma$, corresponds to time domain behaviour of the shape:

$$e^{\sigma t}$$

This is an exponentially increasing (or decreasing) process.

The eigenvalues of **A** thus give an indication of the 'stability' of S; if only exponentially decaying natural modes are present ($\sigma < 0$), then S is stable, while if any non-decreasing mode exists ($\sigma \geqslant 0$), S is unstable.

In slightly more detail, two important types of stability can be defined.

(a) S is 'asymptotically stable' if $\mathbf{x}(t) \to \mathbf{0}$, $t \to \infty$, $\mathbf{u}(t) = \mathbf{0}$, $\mathbf{x}(t_0) \neq \mathbf{0}$.

(b) S is 'bounded-input bounded-output' (BIBO) stable if $\mathbf{y}(t)$ is bounded (i.e. no element tends to ∞) for every $\mathbf{u}(t)$ bounded.

It can be shown that if all the eigenvalues of **A** lie in the open left-hand plane ($\sigma < 0$), S is both asymptotically and BIBO stable.

Example

Consider the example of Eqns (A2.9) and (A2.10).

From Eqn (A2.11), the system transfer function is:

$$\mathbf{G}(s) = (1 \quad 0)\begin{pmatrix} s+2 & -1 \\ 101 & s \end{pmatrix}^{-1}\begin{pmatrix} 0 \\ 101 \end{pmatrix} + 0$$

$$= \frac{101}{s^2 + 2s + 101}$$

The characteristic equation of S is (Eqn (A2.12)):

$$s^2 + 2s + 101 = 0$$

This yields the poles of S:

$$s = -1 \pm j10$$

The eigenvalues of the state matrix are (Eqn (A2.13)):

$$\lambda = -1 \pm j10$$

The natural modes of S are thus of the shape (Eqn (A2.14)):

$$2e^{-t}\cos 10t$$

This accords with Figure A2.1.

S is asymptotically and BIBO stable: its natural mode is a lightly damped oscillation, frequency 10 rad per unit time.

A2.3 Controllability and observability

(a) A system S is 'controllable' if an input $\mathbf{u}(t)$ can be specified which drives the state, $\mathbf{x}(t)$, from any initial state to any other specified state in a finite time. This implies that all the system's natural modes are affected by the input, $\mathbf{u}(t)$.

In the LTI case, it can be shown that S is controllable if and only if the 'controllability matrix', $\mathbf{P}$, has rank m:

$$\mathbf{P} = (\mathbf{B} \quad \mathbf{AB} \quad \mathbf{A}^2\mathbf{B} \quad \ldots \mathbf{A}^{m-1}\mathbf{B}) \tag{A2.15}$$

Equivalently, S is controllable if and only if $\mathbf{B}$ is right prime WRT $(s\mathbf{I} - \mathbf{A})^{-1}$ (Eqn (A2.11)).

An example of an uncontrollable system illustrates the meaning of this.

Example

Consider an LTI system, S, comprising two subsystems connected in series as shown in Figure A2.2.

The system S is second order, and can be modelled by the equations (Eqns (A2.5) and (A2.6)):

$$\dot{\mathbf{x}} = \begin{pmatrix} -3 & 1 \\ -2 & 0 \end{pmatrix}\mathbf{x} + \begin{pmatrix} 1 \\ 1 \end{pmatrix}u$$

$$y = (1 \quad 0)\mathbf{x}$$

where

$$x_1 = y$$

$$x_2 = \dot{y} + 3y$$

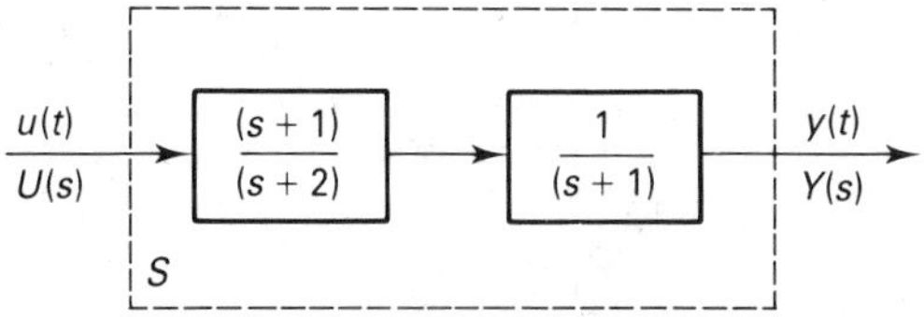

Figure A2.2 Uncontrollable system.

In this case (Eqn (A2.15)):

$$\mathbf{P} = \begin{pmatrix} 1 & -2 \\ 1 & -2 \end{pmatrix}$$

This has rank 1, but $m = 2$, so S is uncontrollable.

Moreover, postmultiplying $(s\mathbf{I} - \mathbf{A})^{-1}$ by $\mathbf{B}$ (Eqn (A2.11)) leads to the 'disappearance' of the term in $(s + 1)$:

$$\begin{aligned}(s\mathbf{I} - \mathbf{A})^{-1}\mathbf{B} &= \begin{pmatrix} s+3 & -1 \\ 2 & s \end{pmatrix}^{-1} \begin{pmatrix} 1 \\ 1 \end{pmatrix} \\ &= \frac{1}{(s+2)(s+1)} \begin{pmatrix} (s+1) \\ (s+1) \end{pmatrix} \\ &= \frac{1}{(s+2)} \begin{pmatrix} 1 \\ 1 \end{pmatrix}\end{aligned}$$

$\mathbf{B}$ is therefore not right prime WRT $(s\mathbf{I} - \mathbf{A})^{-1}$, indicating that S is uncontrollable.

Heuristically, the 'zero' at $s = -1$ filters out the very signal which would affect the natural mode corresponding to the subsequent pole at $s = -1$; this mode is thus uncontrollable from the system input.

(b) A system S is 'observable' if all the states, $\mathbf{x}(t)$, can be deduced from knowledge of the input, $\mathbf{u}(t)$, and output, $\mathbf{y}(t)$. This implies that the system's natural modes all 'appear' in the output, $\mathbf{y}(t)$.

In the LTI case, it can be shown that S is observable if and only if the 'observability matrix', $\mathbf{Q}$, has rank m:

$$\mathbf{Q} = (\mathbf{C}^T \quad (\mathbf{CA})^T \quad (\mathbf{CA}^2)^T \ldots (\mathbf{CA}^{m-1})^T) \tag{A2.16}$$

Equivalently, S is observable if and only if $\mathbf{C}$ is left prime WRT $(s\mathbf{I} - \mathbf{A})^{-1}$ (Eqn (A2.11)).

An example of unobservability illustrates the meaning of this.

Example

Consider the LTI system, S, of Figure A2.3, which comprises the elements of Figure A2.2 connected in reverse order.

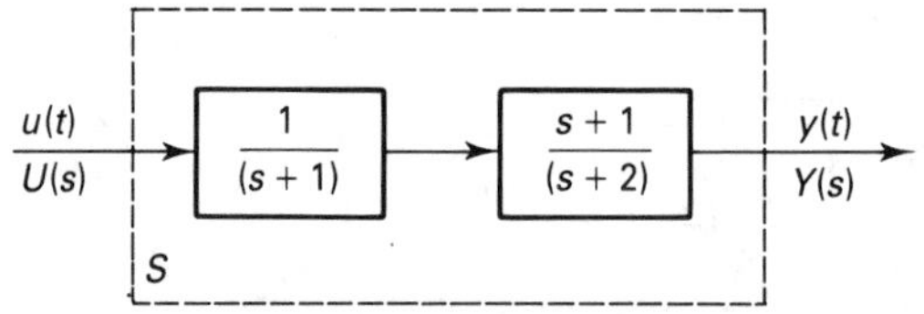

Figure A2.3 Unobservable system.

A system model is:

$$\dot{\mathbf{x}} = \begin{pmatrix} -1 & 0 \\ 0 & -2 \end{pmatrix}\mathbf{x} + \begin{pmatrix} 1 \\ 1 \end{pmatrix}u$$

$$y = (0 \quad 1)\mathbf{x}$$

In this case (Eqn (A2.16)):

$$\mathbf{Q} = \begin{pmatrix} 0 & 0 \\ 1 & -2 \end{pmatrix}$$

This has rank 1, but $m = 2$, so S is unobservable.

Premultiplying $(s\mathbf{I} - \mathbf{A})^{-1}$ by $\mathbf{C}$ leads to the loss of the term in $(s + 1)$, also indicating that S is unobservable. Heuristically, the mode represented by the pole $s = -1$ is filtered out by the zero at $s = -1$. Thus $y(t)$ contains no component corresponding to this mode, which is unobservable at the output.

A2.4 State equation realization from transfer functions (LTI systems)

Realizing an LTI equation model (Eqns (A2.5) and (A2.6)), given a transfer function matrix (Eqn (A2.11)), is generally not easy; it calls for ingenuity or, at worst, the use of algorithms such as the Ho and Kalman algorithm [ref. 3]. For single-input single-output systems, simple formulae exist. Three useful ones are as follows:

Consider the (strictly proper) transfer function (Eqn (A1.8)):

$$G(s) = \frac{Y(s)}{U(s)} = \frac{a_{m-1}s^{m-1} + a_{m-2}s^{m-2} + \cdots + a_0}{s^m + b_{m-1}s^{m-1} + \cdots + b_0} \qquad \text{(A2.17)}$$

where

$$Y(s) = \mathscr{L}(y(t))$$

$$U(s) = \mathscr{L}(u(t))$$

1. Standard-form dynamical equations can be synthesized using:

$$\begin{aligned}
x_1 &= y \\
x_2 &= \dot{x}_1 + b_{m-1}x_1 - a_{m-1}u \\
x_3 &= \dot{x}_2 + b_{m-2}x_1 - a_{m-2}u \\
&\ \ \vdots \\
x_m &= \dot{x}_{m-1} + b_1x_1 - a_1u \\
\dot{x}_m &+ b_0x_1 = a_0u
\end{aligned}$$

These are readily recast in the standard form of Eqns (A2.5) and (A2.6).

$$\dot{\mathbf{x}} = \begin{pmatrix} -b_{m-1} & 1 & 0 & \dots & 0 \\ -b_{m-2} & 0 & 1 & \dots & 0 \\ \vdots & & & & \vdots \\ -b_1 & 0 & 0 & 0 & 1 \\ -b_0 & 0 & 0 & \dots & 0 \end{pmatrix} \mathbf{x} + \begin{pmatrix} a_{m-1} \\ a_{m-2} \\ \vdots \\ a_0 \end{pmatrix} u \tag{A2.18}$$

$$y = (1 \quad 0 \quad 0 \dots 0)\mathbf{x}$$

2. 'Companion form' state equations guaranteed to be controllable, and therefore valid only for controllable systems, are:

$$\dot{\mathbf{x}} = \begin{pmatrix} 0 & 1 & 0 & \dots & 0 \\ 0 & 0 & 1 & & 0 \\ \vdots & & & & \\ 0 & 0 & 0 & & 1 \\ -b_0 & -b_1 & & & -b_{m-1} \end{pmatrix} \mathbf{x} + \begin{pmatrix} 0 \\ 0 \\ \vdots \\ 1 \end{pmatrix} u$$

$$y = (a_0 \quad a_1 \quad a_2 \quad \dots \quad a_{m-1})\mathbf{x} \tag{A2.19}$$

3. Companion form state equations guaranteed to be observable, and therefore valid only for observable systems, are:

$$\dot{\mathbf{x}} = \begin{pmatrix} 0 & 0 & 0 & \dots & -b_0 \\ 1 & 0 & 0 & & -b_1 \\ 0 & 1 & 0 & & -b_2 \\ \vdots & & & & 0 \\ 0 & 0 & 0 & 1 & b_{m-1} \end{pmatrix} \mathbf{x} + \begin{pmatrix} a_0 \\ a_1 \\ a_2 \\ \vdots \\ a_{m-1} \end{pmatrix} u$$

$$y = (0 \quad 0 \quad 0 \quad \dots \quad 0 \quad 1)\mathbf{x} \tag{A2.20}$$

If $G(s)$ (cf. Eqn (A2.17)) is merely proper (as opposed to strictly proper):

$$G(s) = \frac{a_m s^m + a_{m-1}s^{m-1} + \dots + a_0}{s^m + b_{m-1}s^{m-1} + \dots + b_0} \tag{A2.21}$$

then it can be written:

$$G(s) = \frac{Y(s)}{U(s)} = G_R(s) + a_m$$

The above formulae can be applied to $G_R(s)$, which is strictly proper, and the output dynamical equation then augmented to:

$$y = \mathbf{C}_R\mathbf{x} + D$$

where

$\mathbf{C}_\mathrm{R}$ is the output matrix corresponding to $G_\mathrm{R}(s)$, and $D = a_m$.

If $G(s)$ is improper, no state equation realization is possible.

A3 Discrete time linear systems [refs 1, 3, 4]

A3.1 Model equations

The dynamical behaviour of a discrete time linear system, Σ, is modelled by the difference equations: (cf. Eqns (A2.1) and (A2.2)):

$$\mathbf{x}(n+1) = \varphi(n)\mathbf{x}(n) + \psi(n)\mathbf{u}(n) \tag{A3.1}$$

$$\mathbf{y}(n) = \mathbf{C}_d(n)\mathbf{x}(n) + \mathbf{D}_d(n)\mathbf{u}(n) \tag{A3.2}$$

where

$\mathbf{x}(n)$ is the state vector $(x_1 \quad x_2 \quad \dots \quad x_m)^T$, a real variable,
$\mathbf{u}(n)$ is the input vector $(u_1 \quad u_2 \quad \dots \quad u_q)^T$, a real variable,
$\mathbf{y}(n)$ is the output vector $(y_1 \quad y_2 \quad \dots \quad y_r)$, a real variable,
$\varphi(n)$, $\psi(n)$, $\mathbf{C}_d(n)$, $\mathbf{D}_d(n)$ are time-dependent matrices,
φ_n is the state transition matrix,
n is the independent variable, describing time, $n = -\infty, \dots, -1, 0, 1, 2, \dots, +\infty$.

The solution of Eqn (A3.1), given $\mathbf{x}(n_0)$, is:

$$\mathbf{x}(n) = \prod_{i=n_0}^{n-1} \varphi(i)\mathbf{x}(n_0) + \sum_{i=n_0}^{n-1} \left(\prod_{k=i+1}^{n-1} \varphi(k) \right) \psi(i)\mathbf{u}(i) \tag{A3.3}$$

The two components of this represent the zero input response and zero initial state response of Σ respectively.

A large and useful subset of linear systems are linear time invariant (LTI), modelled by the constant parameter equations:

$$\mathbf{x}(n+1) = \varphi\mathbf{x}(n) + \psi\mathbf{u}(n) \tag{A3.4}$$

$$\mathbf{y}(n) = \mathbf{C}_d\mathbf{x}(n) + \mathbf{D}_d\mathbf{u}(n) \tag{A3.5}$$

where φ, ψ, $\mathbf{C}_d$, $\mathbf{D}_d$ are constant matrices of appropriate dimension.

The solution of Eqn (A3.4), given $\mathbf{x}(n_0)$, is:

$$\mathbf{x}(n) = \varphi^{(n-n_0)} \cdot \mathbf{x}(n_0) - \sum_{i=n_0}^{n-1} \varphi^{n-i-1}\psi\mathbf{u}(i) \tag{A3.6}$$

Example

Consider the system:

$$\mathbf{x}(n+1) = \begin{pmatrix} 0.789 & 0.046 \\ -4.606 & 0.880 \end{pmatrix} \mathbf{x}(n) + \begin{pmatrix} 0.120 \\ 4.845 \end{pmatrix} u(n) \tag{A3.7}$$

$$y(n) = (1 \quad 0)\mathbf{x}(n) \tag{A3.8}$$

For $\mathbf{x}(0) = (2 \quad 0)^T$, $u(n) = 0$, Eqn (A3.6) yields the zero input response solution:

$$\mathbf{x}(n) = \begin{pmatrix} 0.789 & 0.046 \\ -4.606 & 0.880 \end{pmatrix}^n \begin{pmatrix} 2 \\ 0 \end{pmatrix}$$

A computer evaluation of this is shown in Figure A3.1(a).

For $x(0) = 0$, $u(n) = 1$, $n \geqslant 0$, the second expression of Eqn (A3.3) yields the zero initial state response solution, a computer evaluation of which is given in Figure A3.1(b).

A3.2 Transfer functions, eigenvalues and stability (LTI systems)

For the LTI case, the transfer function of Σ is found by taking the z transforms of Eqns (A3.4) and (A3.5) and eliminating $\bar{\mathbf{X}}(z)$, the z transform of $\mathbf{x}(n)$ (cf. Eqn (A2.11)) to yield:

$$\bar{\mathbf{Y}}(z) = (\mathbf{C}_d(z\mathbf{I} - \varphi)^{-1}\psi + \mathbf{D}_d)\bar{\mathbf{U}}(z)$$

where

$$\bar{\mathbf{Y}}(z) = Z(\mathbf{y}(n))$$

$$\bar{\mathbf{U}}(z) = Z(\mathbf{u}(n))$$

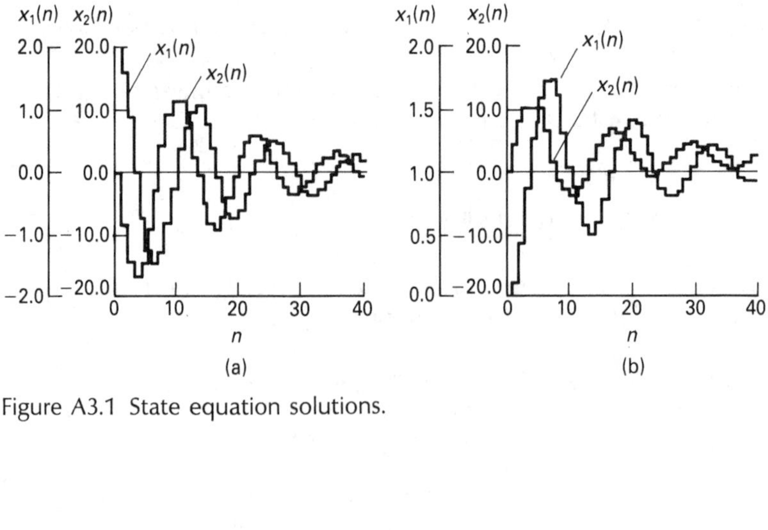

Figure A3.1 State equation solutions.

and the transfer function of Σ is:

$$\bar{\mathbf{G}}(z) = \mathbf{C}_d(z\mathbf{I} - \varphi)^{-1}\psi + \mathbf{D}_d \tag{A3.9}$$

As in the continuous time case, the poles of Σ are the solutions of its characteristic equation:

$$\det(z\mathbf{I} - \varphi) = 0 \tag{A3.10}$$

These are at the eigenvalues of φ, which are given by the solutions of its characteristic equation:

$$\det(\lambda\mathbf{I} - \varphi) = 0 \tag{A3.11}$$

The eigenvalues of φ describes the natural modes of Σ. A complementary pair of eigenvalues at

$$\lambda = re^{\pm j\beta}$$

correspond to time domain behaviour of 'shape':

$$r^n \cos(\beta n) \tag{A3.12}$$

This is an exponentially increasing (or decreasing) sinusoidal process sampled at a rate of $2\pi/\beta$ times per cycle – or, equivalently, with a frequency of β radians per sample. A single real eigenvalue, $\lambda = r$, corresponds to the time-domain behaviour of shape:

$$r^n$$

This is an exponentially increasing (or decreasing) sampled process.

The eigenvalues of φ thus indicate the stability or otherwise of Σ; if only decreasing natural modes are present ($|r| < 1$), Σ is stable, while if any non-decreasing modes are present ($|r| \geqslant 1$), Σ is unstable.

As in the continuous time case, two more detailed types of stability, namely asymptotic and bounded-input bounded-output (BIBO) stability can usefully be defined (Section A2.2).

If all the eigenvalues of Σ lie within the unit circle ($|r| = 1$), Σ is both asymptotically and BIBO stable.

Example

Consider the example of Eqns (A3.7) and (A3.8).

The transfer function of Σ is (Eqn (A3.9)):

$$\bar{G}(z) = (1 \quad 0)\begin{pmatrix} z - 0.789 & -0.046 \\ 4.606 & z - 0.880 \end{pmatrix}^{-1}\begin{pmatrix} 0.120 \\ 4.845 \end{pmatrix}$$

$$= \frac{0.120(z + 0.975}{z^2 - 1.669z + 0.906} \tag{A3.13}$$

The characteristic equation of Σ is (Eqn (A3.10)):

$$z^2 - 1.669z + 0.906 = 0$$

This yields the system poles:

$$z = 0.833 \pm j0.455 = 0.951e$$

The eigenvalues of φ are (Eqn (A3.11)):

$$\lambda = 0.951e \pm j0.500$$

The natural mode of Σ has the shape (Eqn (A3.12)):

$$(0.951)^n \cos\left(0.5n \right)$$

This is an exponentially decreasing sinusoidal signal sampled at a rate of $2\pi/0.5 = 12.6$ samples per cycle. This accords with Figure A3.1.

Σ is asymptotically and BIBO stable; its natural mode is a lightly damped sinusoidal signal, frequency 0.5 radians per sample.

A3.3 Controllability and observability

The properties of controllability and observability follow those for continuous time systems, with isomorphic formulae.

In the LTI case, Σ is controllable if the controllability matrix $\mathbf{P}$ (cf. Eqn (A2.15)) has rank m:

$$\mathbf{P} = (\psi \quad \varphi\psi \quad \varphi^2\psi \quad \ldots \quad \varphi^{m-1}\psi) \tag{A3.14}$$

Equivalently, Σ is controllable if and only if ψ is right prime WRT $(z\mathbf{I} - \varphi)^{-1}$.

Σ is observable if the observability matrix, $\mathbf{Q}$ (cf. Eqn (A2.16)) has rank m:

$$\mathbf{Q} = (\mathbf{C}_d^T \quad (\mathbf{C}_d\varphi)^T \quad (\mathbf{C}_d\varphi^2)^T \quad \ldots \quad (\mathbf{C}_d\varphi^{m-1})^T) \tag{A3.15}$$

Equivalently, Σ is observable if and only if $\mathbf{C}_d$ is left prime WRT $(z\mathbf{I} - \varphi)^{-1}$.

A3.4 Autoregressive moving average (ARMA) models

Consider the mth order system proper transfer function (Eqn (A1.17)):

$$\bar{G}(z) = \frac{\bar{Y}(z)}{\bar{U}(z)} = \frac{a_0 + a_1 z^{-1} + \ldots a_m z^{-m}}{1 - b_1 z^{-1} - \ldots b_m z^{-m}} \tag{A3.16}$$

where

$$\bar{Y}(z) = Z(y(n)), \quad y(0) = 0$$

$$\bar{U}(z) = Z(u(n))$$

Multiplying out and taking inverse transforms yields the 'autogressive moving average' (ARMA) system model relating $y(n)$ to $u(n), u(n-1), \ldots, u(n-m), y(n-1), \ldots, y(n-m)$:

$$y(n) = a_0u(n) + a_1u(n-1) + \cdots + a_mu(n-m) + b_1y(n-1) + \cdots + b_my(n-m) \tag{A3.17}$$

Eqn (A3.17) is a linear, mth order difference equation (cf. Eqn (A1.15)), which models the behaviour of a dynamical system.

Example

Consider the dynamical system, Σ, described by Eqns (A3.7) and (A3.8), whose transfer function is (Eqn (A3.9)):

$$\bar{G}(z) = \frac{\bar{Y}(z)}{\bar{U}(z)} = \frac{0.120 + 0.117z^{-1}}{1 - 1.669z^{-1} + 0.9062z^{-2}}$$

From Eqns (A3.16) and (A3.17), the equivalent ARMA model of Σ is:

$$y(n) = 0.120u(n) + 0.117u(n-1) + 1.669y(n-1) - 0.906y(n-2)$$

A3.5 State equation realization from transfer functions and ARMA models (LTI systems)

ARMA and transfer function models of LTI systems are easily interchangeable using Eqns (A3.16) and (A3.17). To develop a set of standard-form discrete time equations (Eqns (A3.4) and (A3.5)) from a transfer function model (Eqn (A3.16)) is algebraically isomorphic with the equivalent continuous time operation described in Section A2.4. Eqns (A2.17), (A2.18), (A2.19) and (A2.20), are all used directly with z replacing s, and $u(n)$, $\mathbf{x}(n)$, $\mathbf{x}(n+1)$, $y(n)$ replacing $u(t)$, $\mathbf{x}(t)$, $\dot{\mathbf{x}}(t)$, $y(t)$ respectively.

A3.6 Sampled data systems

Perhaps the most important type of discrete time system is that consisting of a continuous time system buffered with digital-to-analog and analog-to-digital converters (DACs and ADCs), as shown in Figure A3.2. This is the 'sampled data' system.

If S is linear, and its model equations (Eqns (A2.1) and (A2.2)) and the sampling period, T, are known, the model equations for Σ are given by Eqns (A3.1) and (A3.2) with (Eqn (A2.3)):

$$\varphi(n) = \varphi((n+1)T, nT) \tag{A3.18}$$

$$\psi(n) = \int_{nT}^{(n+1)T} \varphi(\lambda, nT)\mathbf{B}(\lambda)\, d\lambda \tag{A3.19}$$

$$\mathbf{C}_d(n) = \mathbf{C}(nT) \tag{A3.20}$$

$$\mathbf{D}_d(n) = \mathbf{D}(nT) \tag{A3.21}$$

For LTI systems modelled by Eqns (A2.5) and (A2.6), these relationships simplify to give the parameters of Eqns (A3.4) and (A3.5):

$$\varphi = \mathrm{e}^{\mathbf{A}T}$$

$$= \mathbf{I} + \mathbf{A}T + \frac{1}{2!}\mathbf{A}^2T + \frac{1}{3!}\mathbf{A}^3T + \cdots \tag{A3.22}$$

$$\psi = \int_0^T \mathrm{e}^{\mathbf{A}\lambda}\mathbf{B}\,\mathrm{d}\lambda$$

$$= T\mathbf{B} + \frac{1}{2!}\mathbf{A}\mathbf{B}T^2 + \frac{1}{3!}\mathbf{A}^2\mathbf{B}T^3 + \cdots \tag{A3.23}$$

$$\mathbf{C}_d = \mathbf{C}$$

$$\mathbf{D}_d = \mathbf{D}$$

If T is short compared with the natural periods of S, given by eigenvalues of $\mathbf{A}$, Eqns (A3.22) and (A3.23) simplify to:

$$\varphi \doteqdot \mathbf{I} + \mathbf{A}T \tag{A3.24}$$

$$\psi \doteqdot T\mathbf{B} \tag{A3.25}$$

$$\mathbf{C}_d = \mathbf{C}, \quad \mathbf{D}_d = \mathbf{D}$$

A further point of interest, for LTI systems, is that each eigenvalue of φ, representing a natural mode of Σ, is given by

$$\lambda_d = \mathrm{e}^{\lambda T} \tag{A3.26}$$

where λ is a corresponding eigenvalue of S.

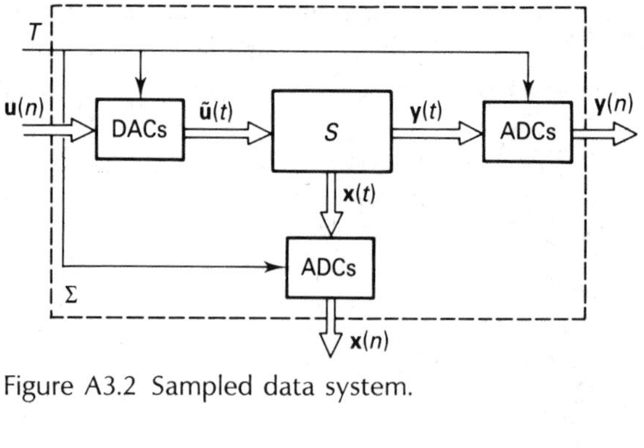

Figure A3.2 Sampled data system.

Example

Consider the system described by Eqns (A2.9) and (A2.10), buffered by DACs and ADCs, $T = 0.050$.

The dynamical equations of Σ are found using Eqns (A3.18)–(A3.21), evaluated where necessary by computer:

$$\mathbf{x}(n+1) = \begin{pmatrix} 0.789 & 0.046 \\ -4.606 & 0.880 \end{pmatrix} \mathbf{x}(n) + \begin{pmatrix} 0.120 \\ 4.845 \end{pmatrix} u(n)$$

$$y(n) = (1 \quad 0)\mathbf{x}(n)$$

Recall that the eigenvalues of $\mathbf{A}$ are at $s = -1 \pm \mathrm{j}10$. Then the eigenvalues of $\boldsymbol{\varphi}$ – and poles of Σ – are (Eqn (A3.26)):

$$z = \mathrm{e}^{(-1 \pm \mathrm{j}10)0.050}$$

$$= 0.951\mathrm{e}^{\pm \mathrm{j}0.500}$$

$$= 0.833 \pm \mathrm{j}0.455$$

This is readily confirmed by finding the solutions of the characteristic equation of Σ (Eqn (A3.10)):

$$\det\begin{pmatrix} z - 0.789 & -0.046 \\ 4.606 & z - 0.880 \end{pmatrix} = 0$$

or

$$z = 0.833 \pm \mathrm{j}0.455$$

APPENDIX B

Low-pass Butterworth filters

The design and properties of filters are outside the scope of this book, but low-pass filters are mentioned in the text and it is useful here to outline the popular Butterworth design in analog and digital forms (Sections B1, B2 respectively). These and other types of Butterworth filter, along with Cauer, Bessel and Chebyshev filters, are described in detail in refs. 5 and 6.

B1 Analog filters [ref. 5]

The Butterworth low-pass filter has a frequency gain characteristic which is optimally flat at zero frequency, its transfer function being based on the expression:

$$G(s)G(-s)|_{s=j\omega} = \frac{k^2}{1 + (\omega^2/\omega_c^2)^m} \tag{B1.1}$$

where

$G(s)$ is the filter transfer function,

ω_c is its 3 dB cut-off frequency, i.e. $|G(j\omega_c)| = \dfrac{1}{\sqrt{2}}$,

m is the order of the filter,
k is chosen so that $G(0) = 1$.

The $2m$ poles of $(G(s)\cdot G(-s))$ lie on the circle $|s| = \omega_c$ in the s-plane Argand diagram, as indicated in Figure B1.1(a). The filter transfer function, $G(s)$, is taken as that corresponding to the stable m poles which lie in the left half s-plane:

$$G(s) = \frac{k}{\prod_{i=1}^{m} (s - s_i)} \tag{B1.2}$$

where:

$$s_i = \exp\left(j\pi\left(0.5 + \frac{2i-1}{2m}\right)\right) \tag{B1.3}$$

and k is chosen so that $G(0) = 1$.

A frequency gain characteristic for a typical filter is shown in Figure B1.1(b). Clearly, the higher in the filter order, the sharper the cut-off characteristic.

m is established from the cut-off frequency and gain (or attenuation) requirement of the filter:

$$m = \frac{\log_{10}(A^2 - 1)}{2\log_{10}(\omega_A/\omega_c)} \tag{B1.4}$$

where a gain of $(1/A)$ is required at $\omega = \omega_A$. Once m is established, the filter transfer function is readily found from Eqns (B1.2) and (B1.3).

Example

A low-pass filter is required to provide an attenuation of 20 at 6 rad s^{-1}, and a 3 dB cut-off at 3 rad s^{-1}.

From Eqn (B1.4):

$$m = \frac{\log_{10}(400 - 1)}{2\log_{10}(6/3)} = 4.32$$

Choose the filter order, $m = 5$.

The filter poles are given by Eqn (B1.3):

$$s_1 = 3\exp(\mathrm{j}\pi(0.5 + 1/10)) = -0.927 + \mathrm{j}1.853$$

$$s_2 = 3\exp(\mathrm{j}\pi(0.5 + 3/10)) = -2.427 + \mathrm{j}1.761$$

$$s_3 = 3\exp(\mathrm{j}\pi(0.5 + 5/10)) = -3$$

$$s_4 = 3\exp(\mathrm{j}\pi(0.5 + 7/10)) = -2.427 - \mathrm{j}1.761$$

$$s_5 = 3\exp(\mathrm{j}\pi(0.5 + 9/10)) = -0.927 - \mathrm{j}1.853$$

k is chosen to give $G(s)$ a gain of 1 at $s = 0$:

$$k = 3^5 = 243$$

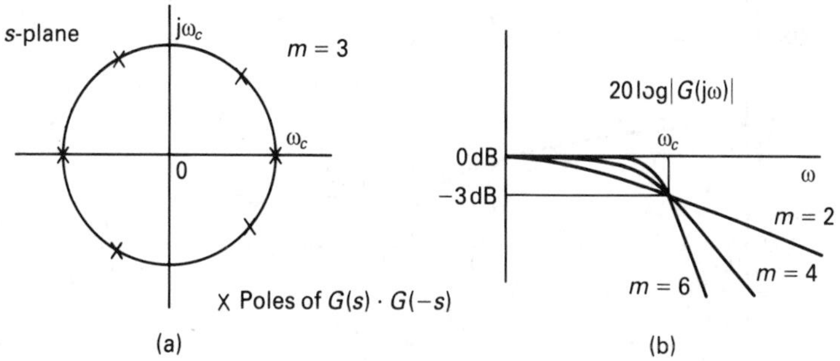

Figure B1.1 Butterworth filter poles and gain characteristics.

Thus, the filter transfer function is:

$$G(s) = \frac{243}{(s+3)(s+0.927 \pm \mathrm{j}1.853)(s+2.427 \pm \mathrm{j}1.761)}$$

The gain frequency characteristic of this, which meets the design requirements, is shown in Figure B1.2.

B2 Digital filters [ref. 6]

The design of the digital filter follows that of the analog filter.

The z-transfer function is based on the expression:

$$\bar{G}(z)\bar{G}(-z)|_{z=\exp(\mathrm{j}\beta)} = \frac{1}{1+\left(\dfrac{\tan(\beta/2)}{\tan(\beta_c/2)}\right)^{2m}} \tag{B2.1}$$

where

$\bar{G}(z)$ is the filter transfer function,
β_c is the filter 3 dB cut-off frequency (rad per sample),
m is the order of the filter.

The $2m$ poles of $\bar{G}(z)\,\bar{G}(-z)$ lie on a circle in the z-plane Argand diagram, as shown in Figure B2.1(a). There are also $2m$ zeros at $z = -1$. Typical frequency characteristics ($m = 3, 4, 6$) are shown in Figure B2.1(b).

Analysis of Eqn (B2.1) shows that the $2m$ poles are at locations given by:

$$z_i = u_i + \mathrm{j}v_i, \quad i = 0, 1, 2, \ldots, (2m-1)$$

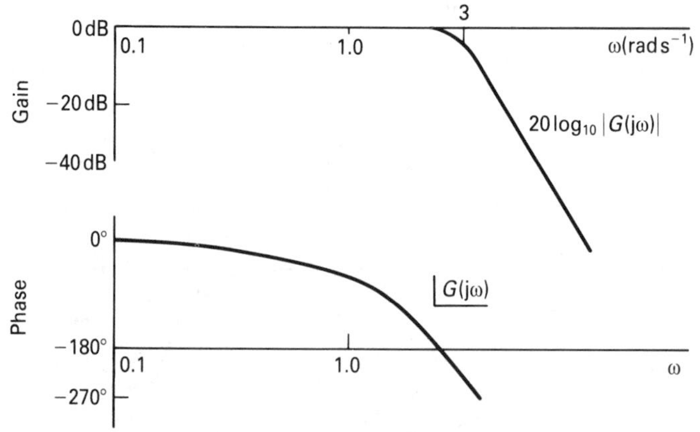

Figure B1.2 Butterworth filter characteristics.

where

$$u_i = \frac{1 - a^2}{d}$$

$$v_i = \frac{2a \sin b}{d}$$

$$d = 1 - 2a \cos b + a^2$$

$$a = \tan\left(\frac{\beta_c}{2}\right)$$

and

$$b = \begin{cases} \dfrac{i\pi}{m}, & m \text{ odd} \\ \left(\dfrac{2i + 1}{2m}\right)\pi, & m \text{ even} \end{cases} \tag{B2.2}$$

The m poles which lie within the unit circle $|z| = 1$ are chosen from the $2m$ poles available, and the filter transfer function is thus:

$$\bar{G}(z) = \frac{k(z + 1)^{m-1}}{\prod_{i=0}^{m-1} (z - z_i)}$$

where k is chosen so that $\bar{G}(1) = 1$.

The design procedure, given β_c and a gain at some higher frequency, is to establish the filter order m from Eqn (B2.1), evaluate the $2m$ poles from Eqn (B2.2), select those inside $|z| = 1$ and convert to algorithm form (Eqns (A1.15) and (A1.17)).

Naturally, if real-time frequencies, ω, and a sampling time, T, are quoted, the relationship $\beta = \omega T$ is used.

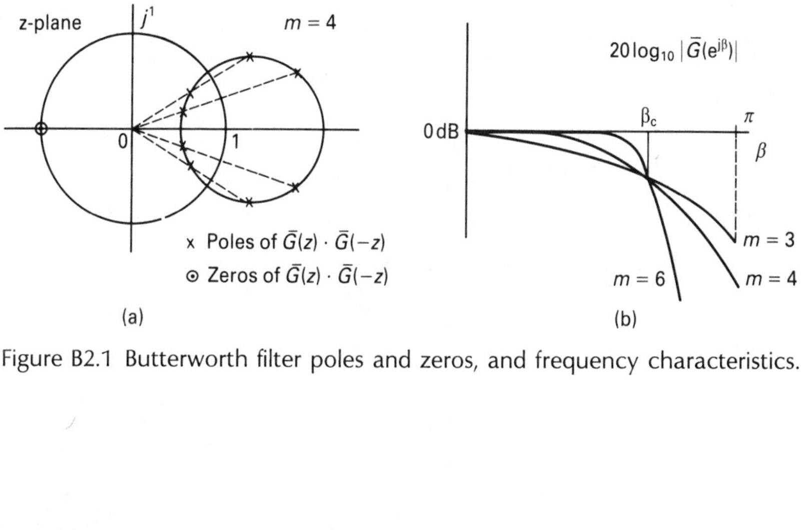

Figure B2.1 Butterworth filter poles and zeros, and frequency characteristics.

Example

A low-pass filter is required to provide attenuation of 10 at 6 rad s^{-1}, and a 3 dB cut-off at 3 rad s^{-1}. A sampling time of $T = 0.2$ s is chosen.

Since $\beta = \omega T$, $\beta_c = 0.60$ rad per sample, and attenuation of 10 at 1.20 rad s^{-1} is specified. From Eqn (B2.1):

$$\left(\frac{\tan(0.6)}{\tan(0.3)}\right)^{2m} = 10^2 - 1 = 99$$

Choose $m = 3$.

In Eqn (B2.2), the six poles are:

$$z_0 = 1.896$$

$$z_1 = 1.393 + j0.681$$

$$z_2 = 0.780 + j0.381$$

$$z_3 = 0.528$$

$$z_4 = 0.780 - j0.381$$

$$z_5 = 1.393 - j0.681$$

The three stable poles inside $|z| = 1$ are chosen, and k is selected to give:

$$\bar{G}(z) = \frac{1.139 \times 10^{-2}(z+1)^3}{(z - 0.528)(z - 0.780 \pm j0.381)}$$

The frequency response of this, $\bar{G}(e^{j\beta})$, $0 \leqslant \beta \leqslant \pi$, is shown in Figure B2.2. The design requirements have been met.

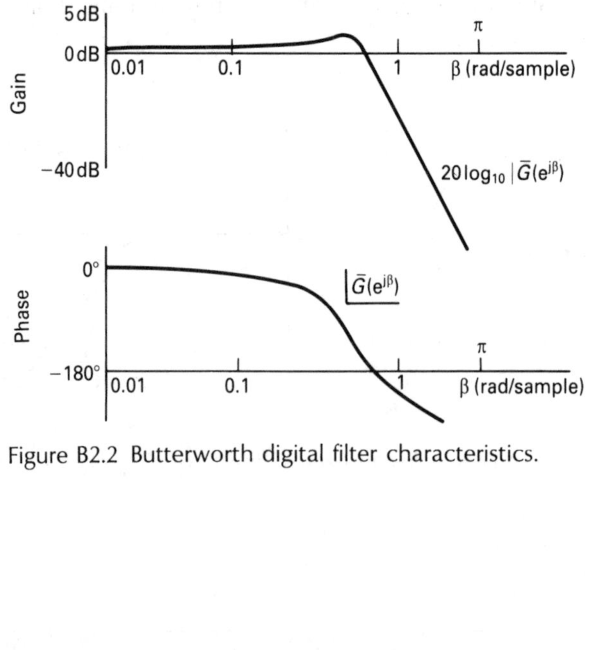

Figure B2.2 Butterworth digital filter characteristics.

The filter is realized in algorithm form using Eqns (A1.15) and (A1.17):

$$y(n) = 1.139 \times 10^{-2}u(n) + 3.417 \times 10^{-2}u(n-1) + 3.417 \times 10^{-2}u(n-2)$$
$$+ 1.139 \times 10^{-2}u(n-3) + 2.088y(n-1) - 1.577y(n-2) + 0.397y(n-3)$$

This is very easily realized in an on-line microprocessor.

APPENDIX C

Counting and statistics

C1 Permutations and combinations [ref. 2]

C1.1 Permutations

The number of ways of arranging r elements from a set of n elements is:

$$n(n-1)(n-2)\ldots(n-r+1)$$

This is written:

$$^{n}P_{r} = \frac{n!}{(n-r)!} \qquad \text{(C1.1)}$$

Example

Consider the number of ways of arranging four cards taken from a pack of fifty-two cards:

$$^{52}P_{4} = \frac{52!}{48!} = 52 \times 51 \times 50 \times 49 = 6\,497\,400$$

C1.2 Combinations

There are $r!$ ways of arranging r elements (Eqn (C1.1)). Thus, if there are $\binom{n}{r}$ ways of choosing r elements from a set of n elements in any order, it follows that there are

$$^{n}P_{r} = \binom{n}{r} r!$$

ways of choosing and arranging r elements from n elements. From Eqn (C1.1):

$$\binom{n}{r} = \frac{n!}{r!(n-r)!} \qquad \text{(C1.2)}$$

Example

The number of ways of choosing four cards from fifty-two in any order is:

$$\binom{52}{4} = \frac{52!}{4!48!} = \frac{52 \times 51 \times 50 \times 49}{1 \times 2 \times 3 \times 4} = 270\,725$$

It is worthwhile extending the definition of $n!$ to include $n = 0$ so that Eqn (C1.2) is valid for the case in which n elements are chosen from n elements. Clearly, there is one such combination:

$$1 = \frac{n!}{n!(n-n)!} = \frac{1}{0!}$$

To validate this 0! is defined as 1.

C2 The binomial theorem

It is not difficult to show, by considering the combinations of terms multiplied together, that:

$$(q+p)^n = q^n + \binom{n}{1} q^{n-1} p^1 + \binom{n}{2} q^{n-2} p^2 + \cdots + p^n$$

$$= \sum_{r=0}^{n} \binom{n}{r} q^{n-r} p^r \qquad \text{(C2.1)}$$

where q, p are any numbers, and n is any positive integer. This is the binomial theorem.

C3 Random numbers and standard normal form tables

A table of random numbers evenly distributed between 0 and 99 is given in Figure C3.1 and a standard normal (Gaussian) form table is given in Figure C3.2.

28 89 65 87 08	13 50 63 04 23	25 47 57 91 13	52 62 24 19 94	91 67 48 57 10
30 29 43 65 42	78 66 28 55 80	47 46 41 90 08	55 98 78 10 70	49 92 05 12 07
95 74 62 60 53	51 57 32 22 27	12 72 72 27 77	44 67 32 23 13	67 95 07 76 30
01 85 54 96 72	66 86 65 64 60	56 59 75 36 75	46 44 33 63 71	54 50 06 44 75
10 91 46 96 86	19 83 52 47 53	65 00 51 93 51	30 80 05 19 29	56 23 27 19 03
05 33 18 08 51	51 78 57 26 17	34 87 96 23 95	89 99 93 39 79	11 28 94 15 52
04 43 13 37 00	79 68 96 26 60	70 39 83 66 56	62 03 55 86 57	77 55 33 62 02
05 85 40 25 24	73 52 93 70 50	48 21 47 74 63	17 27 27 51 26	35 96 29 00 45
84 90 90 65 77	63 99 25 69 02	09 04 03 35 78	19 79 95 07 21	02 84 48 51 97
28 55 53 09 48	86 28 30 02 35	71 30 32 06 47	93 74 21 86 33	49 90 21 69 74
89 83 40 69 80	97 96 47 59 97	56 33 24 87 36	17 18 16 90 46	75 27 28 52 13
73 20 96 05 68	93 41 69 96 07	97 50 81 79 59	42 37 13 81 83	82 42 85 04 31
10 89 07 76 21	40 24 74 36 42	40 33 04 46 24	35 63 02 31 61	34 59 43 36 96
91 50 27 78 37	06 06 16 25 98	17 78 80 36 85	26 41 77 63 37	71 63 94 94 33
03 45 44 66 88	97 81 26 03 89	39 46 67 21 17	98 10 39 33 15	61 63 00 25 92
89 41 58 91 63	65 99 59 97 84	90 14 79 61 55	56 16 88 87 60	32 15 99 67 43
13 43 00 97 26	16 91 21 32 41	60 22 66 72 17	31 85 33 69 07	68 49 20 43 29
71 71 00 51 72	62 03 89 26 32	35 27 99 18 25	78 12 03 09 70	50 93 19 35 56
19 28 15 00 41	92 27 73 40 38	37 11 05 75 16	98 81 99 37 29	92 20 32 39 67
56 38 30 92 30	45 51 94 69 04	00 84 14 36 37	95 66 39 01 09	21 68 40 95 79
39 27 52 89 11	00 81 06 28 48	12 08 05 75 26	03 35 63 05 77	13 81 20 67 58
73 13 28 58 01	05 06 42 24 07	60 60 29 99 93	72 93 78 04 36	25 76 01 54 03
81 60 84 51 57	12 68 46 55 89	60 09 71 87 89	70 81 10 95 91	83 79 68 20 66
05 62 98 07 85	07 79 26 69 61	67 85 72 37 41	85 79 76 84 23	61 58 87 08 05
62 97 16 29 18	52 16 16 23 56	62 95 80 97 63	32 25 34 03 36	48 84 60 37 65
31 13 63 21 08	16 01 92 58 21	48 79 74 73 72	08 64 80 91 38	07 28 66 61 59
97 38 35 34 19	89 84 05 34 47	88 09 31 54 88	97 96 86 01 69	46 13 95 65 96
32 11 78 33 82	51 99 98 44 39	12 75 10 60 36	80 66 39 94 97	42 36 31 16 59
81 99 13 37 05	08 12 60 39 23	61 73 84 89 18	26 02 04 37 95	96 18 69 06 30
45 74 00 03 05	69 99 47 26 52	48 06 30 00 18	03 30 28 55 59	66 10 71 44 05
11 84 13 69 01	88 91 28 79 50	71 42 14 96 55	98 59 96 01 36	88 77 90 45 59
14 66 12 87 22	59 45 27 08 51	85 64 23 85 41	64 72 08 59 44	67 98 56 65 56
40 25 67 87 82	84 27 17 30 37	48 69 49 02 58	98 02 50 58 11	95 39 06 35 63
44 48 97 49 43	65 45 53 41 07	14 83 46 74 11	76 66 63 60 08	90 54 33 65 84
41 94 54 06 57	48 28 01 83 84	09 11 21 91 73	97 28 44 74 06	22 30 95 69 72
07 12 15 58 84	93 18 31 83 45	54 52 62 29 91	53 58 54 66 05	47 19 63 92 75
64 27 90 43 52	18 26 32 96 83	50 58 45 27 57	14 96 39 64 85	73 87 96 76 23
80 71 86 41 03	45 62 63 40 88	35 69 34 10 94	32 22 52 04 74	69 63 21 83 41
27 06 08 09 92	26 22 59 28 27	38 58 22 14 79	24 32 12 38 42	33 56 90 92 57
54 68 97 20 54	33 26 74 03 30	74 22 19 13 48	30 28 01 92 49	58 61 52 27 03
02 92 65 68 99	05 53 15 26 70	04 69 22 64 07	04 73 25 74 82	78 35 22 21 88
83 52 57 78 62	98 61 70 48 22	68 50 64 55 75	42 70 32 09 60	58 70 61 43 97
82 82 76 31 33	85 13 41 38 10	16 47 61 43 77	83 27 19 70 41	34 78 77 60 25
38 61 34 09 49	04 41 66 09 76	20 50 73 40 95	24 77 95 73 20	47 42 80 61 03
01 01 11 88 38	03 10 16 82 24	39 58 20 12 39	82 77 02 18 88	33 11 49 15 16
21 66 14 38 28	54 08 18 07 04	92 17 63 36 75	33 14 11 11 78	97 30 53 62 38
32 29 30 69 59	68 50 33 31 47	15 64 88 75 27	04 51 41 61 96	86 62 93 66 71
04 59 21 65 47	39 90 89 86 77	46 86 86 88 86	50 09 13 24 91	54 80 67 78 66
38 64 50 07 36	56 50 45 94 25	48 28 48 30 51	60 73 73 03 87	68 47 37 10 84
48 33 50 83 53	59 77 64 59 90	58 92 62 50 18	93 09 45 89 06	13 26 98 86 29

Figure C3.1 Random numbers evenly distributed between 0 and 99. (Taken from Table XXXIII of Fisher and Yates, *Statistical Tables for Biological, Agricultural and Medical Research*, published by Longman Group UK Ltd, London (previously published by Oliver and Boyd Ltd, Edinburgh), by permission of the authors and publishers.)

u	.00	.01	.02	.03	.04	.05	.06	.07	.08	.09
0.0	.5000	.4960	.4920	.4880	.4840	.4801	.4761	.4721	.4681	.4641
0.1	.4602	.4562	.4522	.4483	.4443	.4404	.4364	.4325	.4286	.4247
0.2	.4207	.4168	.4129	.4090	.4052	.4013	.3974	.3936	.3897	.3859
0.3	.3821	.3783	.3745	.3707	.3669	.3632	.3594	.3557	.3520	.3483
0.4	.3446	.3409	.3372	.3336	.3300	.3264	.3228	.3192	.3156	.3121
0.5	.3085	.3050	.3015	.2981	.2946	.2912	.2877	.2843	.2810	.2776
0.6	.2743	.2709	.2676	.2643	.2611	.2578	.2546	.2514	.2483	.2451
0.7	.2420	.2389	.2358	.2327	.2296	.2266	.2236	.2206	.2177	.2148
0.8	.2119	.2090	.2061	.2033	.2005	.1977	.1949	.1922	.1894	.1867
0.9	.1841	.1814	.1788	.1762	.1736	.1711	.1685	.1660	.1635	.1611
1.0	.1587	.1562	.1539	.1515	.1492	.1469	.1446	.1423	.1401	.1379
1.1	.1357	.1335	.1314	.1292	.1271	.1251	.1230	.1210	.1190	.1170
1.2	.1151	.1131	.1112	.1093	.1075	.1056	.1038	.1020	.1003	.0985
1.3	.0968	.0951	.0934	.0918	.0901	.0885	.0869	.0853	.0838	.0823
1.4	.0808	.0793	.0778	.0764	.0749	.0735	.0721	.0708	.0694	.0681
1.5	.0668	.0655	.0643	.0630	.0618	.0606	.0594	.0582	.0571	.0559
1.6	.0548	.0537	.0526	.0516	.0505	.0495	.0485	.0475	.0465	.0455
1.7	.0446	.0436	.0427	.0418	.0409	.0401	.0392	.0384	.0375	.0367
1.8	.0359	.0351	.0344	.0336	.0329	.0322	.0314	.0307	.0301	.0294
1.9	.0287	.0281	.0274	.0268	.0262	.0256	.0250	.0244	.0239	.0233
2.0	.02275	.02222	.02169	.02118	.02068	.02018	.01970	.01923	.01876	.01831
2.1	.01786	.01743	.01700	.01659	.01618	.01578	.01539	.01500	.01463	.01426
2.2	.01390	.01355	.01321	.01287	.01255	.01222	.01191	.01160	.01130	.01101
2.3	.01072	.01044	.01017	.00990	.00964	.00939	.00914	.00889	.00866	.00842
2.4	.00820	.00798	.00776	.00755	.00734	.00714	.00695	.00676	.00657	.00639
2.5	.00621	.00604	.00587	.00570	.00554	.00539	.00523	.00508	.00494	.00480
2.6	.00466	.00453	.00440	.00427	.00415	.00402	.00391	.00379	.00368	.00357
2.7	.00347	.00336	.00326	.00317	.00307	.00298	.00289	.00280	.00272	.00264
2.8	.00256	.00248	.00240	.00233	.00226	.00219	.00212	.00205	.00199	.00193
2.9	.00187	.00181	.00175	.00169	.00164	.00159	.00154	.00149	.00144	.00139
3.0	.00135									
3.1	.00097									
3.2	.00069									
3.3	.00048									
3.4	.00034									
3.5	.00023									
3.6	.00016									
3.7	.00011									
3.8	.00007									
3.9	.00005									
4.0	.00003									

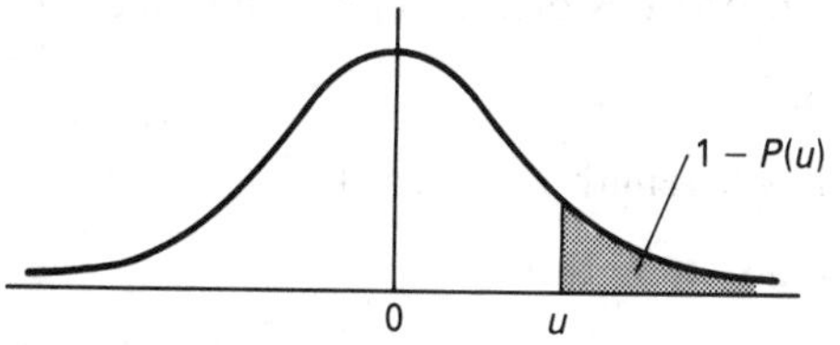

The function tabulated is (1 − $P(u)$ where $P(u)$) is the cumulative distribution function of a standardized normal variable q: $(1 - P(u)) = \frac{1}{\sqrt{2\pi}} \int_u^{\infty} e^{-\frac{u^2}{2}} du$

Figure C3.2 Standard normal form (Gaussian) table of values. (Taken from J. Murdoch and J. A. Barnes, *Statistical Tables*, 1986, Macmillan, and reprinted with the permission of the authors and publishers.)

APPENDIX D

Dirac and Kronecker delta functions [ref. 2]

D1 The Dirac delta function

Consider the function illustrated in Figure D1.1. Define

$$\delta_a(t) = \begin{cases} \dfrac{1}{a}, & 0 \leqslant t \leqslant a \\ 0, & 0 > t, t > a \end{cases}$$

The area under $\delta_a(t)$ is 1, regardless of a. Then a Dirac δ function (or 'impulse function') is:

$$\delta(t) = \lim_{a \to 0} (\delta_a(t))$$

Some properties of the Dirac delta function are as follows.

1. Strength:

   $\delta(t)$ has 'strength', or area, 1

   $k\delta(t)$ has strength k

2. Delay: a Dirac δ function, strength k at time $t = \tau$, is

   $k\delta(t - \tau)$

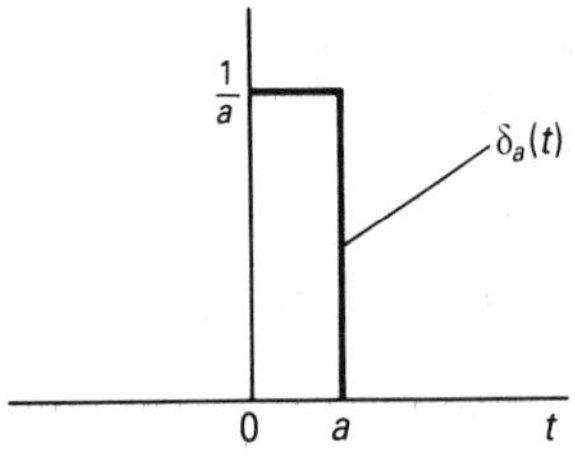

Figure D1.1 Dirac δ function.

3. Integration: if $f(t)$ is continuous at $t = \tau$, then

$$\int_{-\infty}^{+\infty} \delta(t-\tau) f(t)\,\mathrm{d}t = f(\tau)$$

This is the 'sifting' property of $\delta(t)$.

4. Relationship with unit step function: for consistency with (3), for $f(t) = 1$,

$$\frac{\mathrm{d}}{\mathrm{d}t}(u^*(t-\tau)) = \delta(t-\tau)$$

where $u^*(t-\tau)$ represents a unit step function at $t = \tau$. The functions are illustrated in Figure D1.2.

D2 The Kronecker delta function

A Kronecker δ function of the discrete independent variable, n, is

$$\delta(n) = \begin{cases} 1, & n = 0 \\ 0, & n \neq 0 \end{cases}$$

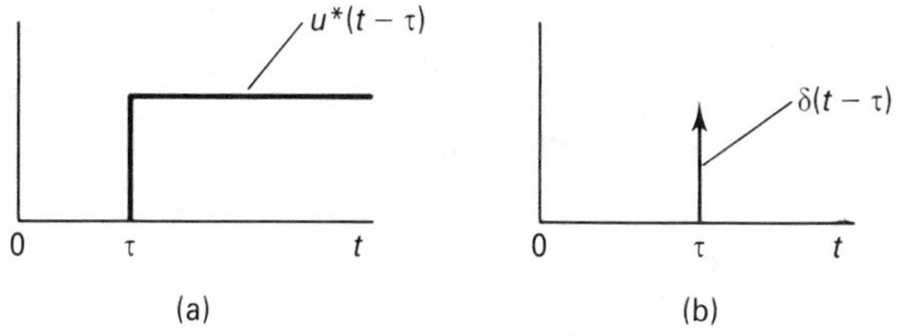

Figure D1.2 Unit step function and Dirac δ function.

APPENDIX E

Riemann and Stieltjes integrals [ref. 2]

E1 The Riemann integral

Consider the function $f(t)$, defined on the interval $a \leqslant t \leqslant b$, and suppose that $[a, b]$ is divided into intervals

$$a = t_0 < t_1 < t_2 < \cdots < t_n = b$$

Consider $f(\tau_k)$ such that $t_{k-1} < \tau_k < t_k$, as shown in Figure E1.1. The area under $f(t)$ is given approximately by:

$$S = \sum_{k=1}^{n} f(\tau_k)(t_k - t_{k-1})$$

For reasonably well-behaved functions, S tends to a limit as $n \to \infty$:

$$\lim_{n \to \infty} (S) = \int_a^b f(t)\,\mathrm{d}t$$

This is the Riemann integral of $f(t)$ WRT t.

E2 The Stieltjes integral

Consider the (not necessarily continuous) non-decreasing function, $f(t)$, and the continuous function $\varphi(t)$, both defined on the interval $a \leqslant t \leqslant b$, and suppose that $[a, b]$ is divided into intervals:

$$a = t_0 < t_1 < t_2 < \cdots < t_n = b$$

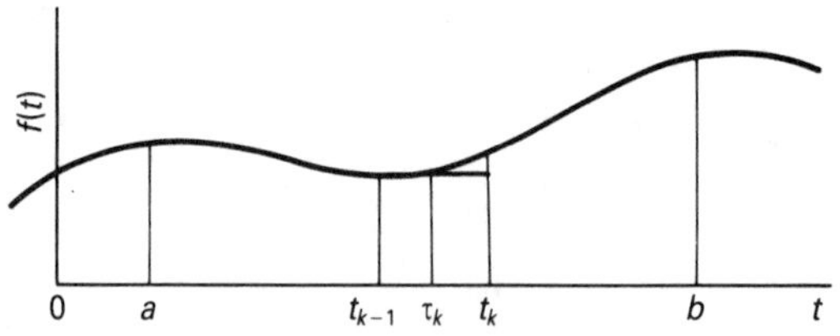

Figure E1.1 The Riemann integral.

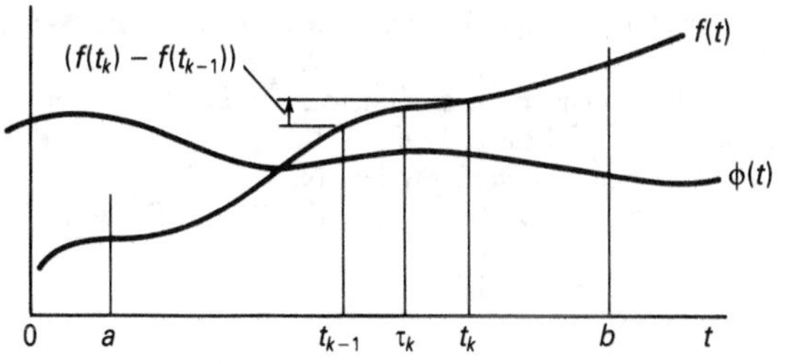

Figure E2.1 The Stieltjes integral.

Consider $\varphi(\tau_k)$ such that $t_{k-1} < \tau_k < t_k$, as indicated in Figure E2.1. It can be shown that the following summation converges as $n \to \infty$:

$$\int_a^b \varphi(t)\,d(f(t)) = \lim_{n\to\infty} \sum_{k=1}^{n} \varphi(\tau_k)(f(t_k) - f(t_{k-1}))$$

This is the Stieltjes integral of $\varphi(t)$ WRT $f(t)$.

In particular, consider the discrete probability function $P_d(x)$, describing the probability of an event of value x (Eqn (1.13)):

$$P_d(x) = \begin{cases} P_d(x_k), & x = x_k \\ 0, & \text{otherwise} \end{cases}$$

Consider the function (Eqn (1.19)):

$$P(\xi) = \text{Prob}(x \leqslant \xi)$$

$P(\xi)$ is clearly a staircase function such that:

$$dP(\xi) = \begin{cases} P_d(x_k), & \xi = x_k \\ 0, & \text{otherwise} \end{cases}$$

or

$$P_d(x_k) = \int_a^b 1\,dP(\xi)$$

where $x_{k-1} < a < x_k < b < x_{k+1}$, and the integral is interpreted as a Stieltjes integral.

REFERENCES

[1] Borrie, J. A., *Modern Control Systems: A manual of design methods*, Prentice Hall International, London, 1986.

[2] Sokolnikoff, I. S. and Redheffer, R. M., *Mathematics of Physics and Modern Engineering*, McGraw-Hill, New York, 1966.

[3] Chen, C. T., *Linear System Theory and Design*, Holt, Rinehart and Winston, New York, 1984.
[4] Franklin, G. F., Powell, J. D. and Emami-Naeini, A., *Feedback Control of Dynamic Systems*, Addison Wesley, Reading, Mass., 1990.
[5] Kuo, F. F., *Network Analysis and Synthesis*, Wiley, New York, 1966.
[6] Bognor, R. E. and Constantinides, A. G., *Introduction to Digital Filtering*, Wiley, New York, 1975.

Solutions to numerical exercises (Chapters 1, 2, and 3)

Chapter 1

1 0.152
2 4.474×10^{-28}
3 25 times
4 (a) 0.037
(b) 7.316×10^{-3}
(c) 0.256
5 (a) 0.651
(b) 0.994
6 $np = 2.800$, $npq = 0.840$
7 $K = 1$
(a) $0.39\dot{3}$
(b) 0.322
$\varepsilon(x) = 1, \quad \varepsilon(x^2) = 2, \quad \sigma_x^2 = 1$
8 $\varepsilon(x) = 0.750$, $\varepsilon(x^2) = 0.600$, $\sigma_x^2 = 0.0375$, $m = 0.794$
9 $8.374\,\mathrm{m}^3$, 16.757 m, 0.207, 0.293
10 $\dfrac{(L - l)}{l}$
11 $K = 1$
$\varepsilon(x) = 0.583$, $\sigma_x^2 = 0.764$, $\varepsilon(y^2) = 0.417$. $\sigma_y^2 = 0.0764$
12 $y = -0.5 \log_e\left(\dfrac{1 - x}{2}\right)$
13 (a) $p(\chi) = \begin{cases} \left|\dfrac{\chi^{-2/3}}{18}\right|, & -27 \leqslant \chi \leqslant +27 \\ 0, & \text{elsewhere} \end{cases}$

(b) $p(\chi) = \begin{cases} \dfrac{1}{\pi\sqrt{1 - \chi^2}}, & -1 \leqslant \chi \leqslant +1 \\ 0, & \text{elsewhere} \end{cases}$

14 0.917

15 $$P(\xi) = \begin{cases} 0.25\left(\dfrac{\xi^2}{2} + 2\xi + 2\right), & -2 \leqslant \xi \leqslant 0 \\ 0.25\left(-\dfrac{\xi^2}{2} + 2\xi + 2\right), & 0 \leqslant \xi \leqslant +2 \\ 0, & \text{elsewhere} \end{cases}$$

$$p(\xi) = \begin{cases} 0.25(-|\xi| + 2), & -2 \leqslant \xi \leqslant +2 \\ 0, & \text{elsewhere} \end{cases}$$

16 $C = \dfrac{3}{\pi}$

$\text{cov}(x, y) = 0$

Chapter 2

1 $$\varepsilon(x(t)) = \frac{1}{t}(e^{-t} - e^{-2t})$$

$$\text{cov}(x(t)) = \frac{1}{t+s}[e^{-(t+s)} - e^{-2(t+s)}] - \frac{1}{ts}[e^{-(t+s)} + e^{-2(t+s)} - e^{-(2t+s)} - e^{-(t+2s)}]$$

2 $$\varepsilon(x(t)) = \frac{1}{\Delta t}\cos\Omega t \sin\Delta t$$

$$\text{cov}(x(t)) = \frac{1}{2\Delta}\left[\frac{\cos\Omega(t+s)\sin\Delta(t+s)}{(t+s)} + \frac{\cos\Omega(t-s)\sin\Delta(t-s)}{(t-s)}\right] - \frac{1}{\Delta^2 ts}\cos\Omega t\sin\Delta t\cos\Omega s\sin\Delta s$$

3 $\text{cov}(x(t)) = \sigma^2\cos(\omega(t-s)) = \sigma^2\cos(\omega\tau)$

4 (a) $S(\omega) = \dfrac{2a}{a^2 + \omega^2}$

(b) $S(\omega) = \dfrac{2a}{a^2 + \omega^2} - \pi(\delta(\omega - b) + \delta(\omega + b))$

(c) $S(\omega) = 12\pi\delta\omega + 4\pi(\delta(\omega - b) + \delta(\omega + b))$

(d) $S(\omega) = \dfrac{a}{a^2 + (\omega + b)^2} + \dfrac{a}{a^2 + (\omega - b)^2}$

5 (a) $\bar{r}(\tau) = 9\delta\tau - \frac{1}{2\pi}\left[\left(\frac{18}{t} - \frac{4}{t^3}\right)\sin 3t + \frac{12}{t^2}\cos 3t\right]$

(b) $\bar{r}(\tau) = \delta\tau + \frac{1}{2\pi}\left(\frac{1}{t^2} - \frac{2\sin 2t}{t} - \frac{\cos 2t}{t^2}\right)$

7 $\frac{\mathrm{d}^n x}{\mathrm{d}x^n}$ exists for all n.

8 (a), (c) are differentiable, (b) is not.

10 $$\operatorname{cov}\left(\int_0^t \beta(t)\,\mathrm{d}t\right) = \begin{cases} \frac{qst^2}{2}, & s \geqslant t \\ \frac{qts^2}{2}, & t \geqslant s \end{cases}$$

11 $$x(t) = \begin{pmatrix} 1 & t \\ 0 & 1 \end{pmatrix}\begin{pmatrix} 1 \\ 1 \end{pmatrix} + \begin{pmatrix} I_1(t) \\ I_2(t) \end{pmatrix}$$

where

$I_1(t)$ is an extended Wiener process, covariance $t^3/3$
$I_2(t)$ is a Wiener process, covariance t.

12 $$\dot{x} = \begin{pmatrix} 0 & 1.0 \\ -1.0 & -0.1 \end{pmatrix} x + \begin{pmatrix} 0 \\ 2 \end{pmatrix} w$$

where

$$x_1 = y, \quad x_2 = \dot{y}$$

$$(\varepsilon(\dot{x})) = \begin{pmatrix} 0 & 1.0 \\ -1.0 & -0.1 \end{pmatrix} \varepsilon(x)$$

$$\dot{R}(t,t) = \begin{pmatrix} 0 & 1.0 \\ -1.0 & -0.1 \end{pmatrix} R(t,t) + R(t,t)\begin{pmatrix} 0 & -1.0 \\ 1.0 & -0.1 \end{pmatrix} + \begin{pmatrix} 0 & 0 \\ 0 & 4 \end{pmatrix}$$

$$R(t,s) = R(t,t)\left[\exp\begin{pmatrix} 0 & t \\ -t & -0.1t \end{pmatrix}\right]^T$$

$$R(\infty,\infty) = \begin{pmatrix} 20 & 0 \\ 0 & 20 \end{pmatrix}$$

13 $$\varepsilon(x(\infty)) = \begin{pmatrix} 0.500 \\ 0.333 \end{pmatrix} u(\infty)$$

$$R(\infty,\infty) = \begin{pmatrix} 1.0 & 1.2 \\ 1.2 & 1.5 \end{pmatrix}$$

14 (a) $F(s) = \dfrac{\sqrt{3}}{s + \sqrt{3}}$

(b) $F(s) = \dfrac{2}{(s+2)(s+3)}$

15 (a) $\dot{x}_1 = x_2$
$\dot{x}_2 = -\sin x_1 - 0.1x_2 - 0.01x_2 \cos x_1 + 0.01w$

where

$x_1 = \theta, \quad x_2 = \dot{\theta},$
w is unit strength white noise.

(b) If $\sin x_1 \doteqdot x_1$, $\cos x \doteqdot 1$,

$$\dot{x} = \begin{pmatrix} 0 & 1.0 \\ -1.0 & -0.11 \end{pmatrix} x + \begin{pmatrix} 0 \\ 0.01 \end{pmatrix} w$$

$$R(\infty, \infty) = \begin{pmatrix} 0.455 \times 10^{-3} & 0 \\ 0 & 0.455 \times 10^{-3} \end{pmatrix}$$

$$\operatorname{cov}(x(t)) = \begin{pmatrix} 0.455 \times 10^{-3} & 0 \\ 0 & 0.455 \times 10^{-3} \end{pmatrix} \left[\exp\begin{pmatrix} 0 & t \\ -t & -0.11t \end{pmatrix} \right]^T$$

Chapter 3

1 $$\varepsilon(x(n)) = \frac{0.5}{n}(1 - e^{-2n})$$

$$\operatorname{cov}(x(n)) = \frac{0.5}{n+l}(1 - e^{-2(n+l)}) - \frac{0.25}{nl}(1 - e^{-2n})(1 - e^{-2l})$$

2 $$\varepsilon(x(n)) = \frac{\sin \Delta n}{\Delta n}$$

$$\operatorname{cov}(x(n)) = \frac{1}{2\Delta}\left[\frac{1}{n+l} \sin \Delta(n+l) + \frac{1}{(n-l)} \sin \Delta(n-l) \right]$$

$$\frac{1}{2\Delta}\left[\frac{\sin \Delta(n+l)}{n+l} + \frac{\sin \Delta(n-l)}{(n-l)} \right] - \frac{\sin \Delta n \sin \Delta l}{\Delta^2 nl}$$

3 $cov(x(n)) = \sigma^2 \cos(\beta(n-l))$

4 (a) $S(\beta) = 2 + 1.47 \cos \beta$

(b) $S(\beta) = 1 + 1.848 \cos \beta + 1.414 \cos 2\beta + 0.766 \cos 3\beta$

5 (a) $y(n) = u(n) + u(n-1) - 0.33y(n-1)$
(b) $y(n) = u(n-1) + 0.5u(n-2) + 1.6y(n-1) - 0.6y(n-2)$

6 $$p(\xi, n) = (0.5)^n \sum_{m=-n}^{+n} \sum_{r=0}^{n} \binom{n}{r} \binom{r}{\frac{m+r}{2}} (0.5)^r \delta(\xi - m\lambda)$$

$$\varepsilon(x(n)) = 0$$
$$\sigma_x^2(n) = 0.5n\lambda^2$$

7 $$\varepsilon(x(n+1)) = \begin{pmatrix} 0.80 & 0.05 \\ -4.00 & 0.90 \end{pmatrix} \varepsilon(x(n)), \quad \varepsilon(x(0)) = \begin{pmatrix} 1 \\ 0 \end{pmatrix}$$

$$R(n+1, n+1) = \begin{pmatrix} 0.80 & 0.05 \\ -4.00 & 0.90 \end{pmatrix} R(n,n) \begin{pmatrix} 0.80 & -4.00 \\ 0.05 & 0.90 \end{pmatrix} + \begin{pmatrix} 0.01 & 0.40 \\ 0.40 & 16.00 \end{pmatrix}$$

Hence

$$R(\infty, \infty) = \begin{pmatrix} 1.253 & 2.255 \\ 2.255 & 104.269 \end{pmatrix}$$

$$R(n, n+k) = \begin{pmatrix} 0.80 & 0.05 \\ -4.00 & 0.09 \end{pmatrix}^k \begin{pmatrix} 1.253 & 2.255 \\ 2.255 & 104.269 \end{pmatrix}, n \to \infty$$

8 $$x(n+1) = \begin{pmatrix} 0.819 & 0 \\ 0 & 0.741 \end{pmatrix} x(n) + \begin{pmatrix} 0.091 \\ 0.086 \end{pmatrix} u(n) + \begin{pmatrix} 0.181 \\ 0.259 \end{pmatrix} w(n)$$

$$\varepsilon(x(n)) = \begin{pmatrix} 0.091 \\ 0.086 \end{pmatrix} u(n)$$

$$R(\infty, \infty) = \begin{pmatrix} 0.010 & 0.012 \\ 0.012 & 0.015 \end{pmatrix}$$

$$R(n, n+k) = \begin{pmatrix} 0.819 & 0 \\ 0 & 0.741 \end{pmatrix}^k \begin{pmatrix} 0.010 & 0.012 \\ 0.012 & 0.015 \end{pmatrix}, n \to \infty$$

Index